AF783648

ASTROBIOLOGY
ORIGINS FROM THE BIG-BANG TO CIVILISATION

THE INTERNATIONAL ORGANISING COMMITTE

HUMBERTO CAMPINS, University of Arizona and Research Corporation, Tucson, USA
IGNASI CASANOVA, Universitat Politecnica de Catalunya, Barcelona, Spain
FRANK DRAKE, Seti Institute, Mountain View USA
ANTONIO LAZCANO, Universidad Nacional Autónoma de México, México
ULISES MOULINES, Institut fuer Philosophie, Logik und Wissenschaftstheorie Ludwig-Maximilians-Universitaet Muenchen, Germany
RAFAEL NAVARRO-GONZALEZ, Universidad Nacional Autónoma de México, México
ALICIA NEGRON-MENDOZA, Universidad Nacional Autónoma de México, México
ADRIANA C. OCAMPO, NASA Headquarters, Washington D.C. USA
YAREMI RIVERO (Coordinator, visit of the astronaut), Lyndon Johnson Space Center, Houston, USA
JUAN G. ROEDERER, University of Alaska-Fairbanks, USA.

LOCAL COMMITTEE

GUSTAVO BRUZUAL, Centro de Investigaciones de Astronomía, Mérida
MARIO I. CAICEDO, Universidad Simón Bolívar, Caracas
ELINOR CALLAROTTI, Universidad Simón Bolívar, Caracas
GREGORIO DRAYER, Universidad Simón Bolívar, Caracas
MARTHA ELENA GALAVÍS, Universidad Metropolitana, Caracas
ERNESTO MAYZ VALLENILLA, CENIT, (IDEA), Caracas
CESAR MENDOZA BRICEÑO, Centro de Astrofísica Teórica, Mérida
HECTOR RAGO, Grupo de Física Teórica/Centro de Astrofísica, ULA, Mérida
TOMAS REVILLA, Escuela de Biología, Facultad de Ciencias, Universidad Central de Venezuela, Caracas
GLORIA VILLEGAS, Centro de Biociencias, (IDEA), Caracas.

SPONSORS

International Centre for Theoretical Physics
International Centre for Genetic Engineering and Biotechnology
Universidad Simon Bolivar
NASA Headquarters,
European Space Agency,
TALVEN Programme, (Delegacion Permanente de Venezuela ante la UNESCO)
The SETI Institute,
Centro Latinoamericano de Física,
The Third World Academy of Sciences,
Academia de Ciencias Físicas, Matemáticas y Naturales,
Red Latinoamericana de Biologia
The Planetary Society,
The Latin American Academy of Sciences (Fondo ACAL).
Alberto Vollmer Foundation, Inc
Fundación J. Oro, Associated to the Catalonian Research Foundation
Red Latinoamericana de Astronomia
Colegio Emil Friedman

ASTROBIOLOGY

ORIGINS FROM THE BIG-BANG TO CIVILISATION

Proceedings of the Iberoamerican School of Astrobiology
Caracas, Venezuela,
28 November- 8 December, 1999

Edited by

JULIÁN CHELA-FLORES,
The Abdus Salam International Centre for Theoretical Physics, Italy and Instituto de Estudios Avanzados (IDEA), Venezuela,

GUILLERMO A. LEMARCHAND,
Instituto Argentino de Radioastronomía (CONICET), and Centro de Estudios Avanzados, Universidad de Buenos Aires, Argentina

and

JOHN ORÓ,
Department of Biochemical and Biophysical Sciences University of Houston, Houston, TX 77204-5934, USA.

SPRINGER SCIENCE+BUSINESS MEDIA, B.V.

Library of Congress Cataloging-in-Publication Data

ISBN 978-94-010-5865-0 ISBN 978-94-011-4313-4 (eBook)
DOI 10.1007/978-94-011-4313-4

Printed on acid-free paper

Cover (illustration European Space Agency.): An artist's impression of "Huygens descend - stages on Titan". Cassini/Huygens is a NASA/ESA mission to the Saturnian System. The spacecraft consists of the NASA Saturn Orbiter and the detachable ESA Huygens Probe designed to explore the atmosphere of Titan, Saturn's largest moon. Within the context of chemical evolution and planetary science the mission, as well as Titan, were discussed during the School. The reader is referred to pp. xxii, 85-87, 158-159, 277 and 307.

Photograph on page (v): Curtsey of The Abdus Salam ICTP Archives.

Dedicated to

Frank Drake

on the occasion of his 70th birthday

FRANK DRAKE

A biographical sketch

Dr. Frank Drake is Chairman of the Board of Trustees of the SETI Institute and provides overall direction for research. In 1960, as a staff member of the National Radio Astronomy Observatory, he conducted the first radio search for extraterrestrial intelligence. The original efforts went under the name of Project Ozma after the Princess in L. Frank Baum's story "Ozma of Oz". Four decades later, with the support by data from many independent projects world-wide and the new evidence for the existence of other planetary systems, the search for extraterrestrial life is an integral part of the new science of astrobiology.

He is a member of the National Academy of Sciences where he chaired the Board of Physics and Astronomy of the National Research Council (1989-92). He is a former president of one of the world's leading astronomical organisations, the Astronomical Society of the Pacific. He was a Professor of Astronomy at Cornell University (1964-84) and served as the Director of the Arecibo Observatory. He is currently a Professor of Astronomy and Astrophysics at the University of California at Santa Cruz where he also served as Dean of Natural Sciences (1984-88).

The proposal of the School was made in 1998 to three institutions, which responded enthusiastically: The Abdus Salam International Centre for Theoretical Physics (ICTP), its main co-sponsor, the International Centre for Genetic Engineering and Biotechnology, both in Trieste, Italy, and the Chancellor's Office, Universidad Simón Bolívar (USB). The secretarial and logistic support was provided in Trieste by the ICTP and in Caracas by USB and the IDEA Convention Center.

In addition the event was generously supported by the following institutes, agencies, foundations and academies: NASA Headquarters, European Space Agency, TALVEN Programme, (Delegación Permanente de Venezuela ante la UNESCO), The SETI Institute, Centro Latinoamericano de Física, The Third World Academy of Sciences, Academia de Ciencias Físicas, Matemáticas y Naturales, Red Latinoamericana de Biología, The Planetary Society, The Latin American Academy of Sciences (Fondo ACAL), Alberto Vollmer Foundation, Inc, Fundación J. Oro, Associated to the Catalonian Research Foundation, Red Latinoamericana de Astronomía and Colegio Emil Friedman.

A total of 36 lectures were delivered by 20 lecturers, of which 14 were from the following countries: Argentina, Mexico, Italy, Spain and the USA. Six lecturers were from the host country. In addition there were 5 chairpersons from the host country that were not participants; two participants acted as chairpersons (Pedro Benítez and Tomás Revilla).

The School brought together 125 participants, which included some of the leading researchers in the subfields of Astrobiology. There were 18 participants form: Argentina, Colombia, Cuba, Mexico, Spain, Uruguay and the USA. The host country was represented by 82 participants registered during the conference, of which 33 had previously registered in Trieste. In addition to the lectures the participants contributed 16 oral presentations of their posters (ten minutes were assigned to each of the participants that presented posters).

The media was represented by 3 participants from the host country. The event was also consistently well represented by the national press. Another important activity of IASA was the following round-table: Music of the Spheres: Would other intelligence also exhibit "artistic creativity?". Juan G. Roederer and Guillermo A. Lemarchand acted as moderators with the participation of: Jacobo Borges, Diana Arismendi, Irene McKinstry de Guinand and Julián Chela-Flores.

Public lectures were delivered by Professor Frank Drake, Professor Juan Oró and Garrett E. Reisman a NASA Astronaut Candidate (Mission Specialist). An event that was particularly appreciated by the participants was a Special Session on Solar System Exploration in collaboration with The Planetary Society (TPS) and Dirección de Servicios Multimedia de la Universidad Simón Bolívar. It was timed so as to make a live Internet link-up with Planetfest in Pasadena, California on the occasion of the Mars lander in the Martian South Pole. The collaboration of Lic. Gregorio Drayer in the coordination of the event is gratefully acknowledged.

In spite of the fact that the lander was lost, the live contact with Planetfest and telephone connection for 45 minutes (due to the generous support of the Planetary Society) with an expert on Mars research (Dr. Christopher McKay) was a very instructive experience for all the School The counselling and very constructive initiatives of the TPS Executive Director, Dr. Louis Friedman were fundamental for this event.

In Caracas we are particularly grateful to Chancellor Freddy Malpica and the Chancellor's Office, without whose support this activity would not have taken place. Our particular thanks to Ms. Nancy Padilla, who acted as the Caracas School Secretary. Her helpful, friendly initiatives contributed significantly to the success of IASA. The whole team of Lic. Rebeca López de Álvarez, namely, Ms. Alicia de Armas, Beatriz Troconis and Mariana Walker, provided essential aspects towards the improvement of the School.

At the IDEA Convention Center, we would like to thank the team headed by Lic. Ramón Garriga; in particular, Mr. Luis González, with the support of Ms Dayana Barreto, acted efficiently as Administrator in Charge of IASA for the whole duration of the School organisation (two years).

In particular, we should highlight the generous and always helpful, critical advice and collaboration of Ms. Pilar Martínez, Secretary of the Latin American Academy of Sciences.

In Italy, special thanks are due to the ICTP administrative staff, particularly Ms. E. Brancaccio, who acted as the School Secretary in Trieste. We would like to acknowledge the valuable support of Mr. Andrej Michelcich, Financial Officer of ICTP, Ms. Manuela Vascotto and Alessandra Ricci for their continuous assistance and wise counselling in financial matters. Our special thanks go to Ms. Dilys Grilli for her generous preparation of part of the manuscript. Finally, we would like to thank Drs. Willem Wamstaker and Jean-Pierre Lebreton for their collaboration in the selection of the cover illustration.

JULIÁN CHELA-FLORES,
The Abdus Salam International Centre for Theoretical Physics, Italy and
Instituto de Estudios Avanzados (IDEA), Venezuela,

GUILLERMO A. LEMARCHAND,
Instituto Argentino de Radioastronomía (CONICET), and
Centro de Estudios Avanzados, Universidad de Buenos Aires, Argentina

and

JUAN ORÓ,
Department of Biochemical and Biophysical Sciences
University of Houston, Houston, TX 77204-5934, USA.

ESCUELA IBEROAMERICANA DE ASTROBIOLOGIA
CARACAS- VENEZUELA
DE NOVIEMBRE AL 8 DE DICIEMBRE DE 1.999

1. DELAYE, Luis
2. LEÓN, Jesús Alberto
3. CHELA-FLORES, Julián
4. FARMER, Josefina
5. ORO, Antonieta de
6. ORÓ, Juan
7. Participante
8. NAVARRO-GONZÁLEZ,Rafael
9. ISLAS GRACIANO, Sara Ernestina
10. GUZMÁN MARMOLEJO, Andrés
11. REVILLA, Tomás Augusto
12. NEGRÓN-MENDOZA, Alicia
13. ROEDERER, Juan
14 ROEDERER, Beatriz
15 DRAKE, Frank
16. ROMERO, Jesús Guillermo
17. SEGURA MOLINA, Antígona
18. DIAZ, Naryttza Namelly
19. HOENICKA, Janet
20. PEREZ CHAVEZ, Itzel
21. Participant
22. MONSALVE DAM, Dorixa D.
23. COLMENARES, Valentina
24. CHANG ROMERO, Ricardo
25. ROJAS, Diego Rafael
26. Colin-Garcia, Maria
27. Participant
28. Participant
29. Participant
30. Participant
31. RAMOS-BERNAL, Sergio
32. MÉNDEZ, Abel
33. CALVA-ALEJO, Leonel
34. ROMERO, Gustavo-Esteban
35. RAMIREZ JIMENEZ, Sandra
36. FALCON RODRIGUEZ, Jersys
37. SUÁREZ MEZA, Luis
38. SANCHEZ, Andrea Leticia
39. TORRES, Diego Fernando
40. GUTIERREZ-CABELLO, Jordi Luis
41. GOMEZ-CHABALA, Sandra María
42. Participant
43. DOPAZO, Hernán
44. BARRAL, José
45. MARTIN-LANDROVE, Miguel
46. MASSARINI, Alicia Isabel
47. RAMIREZ, César Ernesto
48. BENÍTEZ, Pedro
49. CHAURIO, Ricardo Alfredo
50. HERNANDEZ COLMENARES, Javier A.
51. MARCANO, Vicente
52. LAZCANO, Antonio

CONTENTS

A FEW WORDS OF WELCOME

JOHN ORÓ

Department of Biology and Biochemistry
University of Houston
Houston TX 77204-5934, USA

Good morning, distinguished Professor Dr. Ernesto Mayz Vallenilla, former president and founder of the Simon Bolivar University (USB), professor Freddy Malpica, current president of USB, distinguished academic authorities, ladies and gentlemen.

A most hearty welcome to all of you, authorities, sponsors, invited speakers, students and other persons that have come from all around the planet to attend and participate in the first **I**bero**A**merican **S**chool of **A**strobiology (IASA).

First of all I have to congratulate Professor Julian Chela-Flores from the Abdus Salam International Center for theoretical Physics, in Trieste, who has organized this magnificent educative congress called IASA. This scientific meeting is the first School of Astrobiology and promises to be one of the best meetings in this field. The participants are well known scientists from a dozen Spanish speaking countries, including several from North-America and Europe. We have the privilege to have among us Dr. Frank Drake, president of the SETI Institute, pioneer in the search for extraterrestrial intelligent life, and we are looking forward to his usual active participation in the meeting.

I also need to congratulate all the students that attend this meeting, and specially all the institutions that have made possible this important School of Astrobiology, here in Caracas, and its excellent organization by the Institute of Advanced Studies (IDEA) from the University Simon Bolivar.

In addition to the Centro Internacional de Fisica Teorica and the Centro Internacional de Ingenieria Genetica y Biotecnologia from Trieste, Italy, the sponsor institutions of IASA are the following:

Oficina del rector(Universidad Simon Bolivar), NASA Headquarters, European Space Agency, TALVEN Programme (Delegacion Permanente de Venezuela ante la UNESCO), The SETI Institute, Centro Latinoamericano de Fisica, The Third World Academy of Sciences, Academia de Ciencias Fisicas, Matematicas y Naturales, Red Latinoamericana de Biología, The Planetary Society, Oficina de Promoción y Mercadeo(IDEA Convention Center), The Latin American Academy of Sciences (Fondo ACAL), Alberto Vollmer Foundation, Escuela Emil Friedman, Fundación J. Oró, Red Latinoamericana de Astronomia.

It has been said that astrobiology, bioastronomy and exobiology are "sciences" that lack their own subject of study, I mean, that it has not been demonstrated yet that extraterrestrial life exists. Even though it can not be demonstrated mathematically like the relativity theory of Einstein, we can say like Dr. Paul Butler has recently said, echoing and extending the thoughts and statements of Plutarco and Giordano Bruno: "The Universe is too immense for us to be the only intelligent beings."

Briefly the ideas that will be presented in this School of Astrobiology include the following:

(1) Origins of the universe and cosmic evolution
(2) Organic matter, interstellar clouds, comets and meteors
(3) Origin of the solar system, meteorites, planetary atmospheres and catastrophic impacts
(4) Chemical evolution on Earth and Titan
(5) Origin and evolution of life and intelligent life
(6) Search for extraterrestrial civilizations
(7) Evolutionary theory. From the cenancestor to eucariotes and beyond
(8) The origin of neurons, the human brain and language
(9) The Galileo mission
(10) The possibility of life on Mars and Europe
(11) "Mars Express" and beyond.

At the end we will have the honor that an astronaut from Johnson Space Center of Houston, Garrett Reisman will deliver a lecture about "The future of human missions in the exploration of the solar system"

Of course there are a lot of things that we ignore. We hope that from the questions of students or other participants we will clarify or refine our knowledge. So, please ask as many questions as you want to all the speakers. We believe that the dialogue we will excite us and we will contribute in making a better "School" of IASA. It is with meetings like this and future research that we may be able to solve the most fundamental question of mankind "Are we alone in the Universe?"

Thanks for your attention!

CONTEMPORARY RADIO SEARCHES FOR EXTRATERRESTRIAL INTELLIGENCE

FRANK DRAKE

SETI Institute
Mountain View,
California, 94043, USA

For forty years now, radio searches have been conducted in efforts to detect extraterrestrial intelligent radio signals. They exploit the abilities of radio telescopes, as explained beautifully by Guillermo Lemarchand at this School. Although the motivations and abilities of extraterrestrial technologies are not predictable, we nonetheless consider, as we have for forty years, that it is most promising to search for signals at radio wavelengths, and in particular the "microwave" wavelengths of the order of 10-cm.

Why is this? It is not because we are particularly adept at microwave technology, or that we expect the extraterrestrials to be focused on microwaves for communication. Rather, it is because the physics and arrangement of the universe favor microwaves for interstellar communication. The levels of cosmic noise at all relevant electromagnetic wavelengths are now very well known from observation. At frequencies below about 1 Ghz, there is intense cosmic radio noise due to radiation from relativistic electrons orbiting in the magnetic fields of the interstellar medium. At frequencies between about 1 Ghz and 20 Ghz, the primary source of noise is the relic radiation from the Big Bang. This creates a constant radio brightness temperature, all over the sky, of about 2.8 K over this portion of the spectrum. At higher frequencies, the quantum nature of radiation and the randomicity of arrival times of received photons introduce noise into the power level of any signal that is received. This can be represented as an equivalent brightness temperature which exceeds that of the relic radiation, and grows linearly with frequency since the energy per photon increases linearly with frequency. An alternate way of expressing this limitation, should a detector sensitive to individual photons be used (these do not exist on Earth at radio frequencies, but do exist for infra-red and higher frequencies, of course) is that the minimum detectable signal, one photon, requires a minimum received energy which is proportional to frequency. Overall, then, there is a minimum in the noise of a well-designed detection system at microwave frequencies. The second law of thermodynamics states clearly that no technological device can circumvent this noise; thus it limits sensitivity no matter what the expertise of a civilization. Thus it is reasonable to expect that other civilizations will exploit these frequencies for communication channels in deep space, just as we do. Then it is most promising, but not inevitable, that a successful search will be carried out at the microwave frequencies.

Within the microwave band of low noise occur some fundamental lines of the H atom (1420 Mhz) and the OH radical (several lines, the strongest at 1665 and 1667 Mhz). Since these molecules join to make water, a clearly important and perhaps essential material of life, the frequency band containing these frequencies has been dubbed "the water hole" and is considered a prime frequency band for searches for extraterrestrial transmissions. To date, some 60 radio searches have been made since the first modern search, Project Ozma, of 1960. Almost all have been limited by instrumental limitations and choice to frequency bands in the water hole.

The basic SETI radio strategy has been to use radio telescopes of the greatest available collecting area with very sensitive multi-channel receivers. The larger the collecting area, the weaker the signal which can be detected. Indeed, the volume of space from which signals of a given intrinsic power can be detected is just proportional to the diameter of a circular antenna having the same collecting area as the actual collecting area. Multi-channel receivers are desirable because the channel used by the extraterrestrials is totally unknown. The more narrow the transmission, with a given radiated power, the easier it is to detect. This makes it wise to use very narrow channels, in turn demanding that many channels be examined at once if any substantial portion of the total search spectrum is to be covered in a search. We do know that there is a minimum bandwidth to any interstellar signal, a result of multipath propagation through the chaotic electron clouds of the Milky Way. This minimum bandwidth depends on the radio frequency and distance to the source. It is typically about 0.1 Hertz in the Water Hole for sources at a distance of a kiloparsec or so. In order to make progress, most searches make a compromise between minimum bandwidth in the receiving system and total spectral coverage. A typical search bandwidth is about 1 Hertz.

With bandwidths of the order of 1 Hertz, very many channels are called for to provide any substantial overall frequency coverage in a reasonable time. Until recently this was an insurmountable problem. However, the development of low cost computers, and, in particular, special "digital signal processing" computer chips have made possible the construction at affordable cost, perhaps $100,000, of receivers which may monitor as many as 100 million channels at once. This is a remarkable accomplishment, but creates a new challenge, which is to sort through the flood of data from such a system for evidence of intelligent signals. A typical system may be providing 100 million data points per second. To make the challenge even more difficult, the signals may exist in several channels simultaneously, may be pulsed periodically, as is typical of radar, and be drifting in frequency due to a changing Doppler effect or the design of the transmitter system. Other variations are possible. Again, this challenge can be and has been met through the application of special affordable computer systems and, particularly, special algorithms which search for a variety of signal types in the data. For example, the system of Project Phoenix, of the SETI Institute, can detect all the forms of signal just mentioned.

To add a still further challenge, and perhaps the most difficult one to deal with, is radio frequency interference, "RFI", from our own transmitters. All radio telescopes have some sensitivity in all directions; if there is a strong enough signal source visible to the telescope from any direction, the telescope will receive the signal, and it will be identified as an intelligent signal. Our civilization is a very prolific source of radio signals. In the entire developed world there are always present television signals which are detectable by the very sensitive SETI systems, even though the TV signal is so weak as to be unusable by a typical TV set. There may be tens of such signals detectable, even at remote sites. Aircraft provide large number of signals, both through their communication systems but also as a result of the sophisticated air traffic control systems now found throughout the world. There are many other sources of signals, such as cellular phones. Even at the most remote site, a host of strong signals are received from a huge number of transmitters on satellites. Most of these signals are in the Water Hole. The overall result is that every radio spectrum recorded by SETI systems contains many, perhaps fifty or more, human-created signals.

There is no simple solution to this RFI problem. One very simple but powerful step is to keep a catalog of detected signals; when the same signal is seen with the telescope looking in more than one place, it is an interfering signal. Some systems, such as the one at Harvard University, use two feed horns on the telescope, creating two beams. A true extraterrestrial signal will appear in first one beam and then the other, with a precise time delay. This can be used as a criterion to identify an extraterrestrial signal and reject RFI. Other projects, including the Harvard project and Project SERENDIP and its subsidiary, seti@home, observe the same point in the sky and same frequency at quite separate times. If the same signal appears twice or more, it is considered a prime candidate to be a true extraterrestrial signal, and follow-up observations are made.

In the case of Project Phoenix, a more definitive but expensive approach is used. A second telescope, perhaps hundreds or thousands of kilometers from the main search instrument, is fitted with a sensitive receiver and a minimal multi-channel receiver. Any detected signals at the main telescope are examined for a drift in frequency with time as is to be expected from changing Doppler effect, itself a result of the changing velocity component of the telescope along the line-of-site due to the rotation of the Earth. If the easily calculated expected frequency drift is detected, the information about the candidate signal is relayed to the second telescope, and it searches for the signal. If it is not seen, it means the signal was local RFI which mimicked a true signal. If the signal is seen, the second telescope should observe a different radio frequency and frequency drift, which can be precisely calculated. Because of its separation from the main telescope, the velocity component of the second telescope and its derivative will be different. This offers an immediate and conclusive test for the origin of the signal. It even works well with signals from distant spacecraft in the solar system, as has been demonstrated several times. Because of the power of this approach, Project Phoenix has identified the source of all signals it has ever detected. None have been of extraterrestrial intelligent origin.

Recent studies of the next desirable steps in radio SETI have recognized that the greatest weakness in current SETI programs has been the lack of on-going, nearly full-time, access to very large radio telescopes. Project Phoenix, for example, uses the world's largest Arecibo telescope, but only twenty days a year are available to the project. This is very inefficient, and the inefficiency is enhanced by the need to move personnel frequently, restore equipment to proper working order after a long shut down, and to cope with changes in the telescope itself between observing periods. Other projects use only small telescopes, or have to cede control of the telescope pointing to other observers most of the time, in a compromise in which the SETI observers observe in a "parasitic" mode. A further concern is that true SETI signals may be transient, and success in the enterprise may require frequent, or almost constant, monitoring, of many places in the sky and many radio frequencies.

Both of these concerns call for the creation of very large radio telescopes dedicated to the SETI enterprise. Until recently, this was thought impossible because the large funding required, of the order of $100 million, seemed far beyond what might be available. Now there is hope in sight. The many discoveries supporting the idea that there are many habitable planets in the universe, and the improvement in SETI systems, have greatly increased interest in SETI by potential funding sources. Furthermore, the same improvement in computers and decreases in costs which facilitated the solutions to the multi-channel and RFI problems have made the construction of large telescope collecting areas possible at acceptable cost.

This has led to two major projects. The first is the "1hT", or One Hectare Telescope, of the SETI Institute. In this project a radio telescope with an energy collecting area of one hectare (10,000 square meters) will be constructed by utilizing a close-packed array of between 500 and 1000 small antennas, each three to five meters in size. These will be interconnected using fibre optics and digital time delays, all controlled by sophisticated computers, to produce the same effective collecting area as a single one hectare telescope. Not only will this reduce the price to something like $25 million, but also the system offers other important advantages. By using many time delays, it is possible to synthesize a large number of beams on the sky simultaneously, allowing SETI searches to proceed much more quickly, as long as the multi-channel receivers necessary for each beam are available. This capability also makes possible simultaneous SETI and conventional astronomy observations, allowing full-time use by both SETI scientists and those working on such objects as pulsars, quasars, etc. This project is now under development, and it is expected that the 1hT will go into operation in about 2005. It may turn out to be a prototype for a much large instrument, the "One Square Kilometer Array" which is being advocated as the next big step in radio astronomy by radio astronomers worldwide.

The second project is to build a telescope which looks at all the visible sky on all the frequencies of the water hole at all times. The basic principle here is simple. One uses an array of an enormous number of nearly isotropic, therefore very small, antennas. The power collected by these antennas is added together with different time delays to create beams in all possible directions in the sky, with all the frequency information being retained. A huge number of broad-band spectrum analyzers then study the power captured by each of the multitude of synthesized beams. Although simple in principle, the actual system depends for success entirely on enormous computer power. The present estimates is that a computer system which can carry out about ten to the twentieth power calculations per second is required. This is presently beyond both our technical and financial capability. However, if the increases in computer power and decreases in cost follow their historical trends, that is, following "Moore's Law", the required capability at an affordable cost should be available in perhaps a decade. Because of this expected delay in feasibility, for now only preliminary studies are being carried out.

Because of the prospects for major advances, as represented by these new projects, there is a great deal of excitement in the SETI community.

Section 1:
Introduction to Astrobiology

ORIGINS
from the Big-Bang to Civilisation

JULIAN CHELA-FLORES

The Abdus Salam International Centre for Theoretical Physics (ICTP), Office 276, P.O.Box 586; Strada Costiera 11; 34136 Trieste, Italy
and
Instituto de Estudios Avanzados, Apartado 17606, Parque Central, Caracas 1015A, Venezuela.

1. The evolution of the cosmos and life

We may divide astrobiology, which is one of the most remarkable branches of biology, into four parts, to emphasise distinct aspects of the subject: Parts 1 and 2 correspond to the origin and evolution of life in the universe. They are supported by the fields of chemical evolution and Darwin's theory of evolution, respectively. The third aspect of astrobiology, the distribution of life in the universe, still lacks an underlying theory and requires a strong component of the space sciences and missions specific to solar system exploration. The fourth aspect of astrobiology-the destiny of life in the universe-has a common frontier with the time-honoured disciplines of philosophy and theology. They will be discussed separately and briefly in our second lecture. We begin with a review of the space sciences needed for the School.

The American scientist Edwin Hubble discovered in 1929 that the velocity of recession of a galaxy is proportional to its distance from us. (This is known as Hubble's Law.) The constant of proportionality, which is known as the Hubble constant H_0 is, conseqxi, e speed of recession of the galaxy and its distance; H_0 represents quantitatively the current rate of expansion of the universe. At present the value of H_0 in the standard cosmological model (cf., the Friedmann model below) of an expanding universe implies an age of 9-14 thousand million years (Gyr), depending on the particular assumption we may adopt for the matter density present in the universe.

The theory of gravitation formulated by Albert Einstein is known as General Relativity (GR). During the School Hector Rago will discuss the cosmological implications of GR. I will be brief. Cosmological models may be discussed within the context of this theory in terms of a single function R of time t. This function may be referred to as a 'scale factor'. Sometimes, when referring to the particular solution the expression 'radius of the universe' may be preferred for the function R. As the universal expansion sets in, R is found to increase in a model that assumes homogeneity in the distribution of matter (the 'substratum'), as well as isotropy of space.

The dependence of R, as a function of time t, is a smooth increasing function for a specific choice of two free parameters, which have a deep meaning in the GR theory of gravitation, namely, the curvature of space and the cosmological constant. The functional behaviour of the scale factor R was found by the Russian mathematician

J. Chela-Flores et al. (eds.), Astrobiology, 3–12.

Alexander Friedmann in 1922. This solution is also attributed to Howard Robertson and A.G. Walker for their work done in the 1930s. The model is referred to as the Friedmann model. (R is inversely proportional to the substratum temperature T.)

Hence, since R is also found to increase with time t (cf., the previous paragraph), T decreases; this model implies, therefore, that as t tends to zero (the 'zero' of time) the value of the temperature T is large. (The temperature goes to infinity as T tends to zero.) In other words, the Friedmann solution suggests that there was a 'hot' initial condition. As the function R represents a scale of the universe (in the sense we have just explained), the expression 'big bang', due to Sir Fred Hoyle, has been adopted for the Friedmann model. The almost universal acceptance of big bang cosmology is due to its experimental support. The model tells us that as time t increases the universe cools down to a certain temperature, which at present is close to 3° K. The work was performed during the 1960s. It provided solid evidence for the "T = 3° K" radiation, which is a relic from the big bang.

We will mainly attempt to convey the idea of an ongoing transformation in origin of life studies. This progress in understanding our own origins began about three decades ago, triggered by the success in retrieving some key biomolecules in experiments which attempted to simulate prebiotic conditions. Some of the main experiments of the 1950s and early 60s were done by organic chemists. Since that time the field has continued its robust growth. These efforts have led to view the cosmos as a matrix in which organic matter can be inexorably self-organised by the laws of physics and chemistry into what we recognise as living organisms. However, in this context it should be stressed that chemical evolution experiments have been unable to reproduce the complete pathway from inanimate to living matter. In particular, the prebiotic synthesis of all the RNA bases is not clear and cytosine, for instance, may not be prebiotic and it may even have been imported from space.

Thus, at present the physical and chemical bases of life that we have sketched are persuasive, but further research is still needed. Originally the subject began to take shape as a scientific discipline in the early 1920s, when an organist chemist, Alexander Oparin applied the scientific method of conjecture and experiments to the origin of the first cell, thereby allowing scientific enquiry to shed valuable new light on a subject that has traditionally been the focus of philosophy and theology.

Darwin's insights offer some explanation for the existence of life on Earth, and allows for meaningful questions regarding early terrestrial evolution. Jesus Alberto Leon, Hernan Dopazo and Alicia Massarini will discuss this aspect of astrobiology in some detail. With the enormous scope of bacterial evolutionary data available today from an extensive micropaleontological record, Darwin's *Theory of Common Descent* leads back to a single common ancestor, a progenote or "cenancestor", as you will learn form Antonio Lazcano. The characteristics of the cenancestor can be studied today through comparison of macromolecules, allowing researchers to distinguish and recognise early events that led to the divergent genesis of the highest taxa (domains) among microorganisms. In this School we shall discuss certain concepts that may have played a relevant role in the pathway that led to the origin and evolution of the cenancestor. We have argued in favour of the search for extraterrestrial single-celled organisms of a nucleated type. (The word of Greek origin 'eukaryote' is normally assigned to such nucleated cells.) Finally, at the other extreme we have the search for extraterrestrial intelligence (SETI), a time-honoured subject that was pioneered by Frank

Drake, as we have already learned form him in the previous lecture. SETI and its implications will be reviewed by Guillermo Lemarchand in subsequent lectures. To continue we must first return to cosmological models.

In less than one million years after the beginning of the general expansion, the temperature T was already sufficiently low for electrons and protons to be able to form hydrogen atoms. Up to that moment these elementary particles were too energetic to allow atoms to be formed. Once 'recombination' of electrons and protons was possible, due to falling temperatures, thermal motion was no longer able to prevent the electromagnetic interaction from forming hydrogen atoms. This is the 'moment of decoupling' of matter and radiation. At this stage of universal expansion the force of gravity was able to induce the hydrogen gas to coalesce into stars and galaxies.

A series of nuclear reactions in the interior of stars was proposed by Hans Bethe. His aim was to understand the nuclear reactions that are responsible for nucleosynthesis. The underlying phenomenon consists of high energy collisions between atomic nuclei and elementary particles that have been stripped off their corresponding atoms, or even nuclei, due to the presence of the enormous thermal energy in the core of the star. At such high temperatures nuclear fusion may occur; in other words, there can occur nuclear reactions between light atomic nuclei with the release of energy.

In the interior of stars reactions are called thermonuclear when they involve nuclear fusions, in which the reacting bodies have sufficient (kinetic) energy to initiate and sustain the process. One example is provided by a series of nuclear reactions that induce hydrogen nuclei (essentially single 'elementary' particles called protons) to fuse into helium nuclei. A helium nuclei is heavier than the proton; in fact, it consists of two protons and two more massive particles called neutrons.

This process, in addition, releases other particles and some energy. After a long phase (measured in millions of years) dominated by such conversion, or 'burning' of hydrogen into helium, the star evolves: its structure becomes gradually that of a small core, where helium accumulates. To maintain the pressure balance with the gravitational force, both temperature and density of the core increase. The star itself increases in size. It becomes what is normally called a "red giant", because at that stage of their evolution they are changed into a state of high luminosity and red colour. During this long process, from a young star to a red giant, the elements carbon and oxygen, and several others, are formed by fusing helium atoms.

2. The evolution of matter

Stars whose mass is similar to that of the Sun remain at the red giant stage for a few hundred million years. The last stages of burning produce an interesting anomaly: the star pushes off its outer layers forming a large shell of gas; in fact, the shell is much larger than the star itself. This structure is called a planetary nebula. The star itself collapses under its own gravity compressing its matter to a degenerate state, in which the laws of microscopic physics (called quantum mechanics), eventually stabilise the collapse. This is the stage of stellar evolution called a 'white dwarf '.

After the massive star has burnt out its nuclear fuel (in the previous process of nucleosynthesis of most of the lighter elements), a catastrophic explosion follows in which an enormous amount of energy and matter is released.

It is precisely these 'supernovae' explosions that are the source of enrichment of the chemical composition of the interstellar medium. This chemical phenomenon, in turn, provides new raw material for subsequent generations of star formation.

Late in stellar evolution stars are still poor in some of the heavier biogenic elements (such as, for instance, magnesium and phosphorus). Such elements are the product of nucleosynthesis triggered in the extreme physical conditions that are due to the supernova event itself. By this means the newly synthesised elements are disseminated into interstellar space, becoming dust particles after a few generations of stars births and deaths.

As a product of several generations of stellar evolution, a large fraction of the gas within our galaxy is found in the form of clouds of molecular hydrogen, but also many of the heavier elements are present too. The liner dimension of these clouds can be as much as several hundred light years. The mass involved may be something in the range 10^5 to 10^6 solar masses. Images from the orbital Hubble Space Telescope (HST) have shown circumstellar disks surrounding young stars. This supports the old theory of planetary formation from a primeval nebula surrounding the nascent star. In the case of our solar system this gas formation has been called the 'solar nebula'. We have already seen that stars evolve as nuclear reactions convert mass to energy. In fact, stars such as our Sun follow a well known pathway (the *main sequence)* along a *Hertzprung-Russell* (HR) diagram. This diagram was introduced independently by Ejnar Hertzprung and Henry Russell .

They observed many nearby stars, which exhibit a certain regularity; in other words, the stars lie on the same curve. Such stars are called *main sequence* stars. This regularity is evident when we plot luminosity (the total energy of visual light radiated by the star per second) versus its surface temperature or, alternatively, its spectroscopic type. We may ask: *How do stars move on the HR diagram as hydrogen is burnt?* Extensive calculations show that main sequence stars are funnelled into the upper right hand of the HR diagram, where we find red giants of radii that may be 10 to 100 times the Sun radius. Stellar evolution puts a significant constraint on the future of life on Earth, since the radius of the Earth orbit is small. (Since the eccentricity of the Earth orbit is small, we may speak to a good approximation of a circular orbit, instead of an elliptical one.) This brief presentation of astronomy shall be expanded by Willem Wamsteker. We shall continue with some comments on the chemistry of the cosmos.

3. Organic cosmochemistry

The chemistry of the circumstellar zones sets the stage for prebiotic evolution. In other words, a few molecules are sufficient for the synthesis of the key biomolecules of life, namely, the amino acids, the bases, sugars.

By looking at other solar systems in the process of formation, we are led to conjecture what was the nature of our own solar system. It is reasonable to assume that the solar system was formed out of a disk-shaped cloud of gas and dust, which we called the solar nebula. Most of the original matter of which the solar nebula was made of has since been incorporated into its planets and the central star itself, namely, the Sun. The nature of the interstellar dust may be appreciated as the product of condensation of metallic elements, which were themselves produced in stellar interiors (for example, magnesium, silicon and iron). Later, in interstellar space they combined with elements such as oxygen to form particles measuring typically 0.1 microns.

A very significant clue as to the nature of the original solar nebula can still be inferred from the study of comets. In fact, since these small bodies remained far from the star its chemical evolution was insignificant, they remained unchanged. We have to wait for the Rosetta mission to retrieve a piece of a comet early in the first decade of the 21st century. We shall return to this mission again below. Yet, as comets pass in the vicinity of the Sun and the Earth, dust particles are released as the comets heats up. Some of these particles may be retrieved in the next few years by the "Stardust" mission. Such interplanetary dust particles (IDPs) are a major component of our galaxy, including the solar system. In the early 60s Juan Oro was responsible for the conjecture that most of the volatiles on Earth (substances with low boiling point) had been delivered by comets-the most striking example being the water of the oceans. We shall learn form his lecture on "Cosmochemical Evolution" about this currently widely accepted general aspect of the cosmos.

Given the importance of the origin and evolution of the biomolecules, we feel that it is appropriate to comment on the origin of some of the most important biomolecules. They may have been synthesised in the early Earth, or transported here from space. There are also many indications that the origin of life on Earth may not exclude a strong component of extraterrestrial inventories of the precursor molecules that gave rise to the major biomolecules. About 98% of all matter in the universe is made of hydrogen and helium. Besides hydrogen, other biogenic elements, namely, C, N, O, S and P, make up about 1% of the cosmic matter. In fact, some of the more refractory material (i.e., that has gone through chemical reactions at higher temperatures) has remained in the inner solar nebula and have condensed in the form of meteorites, which are called carbonaceous chondrites. These testimonies of the nature of the solar nebula also contain some of the biogenic elements.

The abundance of biogenic elements would suggest that the major part of the molecules in the universe would be based on carbon ('organic'); in fact, out of over a hundred molecules that have been detected, either by microwave or infrared spectroscopy, 75% are organic. Once again, chemical evolution experiments fare well in comparison with the observation of the interstellar medium. Some of the identified molecular species detected by means of radio astronomy are precisely the same as those shown in the laboratory to be precursor biomolecules. Alicia Negron Mendoza and co-workers will discuss the subject of chemical evolution on Earth, whereas Rafael Navarro Gonzalez and co-workers will cover what has been acheived in this field to understand the possible chemical evolution in the solar system, including the interesting cases of Titan and Mars.

The Red Planet, in particular, will play a prominent role in the School, since at this precise time NASA is attempting to land a probe on the Martian South Pole. Willem Wanteker, who is part of the ESA "Mars Express Mission", will help us to put into perspective the future of the exploration of Mars.

4. The origin and evolution of the solar system

Before reviewing chemical evolution itself we should first understand current ideas of how the Earth itself originated. The origin of the solar system goes back to about 4.5 Gyr BP by the gravitational collapse of the solar nebula when a certain critical mass was reached.

Thus, the protosun, the protoplanets, the comets, the parent bodies of meteorites, and other planetesimal bodies were formed as the result of this condensation of interstellar matter. The lunar cratering record demonstrates that during its initial stages, the solar system may have been in a chaotic state with frequent collisions of planetesimals amongst themselves as well as with other larger bodies, including the protoplanets. The composition of the solar nebula must have been analogous to that of the interstellar clouds, namely it may have consisted of hydrogen, helium, as well as carbon compounds. This has been confirmed by astronomical studies of Jupiter, Saturn and their satellites.

To get an deeper insight into the origins of the solar system we need to consider its small objects, a subject that will be covered in the lectures of Humberto Campins. I will restrict myself to comparisons between the densities of carbonaceous chondrites and the rocky planets, or satellites; we know from such comparisons that the terrestrial planets were not formed by a slow process of gradual accumulation of interstellar dust particles. that may have fallen in the solar nebula. Instead, the outline of what may have separated the constituent globules ('chondrules') observed in the carbonaceous chondrites from gases in the nebula is based on simple physical considerations are relics of the accretion episode in the outer nebular disk.

Thus form the combined information provided by the small objects of the inner solar system (chondrites) and messengers form the outer solar system (comets), we can appreciate an outline of the main steps in the formation of the terrestrial planets: We expect heat at the centre of the solar nebula and low temperatures at its periphery. This is today an empirical observation for we can observe in our galaxy nebulae where similar processes of star formation are occurring today. As the thin interstellar medium gets concentrated in some regions in space it forms a protonebula, which eventually collapses onto itself producing a hot interior. The collapse is coupled with rotation of the gas.

Further out form the inner region in the solar nebula, which will eventually give rise to the terrestrial planets, gas is more abundant than IDPs, and cool enough to allow the existence of water ice, which would be more abundant by volume than the dust grains. Without going very deep into the details, we can already appreciate why the terrestrial planets are more dense than the Jovian planets. Besides, since at a distance of about 5-10 AU the temperature was not low enough to discriminate gases form dust, the composition of Jupiter and Saturn, which were formed at these distances, are expected to reflect the composition of the original solar nebula.

5. From chemical to cellular evolution in the solar system

We have only sketched two of the major steps towards life during chemical evolution on Earth. These steps should have taken place from 4.6 - 3.9 Gyr BP, the preliminary interval of geologic time which is known as the Hadean Subera. It should be noticed that impacts by large asteroids in the early Earth do not necessarily exclude the possibility that the period of chemical evolution may have been considerably shorter. Indeed, it should not be ruled out that the Earth may have been continuously habitable by non-photosynthetic ecosystems from a very remote date, possibly over 4 Gyr BP.

The content and the ratios of the two long-lived isotopes of reduced organic carbon in some of the earliest sediments (retrieved from the Isua peninsula, Greenland, some 3.8 Gyr BP), may convey a signal of biological carbon fixation. This reinforces the expectation that chemical evolution may have occurred in a brief fraction of the Hadean Subera, in spite of the considerable destructive potential of large asteroid impacts which took place during the same geologic interval in all the terrestrial planets, the so-called *heavy bombardment period.* In subsequent suberas of the Archean (3.9 - 2.5 Gyr BP) life, as we know it, was present. This is well represented by fossils of the domain Bacteria, which is well documented by many species of cyanobacteria.

In spite of the fundamental work of Darwin that gave rise to modern biology, earlier in the 20th century the fact remained that the Earth biota was still being divided into animals and plants. It was only in the 1930s when taxonomy shifted its emphasis form the multicellular dominated classification to one more oriented towards basic cellular structure. The encapsulation of chromosomes in nuclei was clearly absent in bacteria. This remark led to a division of all living organisms into two groups that went beyond the animal/plant dichotomy.

However, the tree of life could not yet be constructed, since amongst bacteria it was remarked that rapid and random exchange of genes occur. This is a phenomenon called horizontal gene transfer (HGT). This randomness deterred the use of sequences of the biomolecules, in order to construct a tree of life (phylogenetic tree). Yet, Linus Pauling and Emil Zuckerkandl had pointed out that a molecular clock might be identified form the slow mutation rates of some biomolecules.

The question then is to identify the molecule that would not be affected by HGT, as evolution proceeds. Such a molecules are a form of RNA that together with some protein constituents make up ribosomes, the small corpuscules of the cell where proteins are synthesised. This form of RNA is called ribosomal RNA (rRNA). Extensive work with these molecular chronometers have led to a taxonomy of all life on Earth that is deeper than the dichotomy prokaryote/eukaryote.

A classification in which the highest taxa are called domains, instead of Kingdoms, was put forward. In this approach there are three 'branches' in the tree of life:

- Archea, whose cellular membrane differs from the structure of the lipid bilayer of Bacteria and Eucarya.
- Bacteria, encompassing all bacteria.
- Eucarya, including all the truly nucleated cells.

Many 'extremophilic' microorganisms are known: hyperthermophilic archaea have been isolated from deep-sea hydrothermal vents. Besides, today hydrogen sulphide and methane are abundant in the 'black smokers' of the seafloor (fluids of volcanic origin). Chemothrops are microorganisms that obtain the energy they require independent of photosynthesis to fix carbon dioxide and to produce the organic matter that they need. Such environments, the seafloor and deep underground, and microorganisms that thrive in the most extreme conditions, as we have sketched in this section, are likely candidates for the dawn of cellular life on Earth. Several lines of research suggest the absence of current values of oxygen, O_2, for a major part of the history of the Earth. Some arguments militate in favour of Archean atmospheres with values of the partial pressure of atmospheric oxygen O_2 about 10^{-12} of the present atmospheric level (PAL). We have already seen in the Introduction that the growth of atmospheric oxygen was due to the evolutionary success of cyanobacteria, which were able to extract the hydrogen they needed for their photosynthesis directly form water.

One of the chief indicators of the growth of atmospheric oxygen is shale, which is a rock that has played a role in our understanding of biological evolution. The onset of atmospheric oxygen is demonstrated by the presence in the geologic record of red shale, coloured by ferric oxide. The age of such 'red beds' is estimated to be about 2 Gyr. At that time oxygen levels may have reached 1-2% PAL, sufficient for the development of a moderate ozone (O_2) protection from ultraviolet (UV) radiation for microorganisms from the Proterozoic. In fact, UV radiation is able to split the O_2 molecule into the unstable O-atom, which, in turn, reacts with O_2 to produce O_3, which is known to be an efficient filter for the UV radiation.

The paleontological record suggests that the origin of eukaryotes occurred earlier than 1.5 Gyr BP. There are Archean rock formations (which may be found up to 2 Gyr BP) that are significant in the evolution of life. Detailed arguments based on geological evidence supports the conclusion that we had to wait until about 2 Gyr BP for a substantial presence of free O_2. Once the eukaryotes enter the fossil record, its organisation into multicellular organisms followed in a relatively short period (in a geological time scale).

Inevitable random factors driving evolution of life on Earth were the mass extinctions that occurred sporadically in the past. These were due to large impacts on Earth by comets and meteorites. The classic example is the collision in the Yucatan Peninsula, Mexico, at the end of the Cretaceous some 65 million years before the present. It is generally agreed that it was an important factor in the extinction of the dinosaurs.

6. Pathways towards intelligence in the cosmos

6.1. EVOLUTION OF INTELLIGENCE

What is significant for astrobiology is to recognise that natural selection necessarily seeks solutions for the adaptation of evolving organisms to a relatively limited number of possible environments. We have seen in cosmochemistry that the elements used by the macromolecules of life are ubiquitous in the cosmos. We have seen that the formation of solar systems are limited by a set of physical phenomena, which repeat themselves in the Orion nebula; those Jovian planets discovered in our 'cosmic village' can bear satellites in which conditions similar to those in Europa may be replicated.

To sum up, the finite number of environments forces upon natural selection a limited number of options for the evolution of organisms. From these remarks we expect convergent evolution to occur repeatedly, wherever life arises. It will make sense, therefore, to search for the analogues of the attributes that we have learnt to recognise in our own particular planet.We may argue that not only life is a natural consequence of the laws of physics and chemistry, but once the living process has started, then the cellular plans, or blueprints, are also of universal validity: the lowest cellular blueprint (prokaryotic) will lead to the more complex cellular blueprint (eukaryotic). This is a testable hypothesis. Within a decade or two a new generation of space missions may be operational. Juan Roederer will mention some of the important missions that are being planned, or that are currently in operation and in operation by NASA, the American Space Agency. Willem Wamsteker will present the corresponding programme of ESA, the European Space Agency. Some further missions are currently in their planning stages, which are aiming to reach the Jovian satellite Europa in the second decade of next century. Closely related to the above hypothesis (the proposed universality of eukaryogenesis), are different possible conjectures regarding the question of extraterrestrial life: *Is it reasonable to search for Earth-like organisms, such as a eukaryote, or should we be looking for something totally different?*

Intelligence, as we understand it, had to await the emergence of the *Homo* genus some 2 Myr BP, when its first traces arose in our ancestors. A more evident demonstration of the appearance of intelligent life on Earth than the habilines' tools, or the ceremonial burials had to wait till the Magdalenian 'culture'. This group of human beings flourished from about 20,000 to 11,000 years before the present (yr BP). The Magdalenians left some fine works of primitive art as, for instance the 20,000 year old paintings on the walls of caves discovered in various places in Europe.

The problem of communication amongst intelligent beings from different stellar environments has eventually to be faced in truly scientific terms. We consider to the two evolutionary pressures, terrestrial and aquatic, that have given rise to the largely different brains of dolphins and humans (although some general features are common to both of them, such as the modular arrangement of neurons). The intelligence of humans and dolphins are products of a pathway that began from the general prokaryotic blueprint appearing on Earth almost instantaneously (in a geologic time scale), and continued through the sequence of eukaryogenesis, neuron, multicellularity and, finally, brains. The present discussion of universal aspects of communication amongst humans at some point has to be extended to the whole forest of life, and should not remain confined to the species of our single (phylogenetic) tree that we know where cognitive ability has evolved. The presence of extraterrestrial planets, a subject that will be mentioned by Sabatino Sofia, enhances he number of environments where alternative trees of life may develop independently of the single example that is known to us.

6.2. THE ORIGIN OF BRAIN, LANGUAGE AND CIVILIZATION

We have discussed the transition form a simple prokaryotic blueprint of the Archean to the present eukaryotic world. We attempted to show that this transition was probably the most trascendental step in the pathway that led from bacteria to Man. Once eukaryogenesis took place on Earth , the steps leading up to multicellularity, namely, cell signalling and the organisation into cooperative assemblies (tissues) were inevitable. The onset of multicellularity is essentially due to the considerably larger genomes that were compatible with the eukaryotic blueprint.

The densely-packaged chromosomes in the cellular nucleus presented multiple options for opportunistic ways of passing genes to progeny, some of which allowed their carriers to be better adapted to the environment. Mitosis was a more advanced process of cellular division than simple prokaryotic fission. Such variety of options were the raw material for natural selection to improve upon the three billion year old single-cell strategy of life on Earth. The result was that within 30% of the single-cell 'era' a full-organism strategy led to large-brained organisms and intelligence. Juan Roederer will help us to put into perspective the general question of the human brain with what we know about other areas of science.

On the other hand, Raimundo Villegas and Ernesto Palacios Pru with their background form the neurosciences will offer us some reflections on the origin of the nervous system. The first neuron could not arise at the prokaryotic level of develop -

ment. Due to the complex pattern of gene expression that is required for a functional neuron, the first stage of a nervous system had to await for the eukaryotic threshold to be crossed. The superior strategy of translation of the genetic message of nucleated cells permitted a variety of proteins to be inserted into the cellular membrane. This new stage in evolution provided all the channels necessary for the underlying alteration of the ion concentration inside and outside the cell. Indeed, early in the evolution of the eukaryotes we can conjecture that electrochemical imbalances were already being created by a disparity of electric charges on the two sides of the cellular membrane.The question of the origin of the first neuron is of fundamental importance to astrobiology. In particular, it is relevant to the search of other civilisations. It is a difficult subject, one of the reasons being that brains do not fossilise. The origin of language is not only a difficult problem form the point of view of embryological development (the human brain has over 10^{13} neurons), but also the historical perspective is difficult to assess. In our planet natural selection seems to have produced universal characteristics in the first species that has reached cognitive ability. Elinor Callarotti will expand on what I say here, but I need to introduce some concepts that will serve to set the stage for some questions relevant to bioastronomy. Steven Pinker in *The Language Instinct* argues that language is compatible with gradual evolution due to natural selection. Pinker argues his case with a large amount of evidence in favour of a genetic basis for spoken language. This possibility should be seen against the background of the development of a science of language. Noam Chomsky argued that language has underlying structural similarities. Hence, one should conclude that we must be born with knowledge of an innate Universal Grammar (UG). He supplied linguistics with a philosophical foundation (rationalism rather than empiricism). According to Chomsky the structure of language is fixed in the form of innately specified rules: during the process of development, all a child has to do is to turn on a few 'biological switches' to become a fluent speaker of a given language. In his view, children are not learning at all, anymore than birds learn their feathers . Yet this point of view has turned out to be controversial.

Pinker argues convincingly in favour of the adaptationist interpretation of language origin, the 'imperative' by-product of natural selection, rather than a UG. The outline of his argument is as follows: the abrupt increase in brain size may have been due to the evolution of language, instead of language arising only as brain size goes beyond a certain threshold. This faculty of the brain is to be searched as a consequence of natural selection, as any other attribute that characterises a human being (size, skin colour). Language in this evolutionary context, evolved through social interaction in our hominid ancestors, as they adopted collective hunting habits. Better communication by means of rudimentary language would have been favoured by natural selection.

Clearly, if we are not alone in the universe, there are some unavoidable theological and philosophical consequences for which we still have no answer. However, we are convinced that discovering extraterrestrial life would induce a fruitful dialogue between different sectors of culture. For this reason we feel that the problem of extraterrestrial life is an important question. Subjects that are not normally relevant to the scientific discourse are needed within the context of astrobiology. Ernesto Mayz Vallenilla and I will attempt to cover some topics that constrain the age-old questions that are raised in astrobiology. Finally, with the help of Garrett Reisman, astronaut in NASA's Johnson Space Center, we shall explore the future prospects of humans in solar system exploration, and beyond.

DETECTABILITY OF INTELLIGENT LIFE IN THE UNIVERSE: A SEARCH BASED IN OUR KNOWLEDGE OF THE LAWS OF NATURE

GUILLERMO A. LEMARCHAND
Instituto Argentino de Radioastronomía (CONICET) &
Centro de Estudios Avanzados (Universidad de Buenos Aires)
C.C. 8 - Sucursal 25, (1425) Buenos Aires, ARGENTINA
E-Mail: lemar@cea.uba.ar

1. The Scientific Basis of the Search for Extraterrestrial Technological Activities (SETTA)

Almost forty years ago, the Space Science board of the US National Academy of Sciences considered it appropriate to sponsor a preliminary examination of the problem of extraterrestrial intelligent life in the universe. An informal conference was held in November 1961, at the National Radio Astronomy Observatory at Green Bank, West Virginia. The purpose of the discussions was to examine the prospects for the existence of other societies in the galaxy with whom communications might be possible; to attempt an estimate of their number; to consider some of the technical problems involved in the establishment of communication; and to examine ways in which our understanding of the problem might be improved [70]. To facilitate orderly discussion, one of host organizers, Frank Drake[1], sought to formulate the conference's central problem as an equation. The expression presented for this purpose was:

$$N = R_* f_p n_e f_l f_i f_c L \qquad [1]$$

In this equation N is the estimated number of communicative societies in the galaxy at any time, R_* is the rate of star formation, f_p the fraction of stars forming planets, n_e the number of planets per star with environments suitable for life, f_l the fraction of suitable planets on which intelligence appears, f_c the fraction of intelligent cultures which are communicative in an interstellar sense, and L the mean lifetime of such civilizations. Sagan [75, 76, 79] considered an estimate of one million technological civilizations in the galaxy to be conservative. Others are less optimistic [88].

[1] Actually, the first publication of this equation was done by J. P. T. Pearman [70], who was the Green Bank Conference's organizer, representing the Space Science Board of the US National Academy of Sciences. Drake's first publication was done in 1965 [19]. Over the years, different authors proposed several modifications to the equation. Recently, Drake [20] added a new second factor, taking into account, also the fraction of solar type stars.

J. Chela-Flores et al. (eds.), Astrobiology, 13–32.

I believe it really is impossible to realistically estimate those odds because with only a slight difference in the interpretation of the statistics, you can get numbers that are very large, very small, or anywhere in between. Presentations arguing for all of these positions have been delivered at meetings of professional astronomers, which simply reinforces the impossibility of making good *a priori* estimations about the likelihood of intelligent life in our galaxy. The only way to determine this value is by performing a comprehensive observational program.

We can say, that over the years, the *Drake Equation* was taken as the dominant paradigm that justifies the SETI (Search for Extraterrestrial Intelligence) research programs. In this way, SETI proponents operate under a two-pronged hypothesis. The first assumption, known as the *Principle of Mediocrity* [79], is that the development of life is a basic, unexceptional consequence of physical processes taking place in appropriate environments—in this case on Earth-like planets. The idea is that if these processes occur on Earth, they will occur in the same way anywhere else with the same conditions. Because our galaxy has hundreds of billions of stars, and the Universe has billions of galaxies, there should be many "elsewheres," habitable Earth-like planets, and life should be common in the cosmic realm. Of course, the diversity of living systems elsewhere would surpass the imagination, and we cannot even begin to speculate about the abundance of other types of life that might evolve in environments much different from Earth, and which would be based on radically different types of biology.

The second assumption is that on some planets which shelter broods of living creatures, at least one species will develop intelligence and a technological culture that will have an interest in communicating with other sentient creatures elsewhere in the cosmos, and will beam signals out into space with that goal. If, these cultures would, like us, use electromagnetic signals to communicate, and that the signals they produced would have an artificial signature we could recognize, it should be possible to detect these electromagnetic waves, establish contact with these civilizations, and exchange information across interstellar distances.

Under all these assumptions, we may expect a logical progression: the presence of planets with suitable environments will lead to the emergence of life, which will lead to the emergence of intelligence, which will give rise to interstellar communication technology. Viewed from one angle, it may be said that SETI programs are simply attempts to test this theory.

Another basic assumption of SETI researchers is that the physical laws governing the universe are the same everywhere in the cosmos. If this is true, then the basic principles of our science and the science of extraterrestrial beings should be fundamentally the same, and we should be able to communicate with them by referring to those things we share in common: the principles of mathematics, physics, chemistry and so on.

Rescher [72] argues that despite sharing universal laws with us, extraterrestrials are extremely unlikely to have any type of science we would recognize. He contends that they will be very different organisms, with different needs, senses, and behaviors, and that, as Mayr [60] suggests, they may inhabit environments in which neither science nor technology may needed for survival. Even if they do have technology, Rescher argues, the science of an alien civilization would reflect the way they perceive nature, as funneled through the course of their particular evolutionary adjustment to their specific environment, and that will make it impossible for us to distinguish any of their possible intelligent manifestations.

All intelligent problem-solvers are subject to the same ultimate constraints: limitations on space, time and resources. In order for life to evolve powerful ways to deal with such constraints, they must be able to represent the situations they face, and they must have processes for manipulating those representations. Minsky [61] proposes two basic principles for every intelligence:

ECONOMICS: every intelligence must develop symbol-systems for representing objects, causes and goals, and for formulating and remembering the procedures it develops for achieving those goals.

SPARSENESS: every evolving intelligence will eventually encounter certain very special ideas -e.g. about arithmetic, casual reasoning and economics- because these particular ideas are very much simpler than other ideas with similar uses.

Minsky believes that because we and they will have had to develop these principles in order to survive and develop technology, aliens will have evolved thought processes and communication strategies that will match our own to a degree that will enable us to comprehend them. SETI proponents largely agree with him, and go farther to assume that all-galactic civilizations will converge in their interpretations of the physical laws. It is likely, they believe that we will be able to communicate with extraterrestrials about scientific principles, and especially about mathematics.

According to this hypothesis, aliens will have evolved thought processes and languages that will match our own somewhat that will enable us to comprehend them. SETI proponents have the implicit assumption that there is some sort of "convergence" in all the different interpretations of the physical laws, among the galactic civilizations.

Another consideration is that technological civilizations must survive long enough to be discovered. There are obvious advantages to intelligence and technology: once, a mere 100,000 or so hominids dwelt in Africa's Rift; now nearly six billion of their descendants live the world around, an increase in number of nearly five orders of magnitude, a population growth that few other organisms have matched. However, these advantages are counterbalanced by threats when civilizations reach our evolutionary stage—what the late Carl Sagan called "technological adolescence"—when technology brings the potentially civilization-ending threats of ecological catastrophe, the exhaustion of natural resources, and nuclear war. Barring such disasters, the physical environment of the Earth will remain stable for several billion years, because intelligence and technology have developed here about halfway through the stable eight billion-year life-period of the sun.

The only significant test of the existence of extraterrestrial intelligence is an experimental one. No a priori arguments on this subject can be compelling or should be used as a substitute for an observational program.

The acronym SETI is in some sense misnomer. As of now, we have no means for directly detecting *intelligence* over interstellar distances. What we can do is to attempt to detect any manifestation of a technological activity, produced by that intelligence. Imagine that we find, in the middle of the ocean, a message inside a bottle. We may be unable to understand the message, but by sure we can make the abduction that an intelligent being created the bottle. Abductive[2] reasoning accepts a conclusion on the grounds that it explains the available evidence. It addresses a wide range of issues concerning the logic of discovery and the economics of research [6, 7, 44].

Contrary to the traditional statement that the scientific grounds of the SETI project is based on the possible numbers of the Drake Equation; here we proposed a different epistemological approach. Using the *hypothetico-deductive method*[3], we just make the hypothesis that there should be other intelligent beings in the Universe. We just not make any kind of hypothesis about the different evolutionary paths and planetary or other cosmic environments that would let the intelligence to appear. In order to test this hypothesis or more appropriated to falsify[4] it, we must developed a comprehensive search for extraterrestrial technological activities (SETTA). The laws of nature will always place limits to any possible technological manifestation.

It is much simpler to distinguish the artificial origin of a technological activity than to establish a universal criterion to determine how to recognize the signature of life on other worlds[5]. Based in our understanding of the laws of nature, we can systematically explore all the possible artificial signatures. Harwit [35] and Tarter [86] determined how to estimate, in a first approach, the limits of our space of configuration. Our *cosmic haystack* has 10^{29} different "cells". Until now, just a mere fraction of 10^{13} to 10^{14} has been explored and we are still unable to verify or falsify our hypothesis that there are other intelligent beings in the Universe.

2. The Rational Strategies behind SETTA

If we want to find evidence for the existence of extraterrestrial intelligence, we must work out an observational strategy for detecting this evidence in order to establish the various physical quantities in which it involves. This information must be carefully analyzed so that it is neither over-interpreted nor overlooked and can be checked by independent researchers.

For the SETI experiment, as well as in conventional astronomy, the cosmic mean distances are so huge that the researcher can only observe what is received. He or she is entirely dependent on the carriers of information that transmit to him or her all he or she may learn about the Universe.

---

[2] *Abduction or Inference to the best explanation:* a term used by C.S. Peirce (1839-1914) for a special pattern of reasoning by which one infers that a hypothesis is true from the fact that the hypothesis offers the most plausible or satisfactory explanation of evidence. C.S. Peirce (1997), *Collected Papers*, **7**, Thoemmes Press, 89-164.

[3] *Hypothetico-deductive method:* an account of theory confirmation as a process of deducing from hypothesis under test together with appropriate auxiliary statements, predictions whose truths or falsity can be directly observed. Theories are confirmed by their true observational consequences.

[4] *Falsificationism:* the view, advocated by K. Popper (1902-1994), that for a claim to be scientific it must be possible to specify which observation would falsify it or, more generally, the circumstances in which it would be abandoned. Popper also holds that scientific claims can never be verified or positively confirmed to any degree and that science progresses by systematically attempting to falsify previously advanced hypotheses.

[5] The NASA Astrobiology Program set this goal in its roadmap.

However, according to our understanding of the physical laws, information carriers, are not infinite in variety. All information we currently have about the Universe beyond our solar system has been transmitted to us by means of electromagnetic radiation (radio, infrared, optical, ultraviolet, x-rays, and γ-rays), cosmic ray particles (electrons and atomic nuclei), and more recently by neutrinos.

Table 1 summarizes the possible *information carriers* that may let us find the evidence of an ETI, according to our knowledge and interpretation of the laws of physics. The classification of techniques in Table 1 is not intended to be complete in all respects. Thus, only a few fundamental particles have been listed. No attempt has been made to include any antiparticles. This classification, like any such scheme, is also quite arbitrary. Groupings could be made into different *astronomies* [48].

Table 1: The humankind's knowledge of the Universe outside the Earth was obtained by the following list of physical information carriers. According to our present understanding of the laws of physics, some of these information carriers can be used to find evidences of extraterrestrial intelligence beyond our planet.

Boson Astronomy		Particle Astronomy			Direct Techniques
Photon Astronomy	*Graviton Astronomy*	*Atomic Microscopic Particles*		*Macroscopic Particles or Objects*	Space Probes Manned Exploration Discovery of ETI activities in our Solar System?
Radio	Gravity	Fermions	Heavy	Meteors	
Infrared	Waves		Particles	Meteorites	
Optical		Protons	(atomic	Meteoritic Dust	
Ultraviolet		Neutrinos	nuclei)	Asteroids	Pneumatic Astronomy
X-rays		Electrons		Comets	
γ-rays					
		Cosmic Rays			Acoustic waves Magnetohydro-dynamic waves.

The methods of collecting this information as it arrives at the planet Earth make it immediately obvious that it is impossible to gather all of it and measure all its components. Each observation technique acts as an information filter. Only a fraction (usually small) of the complete information can be gathered. The diversity of these filters is considerable. They strongly depend on the available technology at the time.

Using our criteria of *"sparseness"* and of *"economics"* we should select those information carriers that should require a minimum amount of energy to exceed the natural background, to travel at --or close to-- the speed of light; not to be deflected by galactic or stellar fields; be easy to generate, detect and beam and not to be absorbed by the interstellar medium or by planetary atmospheres and ionospheres. Only "photons" survive all these requirements, and so electromagnetic waves of some frequency are the most suitable signal.

Electromagnetic radiation carries virtually all the information on which modern astrophysics is built. The production of electromagnetic radiation is directly related to the physical conditions prevailing in the emitter. The propagation of the information carried by electromagnetic waves is affected by the conditions along its path. The trajectories it follows depend on the local curvature of the Universe, and thus on the local distribution of matter (gravitational lenses), extinction affecting different wavelengths unequally, neutral hydrogen absorbing all radiation below the Lyman limit (912 Angstroms), and absorption and scattering by interstellar dust, which is more severe at short wavelengths.

Interstellar plasma absorbs radio wavelengths of kilometers and above, while the scintillations caused by them become a very important effect for the case of artificial radio transmissions, because of the large temporal variability in the signal amplitude [13]. The inverse Compton effect lifts low-energy photons to high energies in collisions with relativistic electrons, while γ and x-ray photons lose energy by the direct Compton effect. The radiation reaching the observer thus bears the imprint of both the source and the accidents of its passage though space.

The Universe observable with electromagnetic radiation can be characterized as a multi-dimensional phase space. The space of configuration for the transmission and reception of interstellar electromagnetic signals includes a four dimensional coordinates subspace (spatial coordinates and transmission's epoch of the hypothetical partner) and a seven dimensional information subspace (modulation type, transmitting frequency, information rate, frequency width of the transmitting signal, polarization, code and semantics) [17, 49, 51].

In Table 2, we present a summary of the main exotic proposals that use a great variety of information carriers beyond the radio waves. These proposals were published in the scientific literature over the last forty years. A detailed analysis of these ideas as well as a description of some observational programs can be found elsewhere [47, 48, 52]. Different scientists presented several studies and discussions about the physical possibilities of interstellar flights with different outputs [15, 37, 47, 58, 59, 66, 77]. A detailed discussion exceeds the limits of this presentation.

In 1959, Cocconi and Morrison [11], then at Cornell University published an article in the journal *Nature* entitled "Searching for Interstellar Communications," in which they proposed the first realistic strategy for searching for extraterrestrial intelligence, using radio astronomy to scan the nearest sun-like stars for signals at or near the 21-centimeter wavelength of neutral hydrogen.

Independently of Cocconi and Morrison, Frank D. Drake, then an astronomer at the National Radio Astronomy Observatory (NRAO) in Green Bank, West Virginia, was formulating plans to conduct an actual search. Drake was 29 when, on April 8, 1960, he turned the 26-meter Howard Tatel radio telescope toward the nearby solar-type stars τ Ceti and then ε Eridani. Drake dubbed the project "OZMA," because it was searching for distant beings at least as exotic as those in Frank Baum's popular *Wizard of Oz* [18].

Drake's action established that for the first time, man was technologically capable of practical searches to detect evidence of possible ETIs. The low profile, low-budget OZMA system had only one channel with a spectral resolution of 100 Hz and a sensitivity of 10^{-22} Wm^{-2}, and spent only $2,000 on parts. In the end, no signals were found after OZMA's 150 hours of observation. Despite its failure, however, Project OZMA fired the imagination of the public and the scientific community.

Information Carrier	Observational Effect	Proponent(s) - Reference
Radio Waves	Space Vehicles Communication Relays Signals	Vallée & Simard-Normandin [91]
Infrared Radiation	Dyson Sphere at 10 μm	Dyson [22, 23] Jugaku & Nishimura [39] Slysh [80]
Optical Radiation	Discovery of technetium or other short-lived isotope not ordinary found in the typical stellar spectra.	Drake & Shklovskii [79]
	Discovery of Technetium, plutonium, praseodymium or neoymium as a consequence of artificial generated nucleosynthesis to changed the stellar spectra. This would be the case if the central star were used as a repository of radioactive fissile waste material.	Whitmire & Wright [94]
x-rays	The use of an equivalent of the total terrestrial nuclear power for a single space explosion. This could generate a detectable omnidirectional x-ray pulse easily detected over 200 light-years.	Elliot [24]
	X-ray flashes generated by dropping material onto neutron stars.	Fabian [25]
	X-ray binaries as beacons	Corbet [12]
γ-rays	Anomalous γ–rays flashes generated by the artificial annihilation of matter-antimatter processes (e.g. for interstellar propulsion systems)	Harris [32-34] Viewing et al.[92] Zubrin [96]
Neutrinos	Detection of artificial neutrino beams from an ETI	Learned et al [45] Pasachoff & Kutner [67] Saenz et al [73] Subotowicz [81] Überall et al [90]
Matter Exchange	A possible channel for interstellar communication based in a DNA biological coded message capable of self-replication in suitable environments	Nakamura [64] Yokoo & Oshima [95]
Direct Techniques	Small Space Probes (long-delay echoes) Search for Artificial Objects in the Solar System ET objects in the Asteroid's Belt von Neumann machines	Bracewell [8, 9] Freitas -Valdes [27-30] Papagiannis [68-69] Tipler [88]
Exotica: Wormholes, Tachyons, EPR, the use of undiscovered physical laws...		Science Fiction Domain

Another SETI pioneer, Sebastian von Hoerner [36], classified the possible nature of the ETI signals into three general possibilities: local communication on the other planet, interstellar communication with certain distinct partners, and a desire to attract the attention of unknown future partners. Thus, he named them as *local broadcast, long-distance calls,* and contacting signals or *beacons.*

When we finally come to the practical requirements of SETI they can be daunting, and make the old notion of "searching for a needle in a haystack" pale by comparison. An electromagnetic transmission channel always has several variables that must be set in order to establish a communication between two unknown galactic partners. These include a four dimensional sub-group with the location of the extraterrestrial civilization (three dimensions in space), and a temporal dimension that coordinates transmission and reception (you can be looking to the correct place, but at a moment when nobody is transmitting or vice versa). There is also a seven-dimensional sub-group, which includes the frequency, the signal intensity and the cryptographic variables such as polarization, modulation, information rate, code, and semantics (which must be overcome to decode any message). All these variables make up our "cosmic haystack, in which hides the needle—contact with an unknown galactic intelligence.

Considering only the entire microwave region of the electromagnetic spectrum (0.3 to 300 GHz) and avoiding the temporal synchronization between transmission and reception and the cryptographic variables (polarization, modulation, information rate, code and semantics), if we assume that the remaining dimensions of our *cosmic haystack* are all equally probable, then there are roughly three hundred thousand septillion or $3x10^{29}$ places or "cells" in this configuration space. Each cell's dimensions are 0.1 Hz bandwidth per the number of beams that an 300-meter Arecibo-type radio telescope would need to conduct an all-sky survey, assuming a receiver sensitivity of between 10^{-20} and 10^{-30} Wm^{-2}. Assuming one million galactic technological civilizations making random transmissions, this is comparable to a search for an actual five-centimeter sewing needle in an actual haystack with a volume more than 35 times the volume of Earth. The number of cells increase dramatically if we expand our search to other regions of the electromagnetic spectrum or we decide to include the others variables of our space of configuration. So far, only a small fraction of the whole haystack has been explored, a mere 10^{-16} to 10^{-15} of the total possible number of cells.

The success of a radio search for extraterrestrial intelligence depends not only on the unknown abundance of civilizations in our galaxy, but also on their assumed transmission strategies. How can the extraterrestrials announce their presence to his galactic neighbors? The best transmission strategy should set all the variables of the space of configuration following the so-called *Principle of Anti-Cryptography* [17]. In this way we are supposing that the signal will be designed and operated in such a way as to maximize its probability of discovery, both by intentional searches and by accidental observations probably made by all the possible "galactic interlocutors". The parameters should be selected according to the knowledge and interpretation of the astrophysical constraints, choosing those which minimize the number of unknown dimensions to be searched by the recipient.

Even if there are large number of mutually communicative civilizations, the probability that the Earth would be able to eavesdrop on narrow beam ETI signals following a random addressing time's strategy for transmission and reception, is vanishingly small.

Very advanced civilizations might be able to broadcast omnidirectionally over great distances and during large periods, but such civilizations are likely to develop other interests than contacting emerging societies.

The real difficulty in making the first contact stems from the requirement that the transmission should be made in such a way that the target civilization should be pointed their receiver towards the "unknown transmitter" when signals are arriving at their home planet. Probably this is the most difficult element (synchronization in time) of our search strategy. It is possible to select and search rationally all the other elements (e.g. frequency, bandwidth, polarization, space directions, etc.) but without any doubt you should be extremely lucky to catch any extraterrestrial signal, aiming your radio telescope to a place in the sky just at the precisely moment that the message in passing through the Earth. Over the years, a number of original proposals based in the use of several cosmic phenomena were formulated in order to set-up the synchronization requirements [47, 49, 51].

3. Relations for a Radio Astronomical Search of Extraterrestrial Artificial Signals

Suppose a transmitter radiates power P isotropically, then at a distance R an antenna of effective collecting area A_r will receive a power

$$P_r = \frac{PA_r}{4\pi R_d^{\ 2}} \qquad [2]$$

We define the detection range limit R_d, between two civilizations, as the value of R that makes $P_r = mkTb$, where T is the system noise temperature, k the Boltzmann constant, b the spectral resolution and m is the factor by which the detection threshold must exceed the mean noise power in order for a tolerable false alarm probability (say 10^{-12}) to be achieved [21]. Thus

$$R_d = \sqrt{\frac{A_r P}{4\pi mkTb}} \qquad [3]$$

Since the effective area of a circular antenna is $A_r = \eta\pi d^2/4$, where d is the antenna diameter and η is the aperture efficiency, we have

$$R_d = \left(\frac{d}{4}\right)\sqrt{\left(\frac{P\eta}{mkTb}\right)} \qquad [4]$$

If the transmitting antenna radiates only into a solid angle Ω, the transmitter power P_t need be only $P_t = P\Omega/4\pi$. Assuming $\eta=1$, the antenna gain is $g = P/P_t = 4\pi/\Omega$. But $4\pi/\Omega$ is the number of directions, n, in which the antenna must be pointed to cover the sky, so $n=g$. The collecting area, A_0, of an isotropic antenna is $A_0 = \lambda^2/4\pi$. On axis, an antenna of effective area A collects A/A_0 times as much power as an isotropic antenna and so has a gain $g=A/A_0=4\pi A/\lambda^2=(\pi d/\lambda)^2$, here λ is the electromagnetic wavelength. Thus gain (g) and directivity (n) are equal and are proportional to the antenna area in square wavelengths. Small modifications are necessary in the result when $\eta<1$ [43]. Using all these definitions, we are able to re-write the communication range equation in several ways like:

$$R_d=\left(\frac{1}{\lambda}\right)\sqrt{\left[A_t A_r\left(\frac{P_t}{P_r}\right)\right]}=\left(\frac{\lambda}{4\pi}\right)\sqrt{\left(\frac{P_t}{P_r}\right)} \qquad [5]$$

As an example of the application of this relation, we can see at which distance a signal emitted by the radar of Arecibo can be detected by a similar radiotelescope in the galaxy. The equivalent isotropic radiated power[6] (EIRP) of Arecibo is $\cong 10^{14}$ watts, using a spectrometer with a resolution of 1 Hz per channel, a typical system temperature of $\cong 16$ K, an aperture efficiency of $\eta \cong 80$ % and $m = 1$, this signal would be able to be detected at a distance of $\cong$ 4800 light years, only after a few seconds of integration time.

Sullivan et al [84] considered the possibility of eavesdropping on radio emissions inadvertently "leaking" from other technical civilizations. To better understand the information, which might be derived from radio leakage, the case of the planet Earth was analyzed. For example, the US Naval Space Surveillance System had an EIRP of 1.4×10^{10} watts into a bandwidth of only ≤ 0.1 Hz. Its beam is such that any eavesdropper in the directions of the declination range of 0° to 33° (28% of the sky) will be illuminated daily for a period of ≥ 7 seconds. Using the previous equations, it is very simple to show that this radar has a detectability range of leaking terrestrial signals to ≤ 60 light years using an Arecibo-type antenna (d =305-m), or ≤ 600 light years for a Cyclops type array of 1000 dishes with diameters of 100-meter each one [4].

Billingham and Tarter [5] show how sensible are the terrestrial new available technologies for the detection of leakage signals similar to the ones radiated by planet Earth [84] from other nearby stars.

Following this analysis on how to establish an electromagnetic link between two galactic partners, what kind of signal might one expect from a distant civilization? If it is send intentionally it will most likely be narrow-band (~1 Hz or less), ideally a single wavelength, something high monochromatic. This is so for two main reasons (1) to distinguish between natural and artificial signals (e.g. the most monochromatic natural sources are the cosmic masers with bandwidths ~500 Hz) and (2) such a signal travels furthest for a given transmitting power.

We can make several assumptions about the hypothetical distribution of galactic transmitting civilizations [4, 14, 21, 31, 36, 65, 70, 79, 83]. Here on, for their simplicity and generality we will follow Cordes et al. [14] analysis. If we define N_D as the number of transmitting technological societies in the galaxy, we can estimate the number density of ETI sources as

$$\delta(N_D)=\frac{N_D}{2\pi R_G^2 H_G} \qquad [6]$$

[6] The EIRP is defined as the product of the transmitted power and the directive antenna gain.

Assuming that all the civilizations reside in a galactic disk of radius R_G and thickness $2H_G$. The volume through the galactic disk sampled by a radio telescope of beam solid angle Ω_b is $V_b=\Omega_b D^3/3$, where D is the characteristic distance through the disk, $D \sim min[H_G/sin(b), R_G]$, where b is the galactic latitude, D is direction-dependent, but observations in the plane will yield $D \sim R_G$. The number of sources in a typical beam is

$$N_b = \delta(N_D)V_b \approx 10^{-2.7} N_D \left(\frac{D}{R_G}\right)^3 \left(\frac{R_G}{150H_G}\right)\left(\frac{\theta_b}{1^\circ}\right)^2 \qquad [7]$$

Where θ_b is the one-dimensional beam width (FWHM) and we assume $R_G/H_G = 150$. If there are more than 10^3 ETI transmitters homogeneously distributed in the Milky Way, every telescope beam (of 1° size) will contain at least one source, on average. It is convenient to rewrite N_b as

$$N_b = \left(\frac{D}{D_{mfp}}\right)^3 \qquad [8]$$

Where D_{mfp} is the mean free path for the line of sight intersecting a source in the beam and can be estimated as

$$D_{mfp} = \left(\frac{3}{\delta(N_D)\Omega_b}\right)^{\frac{1}{3}} \approx R_G \left(\frac{501}{N_D}\right)^{\frac{1}{3}} \left(\frac{150H_G}{R_G}\right)^{\frac{1}{3}} \left(\frac{1^\circ}{\theta_b}\right)^{\frac{1}{3}} \qquad [9]$$

The typical distances between sources is

$$R_D \approx \begin{cases} H_G \left(\frac{2}{3}\right)^{\frac{1}{2}} \left(\frac{\tilde{N}_D}{N_D}\right)^{\frac{1}{2}} \left(1+\frac{N_D}{2\tilde{N}_D}\right)^{\frac{1}{2}} & N_D \le \tilde{N}_D \\ H_G \left(\frac{\tilde{N}_D}{N_D}\right)^{\frac{1}{3}} & N_D \ge \tilde{N}_D \end{cases} \qquad [10]$$

Where $\tilde{N}_D$ is the number of sources such that R_D equals one scale height of the galactic disk (H_G):

$$\tilde{N}_D = \frac{3}{2}\left(\frac{R_G}{H_G}\right)^2 = 10^{4.5}\left(\frac{R_G}{150H}\right)^2 \qquad [11]$$

Analogous expressions can be obtained for a spherical distribution of ETI Sources. However, the concentration of some candidate signals detected by the META programs in Harvard and Buenos Aires, motivated the disk population analysis. We will continue our study making an estimation of the signal strengths that may be detected by our terrestrial systems.

To simplify our analysis we can start considering a galactic population of extraterrestrial transmitters as a homogeneous distribution of source strengths of standard candle transmitters. If each transmitter radiates with effective isotropic radiation power (EIRP) χ, will show maximum and minimum fluxes as

$$S_1 = \frac{\chi}{4\pi R_{min}^2} \qquad S_2 = \frac{\chi}{4\pi R_{max}^2} \qquad [12]$$

from the nearest and furthest sources, respectively. Here R_{max} and R_{min} are the maximum and minimum distances of galactic transmitters that are detectable with fluxes S_1 and S_2. For simplicity, we ignore transmitting beaming in this analysis. Beaming away from our direction clearly extends the minimum flux S_0 to zero. The net effect is to decrease the probability of detecting strong signals when there is a fixed number of transmitters in the Milky Way. Within these bounds, the probability density function, ϕ_{limit}, for S is

$$\phi_{limit} = \left(\frac{\alpha - 1}{S_0}\right) \bullet \left[1 - \left(\frac{S_0}{S_1}\right)^{\alpha - 1}\right]^{-1} \bullet \left(\frac{S}{S_0}\right)^{-\alpha} \qquad [13]$$

Here α is the exponent of the power law distribution of transmitters (log N – log S). Disk and spherical populations are described by $\alpha = 2$ and $\alpha = 5/2$, respectively, as is well known in studies of natural radio sources and γ-ray.

Taking $R_{min} \sim R_D$ and $R_{max} \sim R_G$, we find that $S_1/S_0 \sim N_D$ for $N_D \leq \tilde{N}_D \sim 10^{4.5}$. In the large civilization limit $(N_D \geq \tilde{N}_D)$, the scaling is $S_1/S_0 \sim (R_G / H_G)^2 (N_D / \tilde{N}_D)^{2/3} \sim 10^{4.4} (N_D / \tilde{N}_D)^{2/3}$ [14]. Depending on the abundance of transmitting civilizations in the Galaxy, the dynamic range of source fluxes may be modest or extremely large. Obviously, radio transmission powers from ETI will be described by a luminosity function for χ that incorporates unknown extraterrestrial technological and sociological constraints.

Drake [75] showed that if the distribution of galactic transmissions from intelligent beings follows a "standard candle" luminosity function, there would be far more civilizations with intrinsically faint signals than civilizations with intrinsically bright ones. Consequently, as with the stars and cosmic radio-sources, the easiest civilizations to detect will not be the closest ones, but the intrinsically brightest and farthest ones. Using this argument, we can cover all the possibilities at once if we make a full-sky survey searching for the strongest artificial galactic signals: this is in fact the strategy followed by projects BETA, META II and all the SERENDIP versions (see Table 3). A different approach is used by Project Phoenix, which is trying to detect signals from several thousand nearby sun-like stars[7].

---

[7] Drake, elsewhere in this proceeding, will make a complete description of the technical characteristics and main results of the target search for artificial electromagnetic signals (between 1-3 GHz) of Project Phoenix

4. Observations

During the last 40 years more than 90 different professional SETI projects have been carried out at observatories in Australia, Argentina, Canada, France, Germany, Italy, Russia, The Netherlands, and the USA. Altogether, they have accumulated more than 320,000 observing hours. Most of these have been in the so-called "magic frequencies" of the microwave region—where we believe ETIs would broadcast— like the 21-centimeter hydrogen line proposed by Cocconi and Morrison. A few of them took a radically different approach, searching for laser signals in the infrared, optical and ultraviolet regions of the spectrum. Unfortunately, none provided any conclusive evidence for the detection of an intelligent extraterrestrial signal.

In recent years there have been important developments in SETI research, in the growth in private financial support for such programs (which allows for much more powerful projects), in the further exploitation of electronic and computer technology to produce powerful systems at low cost, and in the development of effective means to identify and eliminate man-made interference from the signal data.

Today, the private financial support for SETI comes from individual donations to non-profit organizations like the SETI Institute, The Planetary Society and Friends of SERENDIP, with the logistic support from institutions like Harvard University, the University of California at Berkeley, the National Astronomical and Ionospheric Center (NAIC), NRAO, the University of South Wales and the Commonwealth Scientific and Industrial Research Organization (CSIRO) in Australia, the National Research Council of Argentina (CONICET) and the Consiglio Nazionale delle Ricerche (CNR) of Italy.

Project Phoenix at the SETI Institute has a mobile facility with a 56 million channel system; the University of California at Berkeley Project SERENDIP IV has a 168 channel system, and The Planetary Society-Harvard's BETA has a 250 million channel system. If we consider the research and development costs, those systems goes from some hundred thousands dollars to several millions.

Project Phoenix, uses a second antenna hundreds of kilometers from the main radio telescope to identify terrestrial radio interference with received signals. If a signal is detected in both antennas, it will show a specific Doppler signature due to the rotation of Earth, which reveals it as terrestrial in origin. They have already finished two observation campaigns at the CSIRO 64-m antenna in Parkes, Australia and the 43-m antenna of the NRAO in Green Bank. They are now extending their observations to the observatories of Arecibo and Jodrell Bank in the UK.

Another public initiative called SETI@Home is using the Internet to coordinate massive parallel computation on desktop computers. In nine months, since they have started in May 1999, more than 1,760,000 individuals from 220 different countries have downloaded a screen saver program that not only provides the usual graphics, but also performs sophisticated analyses of SETI data using the host computer. The original data is tapped from project SERENDIP IV's receiver operating in Arecibo [1]. This is the biggest computer-demanding project ever done in the human history. The equivalent of 200,000

from the SETI Institute. This program looks for signals around almost 1000 nearby stars, with the highest sensitivity and RFI rejection system. Other details are described by Cullers [10] and Tarter and Chyba [87].

years of CPU time was already reached in February 2000. A total of $1.6x10^{20}$ floating point operations were done during the same period.

Current SETI detection devices simultaneously analyze several million or billion narrow band spectral channels. The computers check whether for any strong narrow band signals among the cacophony of cosmic noises, and other instrumentation eliminates all human-made terrestrial and space radio interference. Of course, if the intensity of an extraterrestrial signal is below the sensitivity of the detection system, we will be incapable of distinguishing it in the ocean of cosmic noise. After years of observation, and the analysis of hundreds of billions of signals, most have been recognized as terrestrial interference or system noises. Fewer than 100 of those hundred billion signals have looked like potential extraterrestrial signals and were defined as extra-statistical events, but none of them could be detected in subsequent observations of the same sky-region [53].

Table 3: Characteristics of the main full-sky survey SETI projects around the world.

Characteristics	BETA	META II	SERENDIP IV	Southern SERENDIP	Italian SERENDIP
Observatory	Agassis	IAR	Arecibo	Parkes	Medicina
Site	Oak Ridge	Bs. As.		NSW	Bologna
Country	USA	Argentina	Puerto Rico	Australia	Italy
Number of channels [Million]	250 x 8	8.4	168	4.2 x 2	4.2 x2
Antenna diameter [m]	26	30	305	64	32
Spectral Res. [Hz]	0.5	0.05	0.6, 1.2, 2.4,..600	0.6	1.2
Operation Freq. [GHz]	1.4-1.7	1.4; 1.6, 3.3	1.4	1.4	0.4, 1.4, 1.6
Instantaneous Bandwidth [MHz]	40	0.4 – 2	100	2.4	5
Total Bandwidth Coverage [MHz]	320	1.2 – 6	180	2.4	5
Sensitivity [$W.m^{-2}$]	$\sim 3x10^{-24}$	$\sim 8x10^{-24}$	$\sim 10^{-24}$	$\sim 10^{-25}$	$\sim 10^{-25}$
Sky Coverage [Percentage of 4 π]	~70	~50	~30	~75	~75
Declination range [Degrees]	-30 to +60	-90 to –10	2 to +38	-90 to + 26	-30 to +90
Types of Signals	C, SC, P	C, CH	C, CH, P	C, CH, P	C, CH, P
References	[46]	[53,56]	[93]	[85]	[62]

Cordes et al. [14] derived Bayesian tests to analyze these unexplained signals detected by the META projects from Harvard between 1986 and 1991 and from Argentina between 1990 and 1995. The results showed were unable to rule out these signals as real alien transmissions that could be originated near the galactic plane and were modulated by interstellar scintillations. The second and more plausible explanation is that these signals are originated by unidentified terrestrial radio frequency interferences.

Due to this finding, new observation strategies were designed to check the origin of these type of signals. For example, Paul Horowitz's Project BETA at Harvard uses a billion channel analyzer and three different antennas beams in order to avoid any possible terrestrial interference [46]. Project Phoenix uses what they call a FUDD (Follow Up Detection Device), that works like a pseudo-interferometer located hundreds kilometers from the main antenna [10]. All the SETI groups worldwide have developed an extraordinary army of software and hardware tools to avoid local interference and have increased the sensitivity of the equipment.

The success of the search depends not only on the unknown number of civilizations in our galaxy, but also on their assumed transmission strategies. Historically, the basic assumption has been that there should exist some "supercivilizations" that are able to make ommidirectional transmissions strong enough to be detected by our full-sky surveys [36, 40, 75, 79]. Recent full-sky surveys made by the Harvard, Arecibo, Ohio and Buenos Aires SETI projects were unable to find any evidence of omnidirectional supercivilization transmissions at a distance of 22 Mpc ($7x10^7$ light years). The survey assumed such signals would have a radiated power equivalent to the entire energy output of a solar type star, or about 10^{26} watts.

One year after Project OZMA, Schwartz and Townes [78], proposed using lasers for interplanetary and interstellar communication[8]. Because laser technology was still new, however, low-power laser devices had not yet been developed to the point of radio technology, which had already been used for decades. Consequently, most SETI efforts took advantage of the microwave region of the spectrum. In 1965, Monte Ross [50] brought the attention of the SETI community that narrow-pulse, low duty-cycle optical systems for SETI could contain more information per received signal photo.

Other proposals have been made to search within the infrared region for beacon signals beamed toward Earth. The higher gain that is available from antennas at shorter wavelengths (up to 10^{14} Hz) compensates for the higher quantum noise in the receiver and wider noise bandwidth at higher frequencies [3]. One concludes that for the same transmitter powers and directed transmission that takes advantage of the high gain, the detectable SNR is comparable at 10 millionths of a meter and 21 centimeters [89].

Ross [41, 42], along with Victor Shvartsman and Gregory Beskin [2] of the Special Astrophysical Observatory (SAO) in Russia, as well as Stuart Kingsley [41,42] from the Columbus Optical SETI Observatory, were some of the first modern optical SETI advocates and researchers. They demonstrated that very low duty cycle, short-pulse, high-peak power laser modulation enables sending signals great distances using less power while maximizing efficiency. Significant information can be sent within each pulse in digital pulse position modulation, when the amounts of pulse intervals are large. To perform this, a search for nanosecond and sub-nanosecond pulses is necessary.

Astrophysics on its shortest time scales is now entering the previously unexplored domains of milli, micro-, and nanosecond variability. At very high time resolution, data rates are very high (megabytes per second), and classical light curves are of little use. Measurements thus have to be of autocorrelations, power spectra, or other statistical properties of the arriving photon stream. All such statistical functions depend on a power

---

[8] On , January 27, 1920, Albert Einstein is quoted by the *New York Times* saying that "If intelligence on other planets attempted communication with Earth, I would expect them to use rays of light, which are more easily controlled."

of average intensity that is higher than one. For example, an autocorrelation (which is obtained by multiplying the intensity signal by itself, shifted by a time lag) is proportional to the square of the intensity. Due to this dependence, very large telescopes are much more sensitive for the detection of rapid variability than the small ones. For faint sources, one wants to study variability also on time scales shorter than the typical intervals between successive photons. While not possible with conventional light-curves, it is enable through a statistical analysis of photon arrival times, testing for deviations from randomness.

In the last years, three innovative optical SETI programs tap into light data from studies of Sun-like stars, searching for possible light pulse or color band signals. Two programs are operating at the University of California and one at Harvard University [38, 50]. The experiment's photometers can detect light as short as a few billionths of a second. This light pulse search uses a relatively simple high-speed optical detector to search for multiple "coincidences" of light pulses of a particular width and intensity. A real laser pulse, if it is delivered at a rate of at least 100 photons per pulse, will always generate a coincidence. Detected coincidences will trigger a measurement of optical pulse, width and intensity, which, along with absolute time and target coordinates, is logged onto a computer.

In my personal point of view, extraterrestrial civilizations will probably not use laser as a source of their beacon signals. In sending SETI messages without know your candidate partner and with limited energy to make your transmission, microwave wavelengths are much more efficient because they can illuminate a larger region of the sky using the same amount of energy. However, if you know the location of your cosmic neighbor, the laser nanosecond pulsed signal can be much more efficient than radio waves because a greater amount of information can be exchanged. If this hypothesis is correct, we should expect laser signals from nearby cosmic civilizations that have already detected our terrestrial technological activities through radiowave (military and planetary radars, television carriers, and other radio transmissions).

Did we already have received a contact signal?, Probably we did, but our detectors didn't have enough sensitivity to distinguish it among the cosmic noise, or perhaps we have never aimed our antenna to the "correct" place at the "correct" moment when the signal were passing through the Earth, or probably we are searching in an incorrect frequency or observing strategy [47, 49, 51]. Perhaps, interstellar scintillation processes faded some signals, or perhaps the extraterrestrial civilizations are using an unknown physical process for interstellar communication than electromagnetic waves [48, 52, 54, 79, 82]. Probably, the first evidence from an ETI will come in a serendipity way, when a traditional astronomer, unable to explain some anomalous observation will realize that their data could be explain only as a consequence of an extraterrestrial domestic technological activity, not necessary a specific contact message. However, as Louis Pasteur (1822-1895) said *"unexpected discoveries are favored by prepared minds"*.

We are living in an era, where it is possible for us to detect evidences of extraterrestrial technological activities or, SETTA, from civilizations no more advanced than we are, over a distance of at least many thousands of light years, using the current astronomical technology. The results of the present SETI programs –whether positive or negative— would have profound implications for our view of our Universe and ourselves.

5. Acknowledgments

The author wants to thank Prof. Julián Chela-Flores (ICTP) and Prof. Joan Oró (University of Houston) for giving him the possibility to participate in the organization of the First Ibero American School on Astrobiology. He wants to extend his acknowledgments to the sponsoring organizations as well as the local hosts at the IDEA Convention Center in Caracas. The SETI activities in Argentina have been possible thanks to the financial support of The Planetary Society.

6. References

1. Anderson, D.P.; Werthimer, D.; Cobb, J.; Korpela, E.; Lebofsky, M.; Gedye, D. and Sullivan, W. (2000), SETI@Home: Internet Distributed Computing for SETI, in G.A. Lemarchand and K. Meech (eds.), *Bioastronomy 99: A New Era in the Search for Life in the Universe*, ASP Conference Series, San Francisco.
2. Beskin, G.M.; N. Borisov; V. Komarova; S. Mitronova, S. Neizvestny, V. Plokhotnichenko and Marina Popova (1997); Methods and Results of an Optical Search for Extraterrestrial Civilizations, *Astrophysics and Space Science*, **252**, 51-57.
3. Betz, A. (1986), A Direct Search for Extraterrestrial Laser Signals, *Acta Astronautica,* **13**, 623-629.
4. Billingham, J. and Oliver, B. (eds.) (1973), *Project Cyclops: A Design Study of a System for Detecting Extraterrestrial Intelligent Life*, NASA CR 114445.
5. Billingham, J. and Tarter, J.C. (1992), Detection of the Earth with the SETI Microwave Observing System Assumed to be Operating out of the Galaxy, *Acta Astronautica*, **26**, 185-188.
6. Boyd, R. (1983), On the Current Status of Scientific Realism, *Erkenntnis*, **19**, 45-90.
7. Boyd, R.; Gasper, P. and Trout, J.D. (eds.) (1991), *The Philosophy of Science*, MIT Press, Cambridge.
8. Bracewell, R.N. (1960), Communications from Superior Galactic Communities, *Nature,* **186**, 670-671.
9. Bracewell, R.N. (1975), *The Galactic Club: Intelligent Life in Outer Space*, H. Freeman & Co., San Francisco.
10. Cullers, K. (2000), Project Phoenix and Beyond, in G.A. Lemarchand and K. Meech (eds.), *Bioastronomy 99: A New Era in the Search for Life in the Universe*, ASP Conference Series, San Francisco.
11. Cocconi, G. and Morrison, P. (1959), Searching for Interstellar Communications, *Nature*, **184**, 844-847.
12. Corbet, R.H.D. (1997), SETI at X-Ray Energies: Parasitic Searches from Astrophysical Observations, *J. of the British Interplanetary Soc.*, **50**, 253-257.
13. Cordes, J. M. and Lazio, T. J. (1991), Interstellar Scattering Effects on the Detection of Narrow-Band Signals, *Astrophysical Journal*, **376**, 123-134.
14. Cordes, J.M., Lazio, J.W. and Sagan C. (1997), Scintillation Induced Intermittency in SETI, *Astrophysical Journal*, 487, 782-808.
15. Crawford, I.A. (1990), Interstellar Travel, A Review for Astronomers, *Q. Jl. Royal astr. Soc.*, **31**, 377-400.
16. Demming, D. and Mumma, M.J. (1983), Modeling of the 10-micrometer Natural Laser Emission from the Atmospheres of Venus and Mars, *Icarus*, **55**, 356-368.
17. Dixon, R.S. (1973), A Search Strategy for Finding Extraterrestrial radio Beacons, *Icarus*, **20**, 187-199.
18. Drake, F.D. (1960), How Can we Detect Radio Transmissions from Distant Planetary Systems?, *Sky and Telescope*, **19**, 140.
19. Drake, F.D. (1965), The Radio Search for Intelligent Extraterrestrial Life, in G. Mamikunian and M.H. Briggs (eds.), *Current Aspects of Exobiology*, Pergamon Press, London.
20. Drake, F.D. (1999), Nuove Prospettive per la Galassia: Il Progetto SETI in Colombo, R.; Giorello, G. and Sindoni, E. (eds.), *L'Intelligenza Dell'Universo*, Edizioni Piemme, Spa, Italy.
21. Drake, F., Wolfe, J.H. and Seeger, C.L. (1983), SETI Science Working Group Report, *NASA Technical Paper,* **2244**.
22. Dyson, F.J. (1959), Search for Artificial Stellar Sources of Infrared Radiation, *Science*, **131**, 1667-1668.

23. Dyson, F.J. (1966), The Search for Extraterrestrial Technology, in *Perspectives in Modern Physics* (Essays in Honor of Hans Bethe), R.E. Marshak (editor), John Wiley & Sons, New York.
24. Elliot, J.L. (1973), X-Ray Pulses for Interstellar Communication, in *Communication with Extra Terrestrial Intelligence*, C. Sagan (Ed.), 398-402, MIT Press.
25. Fabian, A.C. (1977), Signaling Over Stellar Distances with X-rays, *J. of the British Interplanetary Soc.*, **30**, 112-113.
26. Freitas, R.A. (1977), High Energy Particle Beams for CETI, *Spaceflight*, **19**, 379.
27. Freitas, R.A. (1980); Interstellar Probes a New Approach for SETI, *J. of the British Interplanetary Soc.*, **33**, 95-100.
28. Freitas, R.A. and Valdes, F.(1980); A Search for Natural or Artificial Objects Located at the Earth-Moon Libration Points, *Icarus*, **42**, 442-447.
29. Freitas, R.A. and Valdes, F. (1980), A Search for Objects Near the Earth-Moon Lagrangian Points, *Icarus*, **53**, 453-457.
30. Freitas, R.A. and Valdes, F. (1985), The Search for Extraterrestrial Artifacts (SETA), *Acta Astronautica*, **12**, 1027-1034.
31. Gott, J.R. (1993), Implications of the Copernican Principle for our Future Prospects, *Nature*, **363**, 315-319.
32. Harris, M.J. (1986), On the Detectability of Antimatter Propulsion Spacecraft, *Astrophysics and Space Science*, **123**, 297-303.
33. Harris, M.J. (1991a), A Search for Linear Alignments of Gamma Ray Burst Sources, *J. of the British Interplanetary Soc.*, **43**, 551-555.
34. Harris, M.J. (1991b), SETI Through the Gamma Window: A Search for Interstellar Spacecraft, in *Bioastronomy: the Search for Extraterrestrial Life,* J. Heidemann and M.J. Klein (eds.), Lectures Notes in Physics 390, Springer-Verlag, Berlin.
35. Harwit, M. (1981), *Cosmic Discovery: The Search, cope and Heritage of Astronomy*, Basic Books Inc. Pub., New York.
36. Hoerner, von S. (1961); The search for Signal from Other Civilizations, *Science*, **134**, 1839-1843.
37. Hoerner, von S. (1962); The General Limits of Space Travel, *Science*, **137**, 18-23.
38. Howard, A.; Horowitz, P.; Coldwell, C.; Klein, S.; Sung, A.; Wolff, J.; Caruso, J.; Latham, D.; Papaliolios, C.; Stefanik, R. and Zajac, J. (2000), Optical SETI at Harvard-Smithsonian, in G.A. Lemarchand and K. Meech (eds.), *Bioastronomy 99: A New Era in the Search for Life in the Universe*, ASP Conference Series, San Francisco.
39. Jugaku, J. and Nishimura, S. (2000), A Search for Dyson Spheres around Late-type Stars in the Solar Neighborhood III, in G.A. Lemarchand and K. Meech (eds.), *Bioastronomy 99: A New Era in the Search for Life in the Universe*, ASP Conference Series, San Francisco.
40. Kardashev, N.S. (1964), Transmission of Information by Extraterrestrial Civilizations, *Soviet Astronomy*, **8**, 217-220.
41. Kingsley, S.A. (1993), "The Search for Extraterrestrial Intelligence (SETI) in the Optical Spectrum II", *SPIE Conference Proceedings,* **1867**, International Society for Optical Engineering, Washington.
42. Kingsley, S.A. and Lemarchand, G.A. (1996), "The Search for Extraterrestrial Intelligence (SETI) in the Optical Spectrum II", *SPIE Conference Proceedings,* **2704**, International Society for Optical Engineering, Washington.
43. Kraus, J.D. (1986), *Radio Astronomy*, Cygnus-Quasar Books, Powell, Ohio.
44. Laudan, L. (1981), A Confutation of Convergent Realism, *Philosophy of Science,* **48**, 19-48.
45. Learned, J.G., Pakvasa, S., Simmons, W.A., and Tata, X. (1994), Timing Data Communication with Neutrinos: A New Approach to SETI, *Q.J.R.Astr. Soc.*, **35**, 321-329.
46. Leigh, D. and Horowitz, P. (2000); Strategies; Implementation and Results of BETA, in G.A. Lemarchand and K. Meech (eds.), *Bioastronomy 99: A New Era in the Search for Life in the Universe*, ASP Conference Series, San Francisco.
47. Lemarchand, G.A. (1992), *El Llamado de las Estrellas*, Lugar Científico, Buenos Aires.
48. Lemarchand, G.A. (1994a), Detectability of Extraterrestrial Technological Activities, *SETIQuest*, **1**, 3-13. Also available at http://www.coseti.org/lemarch1.htm
49. Lemarchand, G.A. (1994b), Passive and Active SETI Strategies using the Synchronization of SN1987A, *Astrophysics and Space Science*, **214**, 209-222.

50. Lemarchand, G.A. (1996), Interplanetary and Interstellar Optical Communication Between Intelligent Beings: A Historical Approach, in Kingsley, S.A. and Lemarchand, G.A. (eds.), "The Search for Extraterrestrial Intelligence (SETI) in the Optical Spectrum II", *SPIE Conference Proceedings,* **2704**, International Society for Optical Engineering, Washington.
51. Lemarchand, G.A. (1997a), SETI Synchronization Passive and Active Strategies, *SETIQuest,* **3** (3), 8-14.
52. Lemarchand, G.A. (1997b), SETI Technology: Possible Scenarios for the Detectability of Extraterrestrial Intelligence Evidences, in '17th IAA/IISL Scientific-Legal Round Table on SETI and Society' during the *48th International Astronautical Congress*, IAA-97-IAA.7.1.02, October 6-10, 1997, Turin, Italy.
53. Lemarchand, G.A. (1998a), A Full Sky Survey for Ultra Narrowband Artificial Signals, in J. Chela Flores and F. Raulin (eds.) *Exobiology: Matter, Energy, and Information in the Origin and Evolution of Life in the Universe*, 339-346, Kluwer Academic Pub., Dordrecht.
54. Lemarchand, G.A. (1998b), Is There Intelligent Life Out There?, *Scientific American Presents* (Exploring Intelligence), **9** (4), 96-104.
55. Lemarchand, G.A. (1999), A Second Look at Optical SETI, *Bioastronomy News*, 11, 1-3.
56. Lemarchand, G. A. (2000), Progress in the Search for Ultra-Narrow-Band Extraterrestrial Artificial Signals from Argentina, in G.A. Lemarchand and K. Meech (eds.), *Bioastronomy 99: A New Era in the Search for Life in the Universe*, ASP Conference Series, San Francisco.
57. Livio, M., (1999), How rare are Extraterrestrial Civilizations and When Did They Emerge?, *Astrophysical Journal*, **511**, 429-431.
58. Mallove, E. and Matloff, G. (1989), *The Starflight Book: A Pioneer's Guide to Interstellar Travel*, John Wiley & Sons, New York.
59. Mauldin, J.H. (1992), *Prospects for Interstellar Travel*, American Astronautical Society Publication, Univelt, San Diego, CA.
60. Mayr, E. (1995), A Critique of the Search for Extraterrestrial Intelligence, *Bioastronomy News*, **7**, 2-4.
61. Minsky, M. (1985), Why Intelligent Aliens Will be Intelligible? in E. Regis (ed.), *Extraterrestrials: Science and Alien Intelligence*, Cambridge University Press, Cambridge.
62. Montebugnoli, S.; Cattani, A.; Cecchi, M.; Maccaferri, A.; Monari, J.; Mariotti, S.; Cosmovici, C.B. and Maccone, C. (2000), SETItalia, in G.A. Lemarchand and K. Meech (eds.), *Bioastronomy 99: A New Era in the Search for Life in the Universe*, ASP Conference Series, San Francisco.
63. Morrison, P.; Billingham, J. and Wolfe, J. (eds.) (1977), *The Search for Extraterrestrial Intelligence*, NASA SP-419, Scientific and Technical Information Office, Washington, DC.
64. Nakamura, H. (1986), SV40 DNA: A Message from Epsilon Eridani?, *Acta Astronautica*, **13**, 573-578.
65. Oliver, B.M. (1975), Proximity of Galactic Civilizations, *Icarus*, **25**, 360-367.
66. Oliver, B.M. (1981), Search Strategies, in *Life in the Universe*, J. Billingham (Ed.), NASA Conference publication 2156.
67. Pasachoff, J.M., and Kutner, M.L. (1979), Neutrinos for Interstellar Communication, *Cosmic Search*, **1**(3), 2-8.
68. Papagiannis, M.D. (1978), Are we Alone or Could They be in the Asteroid Belt?, *Q.J.R.Astr. Soc.*, **19**, 277-281.
69. Papagiannis, M.D. (1985), An Infrared Search in Our Solar System as Part of a More Flexible Search Strategy, in *The Search for Extraterrestrial Life: Recent Developments*, M.D. Papagiannis (ed.), Reidel Pub. Co., Boston.
70. Pearman, J.P.T. (1963), Extraterrestrial Intelligent Life and Interstellar Communication: An Informal Discussion, in A.G.W. Cameron (ed.), Interstellar Communication: The Search for Extraterrestrial Life, W.A. Benjamin, Inc., New York.
71. Rees, M. (1999), *Just Six Numbers: The Deep Forces that Shape the Universe*, Weidenfeld & Nicolson, London.
72. Rescher, N. (1984), *The Limits of Science*, University of California Press, Ltd., Berkeley.
73. Sáenz, A.W., Überall, H., Kelly, F.J., Padgett, D.W. and Seeman, N. (1977), Telecommunications with Neutrino Beams, *Science*, 198, 295-297.
74. Sagan, C. (1963), Direct Contact among Galactic Civilizations by Relativistic Interstellar Flight, *Planetary and Space Science*, **11**, 485-498.

75. Sagan, C. (ed.) (1973), *Communication with Extraterrestrial Intelligence (CETI)*, MIT Press, Cambridge.
76. Sagan, C. and Drake, F. (1975), The Search for Extraterrestrial Intelligence, *Scientific American*, **232**, 80-89.
77. Scheffer, L.K. (1994), Machine Intelligence, the Cost of Interstellar Travel and Fermi's Paradox, *Q.J.R. Astr. Soc,* **35**, 157-175.
78. Schwartz, R.N. and Townes, C.H. (1961), Interstellar and Interplanetary Communication by Optical Masers, *Nature*, **190**, 205-208.
79. Shklovskii, I.S. and Sagan, C. (1966), *Intelligent Life in the Universe*, Holden-Day, Inc.; San Francisco.
80. Slysh, V.I. (1985), Search in the Infrared to Microwave for Astroingeneering Activity, in *The Search for Extraterrestrial Life: Recent Developments*, M.D. Papagiannis (ed.), Reidel Pub.Co., Boston.
81. Subotowicz, M. (1979), Interstellar Communication by Neutrino Beams, *Acta Astronautica*, **6**, 213-220.
82. Sullivan, W. (1964); *We Are Not Alone*, McGraw-Hill Co., New York.
83. Sullivan III, W.T. and Mighell, K.J. (1984), A Milky Way Search Strategy for Extraterrestrial Intelligence, *Icarus*, **60**, 675-684.
84. Sullivan III, W.T., Brown, S. and Wetherill, C. (1978), Eavesdropping: The Radio Signature of the Earth, *Science*, **199**, 377-388.
85. Stootman, F.H.; De Horta, A.Y.; Wellington, K.J. and Oliver, C.A. (2000), The Southern SERENDIP Project, in G.A. Lemarchand and K. Meech (eds.), *Bioastronomy 99: A New Era in the Search for Life in the Universe*, ASP Conference Series, San Francisco.
86. Tarter, J.C. (1984), SETI and Serendipity, *Acta Astronautica*, **11**, 387-391.
87. Tarter, J.C. and Chyba, C.F. (1999), Is There Life Elsewhere in the Universe?, *Scientific American*, **281**, 80-85.
88. Tipler, F.J. (1980); Extraterrestrial Intelligent Beings do not Exist, *Q.J.R. Astr. Soc.*, 21, 267.
89. Townes, C.H. (1983); At What Wavelengths Should we Search for Signals from Extraterrestrial Intelligence?, *Proc. Natl. Acad. Sci. USA*, **80**, 1147-1151.
90. Überall, H., Kelly, F.J. and Sáenz, A.W. (1979), Neutrino Beams: a New Concept in Telecommunications, *J. Wash. Acad. Sci.*, **69** (2), 48-54
91. Vallée, J.P. and M. Simard-Normandin (1981), Observational Search for Polarized Emission from Space Vehicles/Communication Relays Near the Galactic Centre, *Astronomy & Astrophysics*, 243, 274-276.
92. Viewing, D.R., Horswell, C.J. and Palmer, E.W. (1977), Detection of Starships, *J.of the British Interplanetary Soc.*, **30**, 99-104.
93. Werthimer, D.; Bowyer, S.; Cobb, J.; Lebofsky, M. and Lampton, M. (2000), The Serendip IV Arecibo Sky Survey, in G.A. Lemarchand and K. Meech (eds.), *Bioastronomy 99: A New Era in the Search for Life in the Universe*, ASP Conference Series, San Francisco.
94. Whitmire, D.P and Wright, D.P. (1980), Nuclear Waste Spectrum as Evidence of Technological Extraterrestrial Civilizations, *Icarus*, **42**, 149-156.
95. Yokoo, H. and Oshima, T. (1979), Is Bacteriophage ΦX 174 DNA a Message from an ETI?, *Icarus*, **38**, 148-153.
96. Zubrin, R. (1996), Detection of Extraterrestrial Civilizations via the Spectral Signature of Advanced Interstellar Spacecraft, *J. of the British Interplanetary Soc.*, **49**, 297-302.

COSMOS AND COSMOLOGY

HÉCTOR RAGO
Grupo de Física Teórica and Centro de Astrofísica Teórica
Departamento de Física, Facultad de Ciencias.
Universidad de Los Andes. Mérida 5101. Venezuela
email: rago@ciens.ula.ve

1. Introduction

...ese objeto secreto y conjetural
cuyo nombre usurpan los hombres,
pero que ningun hombre ha mirado:
el inconcebible universo.
Jorge Luis Borges

Perhaps science is not anything else than an attempt (inescapable) to elucidate our own relationship with nature. Seen in this way, the scientific treatment of our origins is vital to the understanding of ourselves. Cosmology, conceived as an effort to give coherence to the physical world at the largest scale using the methods and tools of physics and astronomy, has much to say about how the proper cosmic conditions for the emergence of life arrived. As a matter of fact, perhaps the most important change in our view of the universe since the birth of modern science is the discovery by twentieth century cosmology that the universe is a dynamical entity, which evolves according to local laws that we can (and must) discover (or invent?). Cosmology is the ultimate historical science, that must understand the present universe as the result of the initial conditions that existed 13×10^9 years ago. The evolutionary process allowed to go from a hostile, hot and nearly uniform universe to another highly hospitable, cold, complex and finely structured at very diverse scales world, capable to develop variety, quasi-static structures out of equilibrium (thanks to the anti-entropic tendency of gravitation) and other conditions absolutely necessary for the emergence of any form of life. Within the multidisciplinary approach of

J. Chela-Flores et al. (eds.), Astrobiology, 33–40.

astrobiology, the view that cosmology offers is important. Specifically, we aim in this work to show in a non-technical way why this perspective is worth to be taken into account, that is to say, why we think that it has scientific value. Later we will present, again in a non-technical manner, the present paradigm and its supports. Finally, we shall comment briefly about the challenges posed by the paradigm, looking forward to the near future.

2. Scientific Understanding of the Universe.

> I do not pretend to understand the universe...,
> its a great deal bigger than I am.
> Thomas Carlyle

As cinema and jazz, cosmology was born in the twentieth century. The article which initiates scientific cosmology is by Einstein, published in 1917. Before that, the theoretical tools, the necessary observational technologies and a minimum understanding of the local astronomical phenomena were not available in order to develop an overview of the subject. At first sight we only distinguish our own galaxy, the Milky Way, which is only one of the 10^{11} that exist in the observable universe. It was only sixty years ago that we understood the origin of the energy in the stars that allows them to shine for thousands of millions of years, while the heavy elements in the periodic table are formed in their interior. During the twentieth century the capacity to detect photons through the use of telescopes has been multiplied by a hundred thousand, and recently we can make observations not only in the optical range of the spectrum, but also in wavelengths corresponding to infrared and ultraviolet light, radio waves, microwaves, and up to 10^{12}e.v. We can avoid the distorting effect of the earth atmosphere putting telescopes in outer space. The experiments done in large accelerators have made it possible to invent (or discover?) the laws of the subatomic world. A great variety of equipments, observatories and accelerators throw up permanently a formidable quantity of high quality data about the physical world. This data is handled and analyzed swiftly through the use of modern computers. The combined effects of all these developments have taken cosmology from a conjectural and speculative stage, without solid data and in which prejudice had a lot of weight, to another much more precise, in which observational results and independent and crossed tests validate or not the proposed models and limit the freedom of theorists in their proposals. In these times of globalization, cosmology is, without doubt, big science, a mature science capable to deal with the real world, to correct itself, to lay off models and to convince us that that, at the largest of scales, the physical world is simple enough to be described in scientific terms.

3. Standard Cosmology

The universe is real but you cant see it.
You have to imagine it
Alexander Calder

Cosmology is at the center of a square whose sides are Einstein's gravitation theory, or general relativity, the standard model for elementary particles, statistical physics and some simplifying assumptions. At great scale the universe is dominated by gravity. Consequently, we must resort to the best description of gravitational phenomena, that is to general relativity. The great moral of general relativity is that gravity is a manifestation of the curvature of space-time, which turns them into main actors, at variance with their previous role as the stage in which matter and the physical fields 'live. The content of energy-matter determines the space geometry, and its evolution in time, according to Einstein's equations, so that observations should make an inventory of the content of matter and energy in the universe. It is important to point out that general relativity has passed successfully all tests done through observations and experiments. It is one of the best corroborated physical theories. On the other hand, the behaviour of matter and energy obeys the laws of the standard model for elementary particles. The standard model includes the quantum mechanical description of matter at different energy scales, from molecular physics to high energy physics, through atomic and nuclear physics. It is the physics that we were able to build through the use of great particle accelerators, and provides us with valuable information about the nature of the fundamental interactions. Its successes in the description of reality include the explanation of the structure and the properties of matter, the hierarchy in the periodic table, the nature of electromagnetic radiation, radioactivity, the nuclear reactions that make possible the shining of the stars as well as predict the results of any experiment done in the large particle accelerators. With general relativity and the standard model in hand, cosmologists introduce some assumptions which they believe are valid in our universe, for instance that at very large scale, about 200 million light years, matter and energy are distributed uniformly, so that the geometry of space must reflect that homogeneity and isotropy. This assumption has been verified by measurements of the background radiation, of which we will say more below. The basic idea is, of course, to build models which try to replicate the salient features of the real universe. As is to be expected, the number of models consistent with theory is very large, so that one has to turn to observation to get input for the values of some parameters. Is through this interplay between theory and observation how it has been possible to design a coherent image, the big-bang model or standard cosmological model, with enough empirical successes (and, equally important, without contrary observations) that has established itself as the accepted paradigm for the community of cosmologists. It is important to remark that this does not mean that we understand every detail of the actual structure of the universe, nor that we can answer all questions (some of them very important). The big-bang model is the framework in which observations must be organized and interpreted and the context in which the details of the relevant cosmological processes that have taken place must be refined.

4. The History of a Hot Universe

In the beginnings is the end.
Thomas Steam Elliot

The general scheme of the big-bang model assumes that the universe that we see today comes from a highly dense and hot stage which began to expand some thirteen billion of years ago. The expansion is gauged by the scale factor which measures the relative size of the universe. The model establishes that the temperature measured through the background radiation is inversely proportional to the scale factor, whose dynamics is ruled by the content of energy-matter that we observe, including the energy of the vacuum, and the curvature of space-time, through Einstein's equations. So, the physical processes that take place depend on the temperature scale considered, which in turn depends on the time elapsed since the beginning of the expansion. The history of the universe is, then, the history of the processes that happen while the universe expands and gets colder. Let us point out some of the relevant episodes of this history. We will begin at 10^{-5} sec, time at which the energy that dominated the expansion of the universe was that of radiation and ultra-relativistic particles. The temperature of the environment was about 5×10^{12}K and it was at this stage that protons and neutrons were constituted from quarks, through a process known as baryogenesis. When $t \sim 1$ sec-3 min, temperature goes down from $10^{10}K$ to $10^{9}K$, which are in the range typical in nuclear physics, the density is about 5×10^{5} gr. cm^{-3}. Physics is now conventional and predictions can be made. In this period neutrinos decouple and nuclear reactions create light nuclei like deuterium, helium, helium 3 and some lithium (primordial nucleogenesis). The process of creation of heavier nuclei does not continue because soon the temperature gets too low. When $t \sim 300,000$ years have elapsed, matter density exceeds that of radiation and it begins to control the evolution of the scale factor. The temperature at this stage is about 4000K and photons do not have enough energy to impede the creation of hydrogen and helium atoms. The predominant physics is atomic physics. The universe ceases to be an opaque ionized plasma because radiation interacts very weakly with neutral matter and photons can travel freely, affected only by the expansion of the universe. At $t \sim 10^9 - 10^{10}$ years, the dynamics of the universe begins to be ruled by the vacuum energy or the cosmological constant. This stage corresponds to long range gravitational physics, in which small fluctuations (ten parts per million) in the average density of matter, begin to collapse gravitationally amplifying the contrast in the density by a factor of 10^7. It is the time of the formation of structures: galaxies, galaxy clusters and superclusters. Finally, when 13×10^9 years have elapsed and the temperature is 3K, complex biochemical structures appear, originated from the heavy elements synthesized at the core of the stars and thrown out into space through supernovas. For times less than 10^{-5}sec, the physics is more uncertain. At $t \sim 10^{-43}$sec, the so called Planck period, the quantum effects of gravity were the most important. Lacking a credible theory we can not say anything about it. It is believed that around $10^{-35} - 10^{-32}$ sec, an exponential inflation happened which gave the universe some of its most important features. Without this theory we could not be able to explain these characteristics, except by assuming bizarre initial conditions. As a result of the inflationary phase, the universe was made uniform, the curvature of space was annulled and, perhaps, the quantum fluctuations of the field that created the inflation also began the fluctuations which gave way to the formation of the cosmic structures.

5. Corpus Delicti

> It is a good rule not to push overmuch
> confidence in the observational results
> that are put forward until they are comfirmed by theory.
> Sir Arthur Eddington

What reasons can be put forward in favor of the big bang? Why cosmologists think that the big bang model is a good representation of the actual universe? Aside of the fact that it is supported by local theories with great solvency, the big bang model is solidly supported by recent observations, which have been made in recent years with increasing precision. Mainly these are related to the expansion of the universe, as shown by the red shift of the spectral lines of hundreds of thousands of distant galaxies. Measurements show that this expansion is larger by the same factor in which the distance of the galaxy is larger. The quantitative relation is $v = Hd$, where the Hubble parameter is taken presently as $H = 65$ Km.sec^{-1}Mpc^{-1}, within an error of 10%. One Mpc = 3.2 millions light-years.

The second observational support of the big bang is related to the synthesis of nuclei. The model allows the calculation of the abundance of light elements (helium 4, deuterium, helium 3 and lithium 7) relative to that of hydrogen. The proportion found ($1 : 0.25 : 3\times10^{-5} : 2\times10^{-5} : 2\times10^{-10}$) is consistent with what is found in primitive samples of the universe. Particularly, observations on the abundance of deuterium, measured with great precision using absorption lines from quasars, indicates that the present density of protons and neutrons (baryons) is about 3×10^{-31}.

The third observational basis for the big bang model is the detection of the cosmic microwave background radiation (CMBR) predicted in the forties by Gamow and collaborators and found by Penzias and Wilson in 1964. Its very existence tells us about a hot phase of the universe, but, moreover, the sophisticated study to which it has been submitted during the 90s provides us with further evidence in favor of the big bang model, as well with valuable information about the universe when it only had 0.02% of its present age. CMBR is a residue from the last moment in which matter and radiation were in thermodynamical equilibrium, that ended when matter became neutral. Its present temperature is $T = (2.725 \pm 0.002)$K with a typical wavelength of about 2 mm, and with the most perfect black body spectrum found in nature (deviations are of three parts in 10000). Besides, CMBR has the same temperature for whichever direction of the sky that we are looking at, within a margin of 10 ppm. This isotropy is a consequence of the uniformity of the expansion and of the homogeneous and featureless quality of the universe when it was 300.000 years old and its temperature was 3000 K, and strongly supports the big bang model. Even more interesting is the finding in 1992, through more precise measurements, of changes in this background temperature for different directions (1 part in 100.000), which evidence the non-uniformity which generated the large structures seen today. Since then cosmic radiation has been examined in great detail, due to the fact that the precise form of the anisotropies or, technically, the power spectrum gives information about important cosmic parameters, as the density of matter, the cosmological constant and space curvature.

6. Solving the Puzzle

A place for every thing
And every thing into its place.
Anonymous

The kind of universe in which we live, its geometry and its special way to expand, depends, in every stage, of its content of energy-matter. These numbers, in turn, limit the possible models for structure formation. On the other hand, inflation in the first moments after the big bang imply some characteristics of our universe. Observations of the cosmic radiation, of the abundance of light elements and of the rate of expansion of distant objects give new data about the universe. In units of the critical density necessary for an Euclidean geometry of space, ordinary matter (baryons) contribute with 5%, cosmic radiation photons with 0.01%, neutrinos freed in the first fractions of a second with 3%, and dark matter detected through gravitational lenses, galaxy dynamics or great scale fluxes with 35% (so that the substance in which we are made is not the most abundant in the universe). Besides, the analysis of the fluctuations of cosmic radiation suggest that the total density must be unity, a result supported by inflation predictions, according to which the curvature of space is null and then the total density must have the critical value. The paradigm that begins to have the greater acceptance holds that the remaining 65% is provided by the vacuum energy, that is to say the cosmological constant. The effect of the vacuum energy is to produce a gravitational repulsion and that is what seem to indicate recent observations of distant supernovaes. Instead of diminishing, the expansion of the universe is accelerating due to the existence of a cosmological constant different from zero. Moreover, the more convincing models for the formation of structures are the ones that include non- relativistic dark matter, as well as a cosmological constant.

7. The Challenges

Physics is too complicated
to leave it to the physicists.
David Hilbert

There is no doubt that a healthy relationship between fundamental physics and valuable observations has played an important part in the great advances in cosmology during the last years. Thanks to it we now have a coherent picture of the evolution of the universe since fractions of seconds after the big bang to our days. However, there still remain a lot of loose ends and many unanswered questions. Some of them will be resolved in the near future, through more and better observations, many of which are been carried out already. But others will have to wait for new physical laws at a deeper level. Specifically, the new measurements will allow to determine with greater accuracy the cosmological parameters (mass-energy densities, Hubble constant, radiation anisotropy, cosmological constant or its equivalent), which will allow to adjust the inflation models and those of structure formation. But it will be necessary to identify the composition of non-baryonic dark matter (neutralinos? axions?), without doubt the remains of an age whose physics we do not know well enough. We do not understand why ordinary matter prevails over antimatter. The physics we know is symmetric with respect to particles and antiparticles. Fortunately for us, a few instants before baryogenesis a small asymmetry of one part in 10^{-9} left a slight excess of particles, which are the ones that we now see. The cosmological constant also creates enigmas related to the most fundamental physics. We must identify accurately the reason for the acceleration of the expansion of the universe if observations confirm its existence. Why theoretical calculations differ in 120 orders of magnitude with astronomical observations? The cosmological constant is a quantum originated term (the energy of virtual vacuum pairs) put into a classic equation. These great disagreements show that we are not using an appropriate description. It is possible that the much sought quantum theory of gravity, or some other 'final theory will offer a better understanding of the problem of the cosmological constant. That presumed theory will also be needed to answer some fundamental questions as, for instance,

Which characteristics of the universe are fossils from the age of quantum gravity? Perhaps the number of dimensions of space and time?

What kind of dynamite propelled the expansion? What is the nature of the big bang?

Why the fundamental constants and the cosmological parameters have values that not only allow but favor the emergence of complexity? Are those values determined by basic principles or got in by the back door of chance through the break up of symmetries, for instance?

The history of science shows how foolhardy it is to dare to predict the way by which our understanding of the world will go. Nobody could have foreseen a few decades ago the giddy development of our understanding of the cosmos. Sometimes an unexpected observation or a new theory can alter the intended route. At this moment we can only assert that the turmoil in which cosmology finds itself promises advances which will elucidate our relation with nature. Is not that the aim of science?

NEW DEVELOPMENTS IN ASTRONOMY RELEVANT TO ASTROBIOLOGY

SABATINO SOFIA
Department of Astronomy
Yale University
New Haven, CT, USA

ABSTRACT

This paper outlines the principal relationships between conventional Astronomy and Astrobiology. In particular, it addresses the issues of the origin of the elements that make life possible, and the origin of the planets which are also an essential requirement for the development of living systems. The paper presents the newest developments in Astronomy that confirm the general ideas on these matters developed over the last 50 years, and points out detailed features that disagree with the existing paradigms.

1. Introduction

The basic link between Astronomy and Biology arises from the very fundamental questions: What are we made of? What is the origin of the stuff we are made of? What processes led to the formation of the Sun and Earth so fundamental for the development of life?, and finally, are we alone in our Galaxy (and the Universe), or are we part of a thriving and large community of living (and often intelligent) beings scattered throughout the cosmos? The primary emphasis of Astrobiology, in my opinion, deals with the early stages of the cosmic voyage, firmly anchored in Cosmology; it proceeds through the formation and evolution of galaxies and stars, specifically those surrounded by planetary companions. It ends where a planet has formed which, like Earth, has a benign environment that allows the formation of the most primitive life forms, and its subsequent evolution into complex living systems. At this point, Evolutionary Biology takes over, and explains how these primitive life forms evolve into complex living organisms. We get back to Astrobiology when we wish to determine the likelihood that a similar process occurred elsewhere in the cosmos. The organization of this paper will be in terms of the questions that Astronomy and Astrophysics can answer.

J. Chela-Flores et al. (eds.), Astrobiology, 41–53.

2. What Are We Made of, and Where Did These Elements Originate?

Our bodies, and indeed the bulk of most living organisms and planetary mass are made up of many complex chemicals. For living organisms, water is the most ubiquitous compound, followed by complex combinations of Carbon, Hydrogen, and Oxygen. Other than Hydrogen, the elemental composition of living organisms is made up of what is known in Astrophysics as "heavy elements", or "metals", which means elements heavier than Hydrogen and Helium. It turns out that this elemental composition has profound implications regarding the origin of the material.

2.1. PRIMORDIAL NUCLEOSYNTHESIS

Following the big bang, approximately 15 billion years ago, the Universe was filled by a primordial soup of strange particles and enormous amounts of energy, and governed by forces and processes that we are still hard at work trying to comprehend. In fact, what occurred during the first few minutes following the big bang is studied by a particular branch of Cosmology very akin to Particle Physics theory, and entire books have been written on the subject. What is really relevant to Astrobiology is that about 15 seconds after the big bang, and following a vast process of annihilation of electrons and positrons, we are left with a small number of unmatched electrons, an equal number of protons, and a lot of photons, and, of course, the mysterious gravitons. At this point the temperature has fallen below 3 billion K, and although it is sufficiently cold to prevent pair formation, the collisions between the various ingredients of the soup are sufficiently violent to prevent any stable nuclei beyond Hydrogen. Remember that a proton is a Hydrogen nucleus, and so at this time the elemental component of the soup, in baryonic terms, is only Hydrogen. At about one minute after the big bang, the temperature has dropped below 1.3 billion K. The collisions are now more gentle than before, so that proton-proton and proton-neutron collisions can form larger nuclei that will not break up. This is the beginning of primordial nucleosynthesis.

The process required for two positively charged particles (e.g. protons) to combine is overcoming the Coulomb barrier, that is, a repulsive force proportional to the product of the charges, and inversely proportional to the square of the distance separating them. This Coulomb force (which is attractive for unlike-charged particles) is very large when the particles are very near each other, and it is the reason why spontaneous nuclear fusion does not occur except under extreme conditions. The extreme conditions arise when the temperature of the gas is very high (although not too high, since then subsequent collisions would break the combined nucleus up). The Coulomb force is effective until the two protons are so close to each other that the nuclear force takes over. This force is both attractive, and very strong, but it decreases rapidly with distance. As a consequence, when the particles are pushed together sufficiently close to each other, and the nuclear force takes over, the two particle will become very strongly bound, and the very large binding energy will be released as a neutrino, a positron, and a very high energy photon called gamma ray. This is the energy that powers stars and generates the destructive energy of an H-bomb.

This primordial nucleosynthesis ceases when the Universe is about 5 minutes old. The reason for this is that as the Universe ages, it expands and cools down. Under the conditions of density and composition found in the early Universe, the nuclear reactions cease when the temperature has decreased to about 600 million K, which occurs at the age of 5 minutes. Calculations indicate that by this time approximately 24 percent of the mass of the Universe will have been converted into He-4, (a nucleus containing two protons and two neutrons, also called alpha particle). There will be very small amounts of Deuterium (heavy Hydrogen, formed by a proton and a neutron), and He-3 (formed by two protons and a neutron). The rest, approximately 76 percent of the mass, will remain Hydrogen. The reason that element buildup stops at Helium has to do with the fact that subsequent reactions towards heavier elements would have to overcome a larger Coulomb barrier (2 for He, vs. 1 for H). However, at the same time, the Universe has been cooling down, and the reactions cannot even continue for H, hence they never start for He.

The previous discussion shows that the stuff of which life and living systems are made up were not present in the early Universe. It had to be made elsewhere.

2.2. STELLAR NUCLEOSYNTHESIS

It was in the 1960's when a clear understanding of this process occurred, in a paper published in the Astrophysical Journal by the wife and husband team of Geoffrey and Margaret Burbidge, by the nobel laureate William Fowler, and by Fred Hoyle. They integrated the modeling of stellar interiors developed in the 1920's by Sir Arthur Eddington, with the nuclear astrophysics concepts first developed by Hans Bethe, and then expanded by William Fowler, using tools pioneered by Martin Schwarzschild, to explain how the building up of heavy elements from the debris of formation of the early Universe occurred in nature. I will summarize hereafter the main points.

A star is formed out of material found in the interstellar medium. This material is made up of discrete clouds immersed in a hotter and less dense background called "the diffuse medium". The clouds persist because of a balance between the dispersive forces of pressure, and the contacting force of self-gravity among the particles in the cloud. The dispersive forces increase with the temperature of the gas, whereas the gravitational forces increase with density. Whenever, for any reason, a region within a cloud becomes a bit denser and cooler than the rest of the cloud, gravitational forces dominate over pressure, and a clump of the cloud begins to contract under the driving force of gravitation.

The force of gravitation is spherically symmetric, so a contracting cloud will tend to be spherical in shape. On the other hand, because of conservation of angular momentum, the contraction along the axis of rotation can proceed unimpeded, whereas contraction in the equatorial plane will be opposed by centrifugal forces. As a consequence, the contracting element of the cloud will start out spherical in shape, and subsequently will begin flattening out like a pancake, with a central near-spherical core made up of low angular momentum material. This rotating disk left behind by the contraction will be the material out of which planets will form. The central core will form the star.

Let us now concentrate our attention on the core material, which we can denote as a "proto-star". When the cloud begins to shrink, according to the virial theorem, half of the gravitational energy released is radiated away, whereas the other half becomes internal energy of the cloud, or, in other words, the cloud begins to heat up. It is to be noted that this hotter cloud will not dissipate because the increase in pressure is only one half as large as the increase in gravitational binding energy of the cloud. As this process proceeds, at an ever faster pace, the radiated energy becomes sufficiently large to cause the contracting cloud to glow just like a star. It is at this point that we call it a proto-star, and it looks like any star, excepting that its energy comes from gravitational sources, instead of nuclear sources as it occurs in normal stars.

The central temperature in the proto-star keeps increasing as a consequence of the contraction, until it reaches approximately 5 million K. When this happens, the density conditions are such that proton-proton reactions can begin. In other words, the proton collisions can overcome the Coulomb barrier, and the conversion of Hydrogen into Helium can proceed. At this point, a very stable dynamical equilibrium sets in, and the contraction stops. We say that the star is just entering the "main sequence", which is the stage when the star gets its energy from processing H into He. The ignition point is known as Zero Age Main Sequence, or ZAMS.

The contraction is halted at the instant when the nuclear processes in the stellar core exactly equal the energy loss by radiation at the stellar surface. The reason the equilibrium is very stable is due to the high sensitivity of the rate of nuclear reactions to temperature. Let us assume that, for a moment, more energy is produced at the core than is dissipated at the surface. The core would quickly expand a bit, which would lower the central temperature, and thus decrease the subsequent rate of energy generation. Similarly, if less energy were replaced through nuclear processes in the core, it would cool down, shrink, heat up sufficiently until just the right amount of energy is produced.

Nuclear fusion processes can yield energy (i.e. be exothermic) only up to Fe. However, the step that is most efficient in yielding energy is the first one, namely, the conversion of H into He, also known as "hydrogen burning". As long as there is any H remaining in the stellar core, the star is in the main sequence stages of evolution. On the other hand, as the proportion of H decreases, the temperature has to increase to provide the same energy output. As a consequence, the central temperature of the star has to increase. For example, for the Sun, approximately 50 percent of the initial H in the core has been converted into He during the nearly 5 billion years of its existence. It will take another 5 billion years to exhaust the hydrogen in the core, at which point the Sun will exit the main sequence. At the present time, the temperature in the solar core is approximately 17 million K.

After the H is exhausted in the core, the temperature is not sufficiently high for the subsequent nuclear step. Since the energy lost by the core to power the star is not replenished, it will cool down, and begin contracting. When this happens the temperature in the core increases until it reaches approximately 100 million K. At this point, the central temperature is high enough to allow overcoming the

Coulomb barrier of He, thus starting the stage of He "burning". At the same time, the temperature in a shell surrounding this hot core has also become sufficiently hot to start H-burning. The consequences are that all this energy (initiated by a shrinking of the core) causes the outer regions of the star to expand and become highly luminous. This stage of evolution is called "red giant", because the star is very large, and the surface temperature is lower than it was on the main sequence. In the case of the Sun, when approximately 6 billion years from now it becomes a red giant, it will gobble up the Earth and the radius will extend up to near the orbit of Mars. The primary reaction in the core is called the triple-α reaction, in which 3 He atoms combine to form Carbon. Also, a part of this reaction will proceed further to Oxygen. This evolved core will be surrounded by a He shell, which in turn is surrounded by a H-rich, unprocessed envelope.

What happens in the stages subsequent to He exhaustion in the core is critically dependent on the mass of the star. In particular, very different processes occur for low mass and for high mass stars. In this regard, the dividing line (not completely well established) is around 4 times the mass of the Sun.

3. Low Mass Stars

For low mass stars, following He-exhaustion in the core, the contraction will not produce sufficiently high temperature to produce C and O burning before the core material becomes "degenerate". Degeneracy arises when the density of the gas is so high that the exclusion principle becomes operative. The exclusion principle states that in a unit of phase space no more than two electrons can fit. This means that, as the density of a gas increases, the Maxwellian distribution of momentum of the particles is misshapen, until it becomes a square box whose upper boundary only dependent on the density. At this point, the equation of state, which relates pressure, temperature, and density, changes from the ideal gas law to a degenerate matter equation of state. In particular, in an ideal gas (which accurately represents what happens inside a normal star), pressure is proportional to the product of density and temperature. That is the reason why a contracting core becomes stabilized as both the temperature and the density increase. By contrast, in a degenerate gas the pressure in proportional only to a power of the density, and totally unrelated to temperature. The power of the density dependence is 5/3 for non-relativistic, and 4/3 for relativistic degeneracy. For a low mass star, then, following He-exhaustion, the core becomes non- relativistically degenerate. At this point, temperature plays no role. However, the density dependence of the pressure is sufficiently high so that a contracting core gains more degenerate pressure than the increase in gravitational binding force. As a consequence, the contracting core can (and does) get stabilized against further contraction by the degeneracy. This prevents the temperature from ever reaching a value in which subsequent nuclear burning, to heavier elements, will take place. As a consequence, the final fate of a low mass star is to have a hot core of C and O surrounded by a He shell, surrounded by a H-rich envelope. This will happen as long as this core is less massive than 1.4 times the mass of the Sun, which is known as the Chandrasekhar limit.

Remember that a star is called "low mass" as long as its mass is below about

4 solar masses. What happened to the difference between 4 and 1.4 solar masses? This material is lost to the interstellar medium both in the form of a strong stellar wind, and in various episodic sheddings of mass shells which produce what is called "planetary nebulae" (shells of gas having nothing to do with planets). These events of mass loss get rid of the major portion of the envelope of the red giant. Having shed most of the envelope, what remains is only the naked core. This naked core is very hot and very dense, and is called a "white dwarf". It is white (or actually blue) because it is hot, and it is dwarf because its size is closer to that of the Earth than that of a normal star. As a consequence, its density is extremely high, about a few million grams per cubic centi-meter. This object has reached the end of the evolution life. It will not change excepting very slowly cooling down until it becomes a black cinder in the sky.

What is important to note in terms of astrobiology is that, even though a good fraction (about 1/3) of the mass of a low mass star has been converted into C and O, elements very important for life, these heavy materials remain trapped inside the white dwarf, and so they are never recycled into the interstellar medium. Thus, they do not help enrich the intestellar medium with the material that a next generation star might convert into planets, and perhaps life.

4. High Mass Stars

Fortunately, for the high mass stars, the more massive contracting core following He-burning will reach temperatures high enough to begin subsequent stages of nuclear burning before the condition of degeneracy is reached. Consequently the next steps follow each other until the core has been converted into Fe. At this point, the structure of the star is as shown in Figure 1.

Remember that Fe cannot generate energy. Hence, as the Fe core begins to cool down, and contract, no new energy source can stop this contraction. The core will become degenerate in the relativistic regime. Because of the low density dependence of relativistic degeneracy, the increase in pressure is only exactly countering the increase in gravitational binding energy. In other words, if the gravitational (contracting) forces exceed the pressure (against contraction) forces at the beginning of degeneracy, the contraction will continue, not only unimpeded, but at an ever increasing pace. The collapse crunches the Fe nuclei so violently that they break up into their component protons and neutrons. At the same time, the protons are forced to combine with the ambient electrons to also be converted into neutrons, and so the core is made up entirely by neutrons. Since the electrons provided some pressure support, their disappearance further fail to inhibit the contraction which now reaches a "free fall" velocity of nearly 10 percent the speed of light. When the density of the core approaches 10^{14} gr/cm^3, the neutrons touch each other and strongly resist further collapse. The core begins bouncing back. Because the core collapse proceeds so quickly (the last steps taking only a fraction of a second), the surrounding envelope at first remains totally unaware of the great catastrophe going on at its center. However, this does not last very long. The gravitational energy released during this final collapse is as large as the energy released by over a hundred suns during their entire lifetimes. A lot of this energy

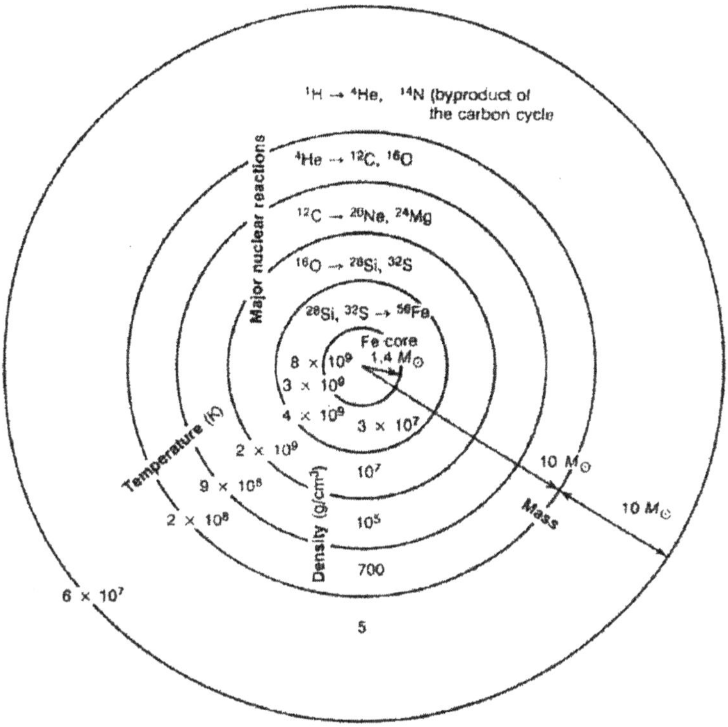

Figure 1. The Structure of a 20 $M_\odot$ star just prior to collapse.

goes into breaking up the iron nuclei into protons and neutrons, and into squeezing the electrons into protons. The latter operation generates energetic neutrinos, which can escape the stellar core quite unimpeded, and manage to carry away a very large share of the collapse energy. The rest of the energy creates a strong shock wave which begins to propagate outwards from the stellar core. This shock wave strongly heats the outer envelope fostering nuclear buildup of elements, primarily by creating neutrons which can enter existing stable nuclei without having to overcome a Coulomb barrier, since they are neutral. This process (and only it) is capable of building up nuclei heavier than iron (for example, Cu, Zn, Au, etc., elements that in low quantities are part of living systems, and part of the planetary masses). Besides building up heavy elements, the shock and the nuclear reactions heat and accelerate all the material in the envelope. Eventually, in a great explosion, the bulk of the stellar envelope is hurdled outwards with speeds exceeding 1/10 c. The entire material acquires momentarily a luminosy rivaling that of an entire dwarf galaxy. This phenomenon is called a SUPERNOVA explosion. Meanwhile, depending on the mass of the stellar core left behind, we have a neutron star, or a black hole, objects which have been detected in recent years both as binary X-ray sources, and as pulsars. These objects have stellar size masses and diameters of only a few kms.

The supernova event, then, in terms of astrobiology, has two main purposes. One, it builds up elements heavier than iron. The other, it takes up a lot of the heavy elements built up in the stellar interior and throws them into the interstellar medium. Subsequent generations of stars, like the Sun, start out with, besides the H and He produced early in the Universe, the heavy elements cooked up in the stellar interiors. It is these elements that form the bulk of the mass of the terrestrial type planets, and of living systems. So, life is primarily made up of heavy elements cooked up by nuclear processes in the interior of massive stars, further built up by neutron enrichment of nuclei during the supernova explosion which marks the death of each massive star. We are made of star debris.

5. Formation of Planets

As stated earlier, when a star begins to form, it leaves behind a disk which persists a good fraction of the time until the star has completed its formation. Young stars, most effectively represented by so-called T-Tauri stars, are surrounded by gaseous disks approximately 50 percent of the time, for periods of time extending up to 5-10 million years. For the other half without disks, (also called naked T-Tauri stars), the pancake left behind during the cloud contraction was presumably disrupted quickly (for example, by the presence of a nearby companion). Similarly, the disk which do remain will also disappear either by forming planets, or by dissipation. Either process must be very efficient because of the short lifetime of the disks.

It is difficult at the present time to establish with any degree of certainty precisely what fraction of the disks is dissipated, and what fraction produces planets. In general terms, we can understand how both processes operate, but the quantitative understanding of the phenomenon (based on some mysterious "disk viscosity" whose specific nature still eludes us) is still not there. However, for a long time, it has been our belief that planetary formation should be a reasonably efficient process, and that planets surrounding low mass stars should be ubiquitous. The problem is that to observationally test this belief has been a difficult and frustrating process, which only recently has started to produce results.

6. How Do We Observe Planets?

The enormous brightness contrast between a star and any companion planet is so extremely large that it is hopeless to actually directly observe the planet in the presence of the nearby star, even for the largest planets, and for the faintest stars. It is a well known limitation of the best current technology that for an unresolved system (as all planetary systems are bound to be), we cannot detect a companion which is less luminous than 1/1000 of the brighter star. This is true even for a binary star system. In the case of Jupiter and the Sun, for example, the brightness ratio (for an observer outside the solar system) is approximately 1 part in 1 billion (at visual wavelengths), and 1 part in a million in the infrared.

The only hope to detect a massive unseen planetary companion to a normal star is by detecting the dynamical effects of the companion on the star. Basically, it is said that a star meanders among its neighbors in a straight line, as required by Newton's first law. However, if a star has a companion, they orbit each other, and it is their common center of mass that moves in a straight line. For two stars of equal mass, the center of mass resides half way between them, and the stars move in space describing figures 8 with respect to the stellar background. Since the center of mass is the balancing point between the two objects, it lies much closer to the more massive than to the least massive object. For example, if the Solar System consisted only of the Sun and Jupiter (not a very inaccurate assumption in terms of mass), the center of mass would be near the surface of the Sun in the direction of Jupiter. Thus, an observer outside the solar system could detect the presence of the unseen Jupiter only by detecting the fact that as the Sun travels through the Galaxy, it wobbles around its center by an amount approximately equal to its diameter over a period of 11.86 years. The orbital velocity of this motion is about 12.5 m/s.

Because of the requirements just described, detection of planets outside the solar system has been a long and frustrating undertaking that has often produced results that later were proven wrong. The situation has changed drastically in recent years when through instrumental and methodology advances,the determination of radial velocities migrated from standard errors of 20 km/s to 1 km/s, to 2-3 m/s.

First of all, we must understand the origin of the radial velocity changes. The orbital motion of the star-planet system occurs in a fixed plane. If the plane were orthogonal to the line of sight, the orbital motion would produce no radial velocity changes. If, on the other hand, the orbital plane has any other inclination, the orbital motion has a component along the line of sight. This causes a shift of the spectral lines of the star or stars because of the doppler effect.

The shift of the spectral line as a consequence of the doppler effect is:

$$d_\lambda = \lambda \frac{v}{c}$$

where v is the component of orbital velocity along the line of sight (radial velocity), and c is the speed of light. The line shift for a typical line ($\lambda = 5000$ Å), and for a velocity of 10 m/s would be

$$d_\lambda = 1.5 \times 10^{-4} \mathring{A}$$

Of course, to measure this is a very difficult undertaking, since the width of a typical spectral line is of the order of 1 Å, and noise on the line make it very difficult to establish a shift as small as 1 in 10,000. The advance in the last few years has consisted in developing absorption cells (in this case filled with iodine vapor) to sample a well established subset of the line, and to detect minute shifts. The current sensitivity is at about ± 3 m/s.

With the above technique, approximately 30 planets have been detected to date. They are all very massive (masses closer to or larger than Jupiter than the Earth), and most are much closer to their parent star than Jupiter is to the Sun. Also, keep in mind that since the orbital inclination is often unknown, we can only

TABLE 1. Masses and Orbital Characteristics of Extrasolar Planets

Star Name	Msin i M_{jup}	Period day	Semimajor Axis AU	Eccentricity	K m/s
HD 187123	0.52	3.097	0.042	0.00	72
τ Bootis	3.64	3.3126	0.042	0.00	469
HD 75289	0.42	3.508	0.047	0.00	54
51 Pegasi	0.44	4.2308	0.051	0.01	56.
υ Andromedae (b)	0.69	4.617	0.059	0.04	73.
υ Andromedae (c)	2.0	241.3	0.82	0.23	54.
υ Andromedae (d)	4.1	1280.6	2.4	0.31	67.
HD 217107	1.28	7.11	0.07	0.14	140
ρ^1 55 Cancri	0.85	14.656	0.12	0.03	75.8
Gliese 86	3.6	15.8	0.11	0.04	379
HD 195019	3.43	18.3	0.14	0.05	268
ρ Corona Borealis	1.1	39.6	0.23	0.1	67
HD 168443	5.04	58	0.28	0.54	330
Gliese 876	2.1	60.9	0.21	0.27	239.
HD 114762	11.0	84	0.41	0.33	619
70 Virginis	7.4	116.7	0.47	0.40	316.8
HD 210277	1.36	437	1.15	0.45	41
16 Cygni B	1.74	802.8	1.70	0.68	52.2
47 Ursae Majoris	2.42	1093	2.08	0.10	47.2
14 Herculis	4	$\approx$2000	$\approx$3	$\approx$0.35	80
ι Horologii	2.26	320	0.91	0.16	80
HD 130322	1.08	10.7	0.103	0.03	115
HD 192263	0.76	24.1	0.18	0.12	68
HD 10697	6.59	1083	2.0	0.12	123
HD 37124	1.04	155	0.585	0.19	43
HD 134987	1.58	260	0.78	0.24	50
HD 177830	1.28	391	1.00	0.43	36
HD 222582	5.4	576	1.35	0.71	187
HD 209458	0.63	3.5238	0.045	0.00	81

get Msin i. The current list of detected planets is presented in Table 1, and an example of the radial velocity curve in which the discovery is based is shown in Fig.2.

These discoveries have consequences that are mixed in terms of astrobiology. On the positive side, it means that the scenario envisioned for planetary formation, that is, as a byproduct of star formation is basically correct. On the negative side, the characteristics of these planetary systems are very different from those envisioned in our theories, which were obviously based on what we see in the

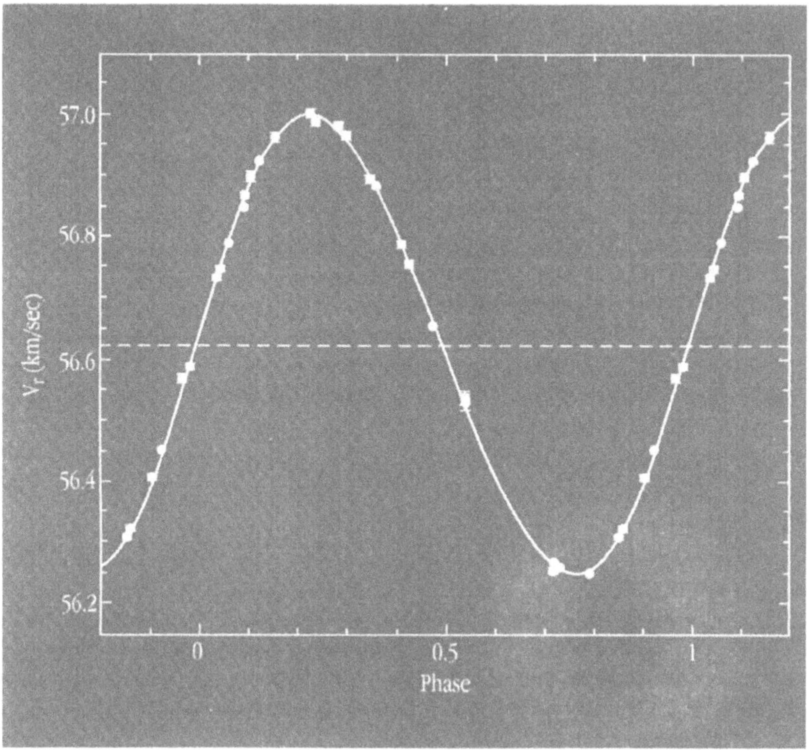

Figure 2. Radial velocity curve from a 3.6 Jupiter Mass Planet Around the Nearby Star Gliese 86

Solar System. Here, Jupiter has a nearly circular orbit at a large distance from the Sun. Both characteristics are critical for the existence of Earth. If the orbit were very eccentric, and/or if its semimajor axis (i.e. distance from the Sun) were much smaller, the perturbations that it would have exerted on the Earth would have thrown the latter either into an orbit where it would have become very inhospitable for the development and maintenance of life, or completely outside the Solar System. Since most of the discovered planets do have a small distance from their parent star, and/or a large eccentricity in their orbit, these systems will very unlikely harbor a planet amenable to life.

Since most of the planets discovered are both too massive, and their orbits have characteristics opposite to those associated with the development and maintenance of life, it would appear that the discoveries do more harm than good in our expectations to find life outside Earth. To some degree, the findings are sufficiently unexpected to cause us to profoundly revise the details of the processes which not only form planets, but also which cause their characteristics to be modified by interactions with each other. This theoretical activity has already started in earnest. However, we must understand that the properties of the planets discovered are affected by the limitations of our measurements. For example, if we were observing the Solar System from outside with the best current instrumentation, we could in principle detect Jupiter, because the wobble that it causes can be as large as 12.5 m/s, substantially above the threshold of detectability. However, because its period is 11.8 years, it might not have been observed long enough to have seen the orbital change. In fact, planets close to the parent star are easier to detect, first because their short period uncovers their nature within a few days of observation. Second, from Kepler's laws it follows that the orbital velocity decreases as the square root of the distance separating the objects, everything else being equal, the closer planets move faster in their orbits and thus produce a more easily detectable signature. However, we could not detect the existence of Earth for two reasons. One, the maximum amplitude of the wobble that it causes on the Sun is 818 times smaller than that due to Jupiter. This is 0.04 m/s, 100 times too small to be detected with current instrumentation. Second, this tiny wobble (with a 1 year period) would be superposed to all the larger wobbles, each of different periods, produced by all the more massive planets. There is no way that could be ever detected with current technology.

7. Summary

As the very name of the discipline suggests, Astrobiology has a key astronomical component, all the way from what made the development of life possible to detection of life outside the Solar System, which as is amply discussed in other papers at this meeting uses the techniques of Radio Astronomy. The rough road from the big bang to planetary formation had been outlined as far back as the 1950's and 1960's, when the detection of the cosmic background radiation confirmed the big bang cosmological ideas, including primordial nucleosynthesis, and the role of stellar interiors in subsequent nucleosynthesis was precisely understood. At that time, also, the first models of cloud collapse in star formation were developed, and the role of conservation of angular momentum in helping to form planets was glimpsed. Prior to this time, for example, one possibility of forming planets was as a byproduct of stellar collisions. Stellar collisions are so rare that, if this were the origin of the Solar System, planetary systems would be exceedingly rare. In the now conventional picture, on the other hand, planets would be ubiquitous.

The main recent events in Astronomy relevant to Astrobiology relate, first to the advances in understanding the details of the big bang cosmology. A very refined picture of the early Universe was obtained by the satellite COBE. On its foundation, the ideas of dark matter were developed, the role of primordial nucleosysnthesis was made more precise, and we are currently obtaining the properties of the current Universe, including the possibility of a non-zero cosmological constant, which could lead to an expanding Universe with accelerating velocity, rather than the conventional slowing down picture. However, the most significant advance has been the undisputed detection of approximately 30 extra solar planets, a number that is increasing on a daily basis. Because the orbital properties of these planets are in marked contrast with previous expectations, these discoveries have provided information to model the process of planetary formation with a degree of sophistication orders of magnitude higher than what was possible before.

The astronomical component of astrobiology is an active and productive field. Besides the topics alluded to above, studying the details of all the planets in the Solar System is a very active and fruitful undertaking greatly enhanced by the continued launching of planetary probes. However, to me the most significant astronomical contribution to the field will be made when and if our radio telescopes detect the first uncontested signal from a civilization outside planet Earth. Arguably, such an event might constitute the most important occurrence in the history of humanity.

REFERENCES

Because this article is primarily didactic in nature, the most appropriate references are secondary sources which provide an up-to-date discussion of the current views of the relevant fields.

For cosmology and primordial nucleosynthesis, see:
Peebles, P.J.E., Schramm, D.N., et al. 1995, The Evolution of the Universe, 271, 10.

For stellar structure and evolution, see the book:
Principles of Stellar Evolution and Nucleosynthesis, by Clayton, D.D., 1968 (McGraw Hill: New York)

For nucleosynthesis in massive stars and supernovae, see:
Woosley, S.E., Hoffman, R.D., et al. 1997, Nucleosynthesis of massive stars and supernovae, Nuclear Phys. A 621, C 397.

For the detection of extrasolar planets, see:
Marcy, G.W., and Butler, R.P., 1998, Detection of Extrasolar Giant Planets, Ann. Rev. Astron. Astrophys., 36, 57.

Section 2:
Chemical Evolution

COSMOCHEMICAL EVOLUTION AND THE ORIGIN OF LIFE ON EARTH

JOHN ORÓ
Deptarment of Biology and Biochemistry
University of Houston, Houston TX 77204-5934, USA

1. Introduction

The 20th century has been characterized, among other things, as the century of science and technology. Aside from important discoveries made during this century, three of the major concepts and areas of knowledge that have acquired singular preeminence are the Universe, Life and Man. In the last century Charles Darwin quite successfully applied the concept of evolution by natural selection to living creatures. In more general terms the concept of evolution by natural causes can be applied to the inanimate world and to the Universe as a whole including all the cosmic bodies, namely galaxies, stars, circumstellar and interstellar clouds, chemical elements, interstellar molecules, planetary systems, planets, comets, asteroids, meteorites, etc. It is within this context that we can speak of cosmochemical evolution and its presumed relation to the origin of life on the planet Earth. More specifically, in this paper, we will focus our attention on the origin and evolution of life and we will briefly cover the pertinent physical and chemical aspects of cosmic evolution which were responsible for the formation of the main biogenic elements in stars, which aside from H include C, N, O, S and P; and the circumstellar and interstellar organic and other molecules which constitute the dust and gas clouds of the interstellar medium. We will discuss the formation of both the Solar and the Earth-moon systems, the role of comets in providing the water and the biogenic precursors necessary for the formation of biochemical compounds on the early Earth, the probable processes of chemical evolution that presumably preceded the emergence of Darwin's ancestral cell from which all other living systems have evolved on the Earth, including this chimerical and probably ephemeral creature we call *Homo sapiens sapiens.*

Moreover, after exploring some of the critical stages of biological evolution, such as the K-T boundary event responsible for the disappearance of dinosaurs and very probably, much later on, for the evolution of hominids, we will also consider briefly the question of life beyond the Earth. Perhaps more importantly, we will discuss one of the major discoveries of this century, namely the recent findings of extrasolar planets around stars other than pulsars. We will consider the possibility of development of technological space-faring civilizations on the basis of our present knowledge and the estimates of existence of extraterrestrial intelligence. Finally, we will end with some ethical unifying conclusions that can be derived from a better knowledge of the evolution of the cosmos and the origin and evolution of life. After all, we are children of the Universe.

J. Chela-Flores et al. (eds.), Astrobiology, 57–69.

2. The nucleosyntheses of carbon and biogenic elements

From a chemist point of view one of the most creative processes in the evolution of the Universe is the nuclear synthesis of the chemical elements of the periodic system. That took place first during the Big Bang explosion, then in the core of the stars and later in supernova explosions. Three minutes after the Big Bang, the free nuclei of hydrogen and helium were generated. However, a few million years had to pass before protogalaxies and stars could be formed from the condensation of clouds of hydrogen and helium. First the light elements were synthesized in the core of stars. Most of the heavy elements were generated much later by supernova explosions. These two processes are continuously going on at the present time. The Universe is made of about 98% H and He and 1% of C, N, O, S and P and about 1% of the rest of the elements.

The formation of carbon and of the biogenic elements (H, C, N, O, S and P) is most important. These elements are needed not only to make the organic compounds that are present in the circumstellar and interstellar medium, in comets and other cosmic bodies, but they are necessary for the formation of the biochemical compounds of the living systems.

The corresponding nuclear reactions for the synthesis of the biogenic elements have been briefly described in a paper from our laboratory (Macià et al., 1997). It is worth to consider two of the first nuclear reactions which lead to the formation of helium and carbon nuclides. In addition to its synthesis during the Big Bang, the stellar formation of the helium nuclide takes place by the proton-proton chain. This occurs in the core of many ordinary stars and specifically in our sun at a temperature of about 15 million degrees. The small deficiency of mass which results from the condensation of four protons into one nuclide of helium is converted into energy in accordance with Einstein's equation, $E=mc^2$. This large amount of energy in the form of heat and radiation has been partly responsible for the appearance of life on our planet and for its evolution during the past 4,5 thousand million years. The next most important nuclear reaction is the one that involves the condensation of three helium nuclides into one carbon-12 nuclide. This nuclear reaction is known as the triple-alpha process. It occurs inside carbon stars at a temperature of 100 million degrees. Obviously, without the triple-alpha process, we would not be able to talk about life in the Universe. In fact we would not be here.

Once carbon-12 is made by the low probability collision of three alpha particles, subsequent alpha-capture processes produced the oxygen and sulfur nuclides. The nitrogen nuclide is catalytically generated by the CON cycle. However, the phosphorus nuclide requires the participation of many complex nuclear reactions. This explains the relatively lower cosmic abundance of this element (Macià et al., 1997).

3. Circumstellar and interstellar molecules

Carbon stars are a rich source of carbon compounds or organic molecules. From the core of these and other stars, the biogenic elements (H, C, N, O, S, P) migrate to the outer and cooler regions of the stars. There, ordinary chemical reactions give rise to the formation of diatomic and triatomic combinations that can be observed in stellar atmospheres. Among the most common species on finds C_2, CN, CO, CH, NH, OH, and H_2O that are present for instance in the atmospheres of ordinary stars such as our

Sun, a third generation star.

At least one hundred chemical species have been identified so far in the interstellar medium (ISM) by their gas-phase molecular spectra (Oró and Cosmovici, 1997; Oró, 2000). All these molecules, ions, and radicals are relatively simple. The ones specifically identified are made from two, three, four, etc., up to about 11 atoms. About 75% of them are organic, or contain carbon.

Since the biogenic elements are the most abundant active elements in the Universe, and since most of the interstellar molecules contain carbon, one could say that the Universe is essentially organic, and prepared for life to emerge wherever and whenever the conditions are right. It should also be pointed out that these one hundred chemical species do not include other organic compounds that have not been individually detected in the interstellar medium but that probably form the bulk of the so called diffuse infrared bands (DIB). According to Allamandola and colleagues (1989), polycyclic aromatic hydrocarbons, or PAHs, are one of the major components of the interstellar dust and gas clouds. In a recent review, Allamandola shows that these compounds are not only abundant in the ISM, but in many other places of the cosmos such as meteorites, interplanetary dust particles, and comet Halley. Indeed, as he has said, PAHs are everywhere including the Martian meteorite ALH 84001. The Martian PAHs are not products of biological origin, since the same PAHs have also been found in the Murchison meteorite (Basile et al., 1984). The most probable sources of cosmic PAHs are the circumstellar clouds. Eventually they become part of the ISM as well as comets, meteorites, and other bodies of the Solar System. On the basis of experimental work on the production of aromatic hydrocarbons cosmic PAHs probably result from the high temperature condensation of C_2 or C_2H_2 in the atmospheres of carbon stars and their circumstellar envelopes.

Among all these species, it is of the interest to point out the following ten molecules: Hydrogen, ammonia, water, formaldehyde, hydrogen cyanide, cyanacetylene, carbon monoxide, hydrogen sulfide, cyanamide and phosphorus nitrile. With them one could synthesize in the laboratory under prebiological conditions, the amino acids, nucleic acid bases, sugars, lipids, and mononucleotides which are part of the biochemical compounds of all living systems.

4. The Solar System

Approximately 5 Ga ago the Solar System was formed by the gravitational collapse of a dusty gaseous nebula of interstellar matter, presumably triggered by the shock-wave of a supernova explosion. Hoppe et al. (1997) have obtained evidence for the presence of SiC grains in the Murchison meteorite where several isotopic ratios indicate they are matter from a type II supernova. Subsequently, during its first several million years, the evolving Solar System was in a state of great upheaval, where the norm was the continuous collision of planetesimals or comets with other major bodies. Thus, at least two phases can be distinguished in the formation of the Solar System, first the protosun by the gravitational collapse of the solar nebula, and then the planets, satellites, and other bodies by accretion of the differentiated nebular matter. As pointed out by Delsemme, the precursors of the terrestrial planets were rocky planetesimals essentially devoid of light molecules due to the relatively high temperature prevailing around the orbit of the Sun within the inner part of the asteroid belt.

5. The Earth-Moon system

The impact record on the Moon is witness to the turbulent state of flux during the early stages of formation of the Solar System. The proto-Earth was obviously subject to many collisions from small and large planetesimals. A relatively recent model proposed by Cameron and other investigators, e.g. Cameron and Benz (1991), has led to a new theory for the formation of the Earth-Moon system which apparently avoids the deficiencies of previous theories. According to this theory, a celestial body with a mass comparable to that of Mars collided with the proto-Earth, injected most of the iron into the nucleus of the Earth, and caused the fusion and ejection into orbit of portions of the mantle which eventually coalesced to form the Earth's only natural satellite, the largest moon of the terrestrial planets. This origin explains most of the similarities and slight differences between the composition of the Earth's mantle and the Moon, for instance the iron-poor nature of our satellite. In the dynamic aspects, it also explains the angular momentum of the Earth-Moon system.

6. Cometary matter captured by the early Earth

The above theory, simply expressed, means that the Earth-Moon system resulted from a single-impact massive collision which led to the loss of most of the volatiles into space. Also, as suggested by Cameron, comets and other small bodies of the evolving Solar System subsequently contributed by late accretion most of the water and biogenic compounds through smaller collisions which occurred during the first 600 million years of Earth's history. On the other hand, due to its low gravitational mass, the Moon could not retain any significant amount of volatile compounds.

TABLE1. Cometary matter trapped by Solar System bodies

	Cometary matter(g)	Time-span (years)
Venus	4.0×10^{20}	$2x\ 10^{9}$
Moon	2.0×10^{20}	Late-accretion
Earth	$1.0 \times 10^{14\text{-}18}$	2×10^{9} years
	$1.0 \times 10^{25\text{-}26}$	Late-accretion
	3.5×10^{21}	Late-accretion
	7.0×10^{23}	4.5×10^{9}
	2.0×10^{22}	4.5×10^{9}
	1.0×10^{23}	2.0×10^{9}
	$1.0 \times 10^{24\text{-}25}$	1.0×10^{9}
	$6.0 \times 10^{24\text{-}25}$	1.0×10^{9}
	$1.0 \times 10^{23\text{-}26}$	4.5×10^{9}

Calculated by several authors, listed in Oró, Mills and Lazcano (1992)

We know that currently the atmosphere of the Moon is 10^{-10} torr. There is no doubt that, while some of the collisions contributed matter to the Earth, others removed water and volatile compounds, because of the dual aspects of impact capture and impact erosion. However, at the end of this bombardment period, as indicated by the Moon cratering record, the Earth came out with a significant increase of its mass, particularly in water, carbon, and other biogenic elements (Table 1), in a similar manner as it probably happened to the major terrestrial planets, to Titan and other bodies of the Solar System.

7. Comets

Comets are aggregates of the interstellar matter at the very low temperatures of interstellar space (Oort cloud) and outer regions of the Solar System (Kuiper belt). The dirty ice model that has prevailed with some modifications is that of Fred Whipple. The composition and many aspects of the relation of comets to the terrestrial planets has been studied in detail by Delsemme (1992), and the model for the condensation of interstellar grains to larger and larger aggregates to generate cometesimals and planetesimals has been developed theoretically and experimentally by Greenberg and Hage (1990). Several interesting studies have been reported on Halley's comet, by Kissel and Kruger on Shoemaker-Levi's 9 comet, by Gautier and other investigators; and on Hyakutake, by Mumma and others (e.g. Mumma, 1997; Oró and Lazcano, 1997). A large number of papers should appear in relation to the spectacular Hale-Bopp comet. So far, more than 33 chemical species have been detected in Hale-Bopp, most of which coincide with the organic and inorganic molecules present in the interstellar medium. Current knowledge of the organic composition of comets is reviewed by Mumma (1997) who applied high resolution infrared spectroscopy for the first time to Hyakutake comet. He detected strong emissions from H_2O, HDO, CO, CH_4, C_2H_2, C_2H_6, CH_3OH, H_2CO, OCS, HCN, OH, and other chemical species. Of particular interest are the large amounts of methane and ethane in Hyakutake's nucleus. The question of comets and life has been discussed recently by Oró and Lazcano (1997), Oró and Cosmovici (1997) and Oró (2000).

Comets can be seen as the bridge that connects life to the Universe at large. Perhaps this is one of the most important and ultimate consequences of the creative processes of cosmochemical evolution. The relevance of comets as the source of the organic molecules necessary for the development of life on our planet was first published by the author (Oró, 1961) under the title "Comets and the Formation of Biochemical compounds on the Primitive Earth". A further elaboration and confirmation of this theory has been recently published by Armand Delsemme (1999) under the title "The Cometary Origin of the Biosphere". A more detailed discussion of the role of cometary organic matter on the origin of life has been published recently (Oró, 2000).

8. Carbonaceous chondrites

In addition to comets, the Earth was and is bombarded by asteroids and meteorites. Thousands of meteorites have been recovered from Antarctica and many parts of the world. Some of them are carbonaceous chondrites, which contain organic compounds.

The Alais carbonaceous chondrite, which fell in France in 1806, was analyzed by Berzelius. In the past few years a large number of meteorites have been found in Antarctica. So far, one of the most interesting carbonaceous chondrites is the one that fell in Murchison, Australia, in 1969. Following the initial findings, in a very systematic research Cronin and his collaborators have analyzed this meteorite for organic compounds, and have recently reviewed this information (Cronin and Chang, 1993). The organic matter is largely macromolecular, and possibly related to the refractory organic mantle of interstellar dust grains of Greenberg's model (Greenberg and Hage, 1990).

The complex mixture of monomeric organic compounds in the Murchison meteorite includes carboxylic acids, amino acids, hydroxy acids, sulfonic acids, phosphonic acids, amides, amines, purines (adenine and guanine), and a pyrimidine (uracil), alcohols, carbonyl compounds, aliphatic, aromatic, and polar hydrocarbons. Some seventy five amino acids have been identified. Of these, eight are common constituents of proteins, such as glycine, alanine, and aspartic acid. A few others are of metabolic interest, such as g-amino butyric (GABA), but many of them are not found in biological systems. With relation to the chirality of the a-carbon, all the amino acids are racemic (equal mixtures of D- and L-isomers). The analyses show the presence of many non-biological amino acids. This, together with other properties such as the higher D/H and $^{13}C/^{12}C$ ratios, suggests that these amino acids were synthesized from extraterrestrial precursors (Cronin and Chang, 1993) predating the formation of the Solar System. Alkyl phosphonates, with the alkyl group (from C_1 to C_4) attached to the phosphorous atom of the phosphonate, are also present in this meteorite. This finding suggest a possible derivation of phosphonic acids from the interstellar molecule CP (Cronin and Chang, 1993). Phosphates are presumably derived from interstellar PO_2 as indicated by the IDPs of cometary origin (Macià et al., 1997).

9. Prebiotic chemical evolution

So far we have discussed a number of physical and chemical creative processes from the initial Big Bang to the formation of the Earth-Moon system. These creative processes include the formation of hydrogen and helium shortly after the Big Bang. Once
(a) hydrogen was formed, it seems that the evolution of the rest of the Universe was followed by
(b) the formation of galaxies and stars,
(c) the nuclear synthesis of elements in stars and supernova explosions,
(d) the formation of atoms and molecules in circumstellar and interstellar clouds, and eventually
(e) the formation of planetary systems like our own.

Once the Earth was formed and acquired water and the cometary precursors for the synthesis of biochemical compounds, there had to be appropriate conditions that could allow the synthesis of biochemical monomers and polymers (polynucleotides, polypeptides, and lipids) that upon self-organization eventually gave rise to the appearance of the first self-reproductive living system. The processes of prebiological chemical evolution have been described in a number of papers from different laboratories, including our own. A review of this subject can be seen in Oró, Miller and Lazcano (1990) and Oró (2000).

10. Early evolution of life

In addition to the chemical evolution approach, the study of the origin of life can be approached from an evolutionary biological point of view. The most ancient fossils are those found in sediments and rocks in western Australia (the Warrawoona formation). These fossils are 3,500 million years old and their morphology is similar to that of cyanobacteria, as shown by Awramik, Schopf, and colleagues (e.g., Schopf, 1983). However, they cannot represent the oldest organisms on Earth because they are too complex or too evolved. Of the three contemporary cell domains (Archaea, Bacteria, and Eucarya), the oldest are probably hyperthermophilic archaebacteria which live at high temperatures in the hot springs at the bottom of oceans. But those organisms are not widely believed to be the ancestral cells of all living beings either. It is to be hoped that by studying the sequences of some of the most primitive enzymes which are most preserved in ancient organisms, we shall be able to establish the root of the philogenetic tree of the three unicellular domains, as is being pursued by several investigators (e.g., Lazcano, 1994). Resolving this problem will not be easy, as the enzymes of any given unicellular organism exiting today have had more than 3,500 million years to evolve and many changes may have occurred in the process, even in sequences that today seem very well preserved. However, recent progress in molecular biology raises hopes regarding the application of this method to the study of the origin of life.

A high concentration of oxygen (approximately 15% of present levels) appeared on Earth about 2,000 million years ago, or earlier. Presumably, this was the accumulation of molecular oxygen produced by the oxygenic cyanobacteria, once most of the exposed divalent iron had been oxidized to the trivalent form. This led to the evolutionary emergence of cells with a highly efficient oxidative phosphorylation system. These were the aerobic bacteria which, in the form of mitochondria and through the endosymbiotic evolution process studied by L. Margulis (1982), gave rise to the emergence of eukaryotes at about the time of the appearance of significant concentrations of molecular oxygen. The newly emerged eukaryotic cells eventually evolved into multicellular organisms at about 800 million years ago and branched into three great classes, fungi, plants, and animals.

11. The K-T boundary impact and other cosmic events

An important inflection in the evolutionary process took place about 65 million years ago when the collision of a comet, or asteroid, in the Yucatan Peninsula, caused a catastrophic darkening of our planet and left as a mark one of the largest craters on Earth, the Chicxulub crater. As is known this was first suggested by Alvarez and coworkers based on the presence of iridium as a cosmic marker in the Cretaceous-Tertiary (K-T) boundary sediments. Such a catastrophe is thought to have been responsible for the disappearance of dinosaurs, as well as, many other living species, and it must have opened a niche which allowed the evolutionary development of mammals and led to the eventual appearance of the ancestors of primates and man several million years ago. Had this not occurred, the intelligent beings that now populate Earth would probably have been descendants of the dinosaurs. Apparently, some little mammals were able to survive the K-T boundary explosion, presumably because they were omnivorous and living in caves.

With the new vital space created by the disappearance of the dinosaurs, the little mammals were able to evolve freely giving rise to a multitude of species. Some of them were prosimians which evolved in arboreal habitats in a parallel manner with fruit producing flowering trees. Some of these arboreal mammals belonged to the suborder *Antropoidea* and to the family *Hominidae*. The evolution of our ancestors, *Australophitecus* and hominid species leading to *Homo sapiens sapiens*, is briefly described elsewhere (e.g., Oró 1996).

A similar cosmic event occurred not long ago. In July 1994 we witnessed a relatively large collision of comet Shoemaker-Levy 9 with Jupiter. Had this collision occurred with the Earth, it may have wiped out most of the biosphere, including probably also humans, with the possible exception of microorganisms and their symbionts living at the bottom of seas. It has recently been pointed out that the comet Swift-Tutle, responsible for the Perseid meteors, will pass again close to the orbit of the Earth. Brian Marsden indicated that coincidence will only be 15 days apart, which is less than the margin of error in these calculations. Let's hope that we do not have a collision!

I believe the most important practical lesson of Solar System space science is to know and decide what can be done to avoid any such cosmic catastrophic impact with the Earth in the future. The technological means for detection and control of incoming cosmic bodies are already available to us. They only need to be organized into a global cooperative network of vigilance, action and reaction. There is no question that a unified policy should be established by all countries of the world through the United Nations Assembly. The more developed countries of the world should be sponsoring the educational efforts of UNESCO in this area of space science and other areas of global interest which are major of concern for the future of humankind.

12. Exploration of the Solar System: The Apollo mission

On July 20, 1969, Neil Armstrong became the first man to set foot on a heavenly body other than Earth, the Moon. Upon the return of the lunar samples, the moon dust was found to be blackish and the first reports to appear in Houston newspapers, when the crates containing the lunar samples were opened, suggested that it could well be graphite, rekindling hopes that life could have existed on the Moon some time in the past. After analysis of the lunar matter, we know today that life could never have existed on the Earth's satellite, and that the lunar dust dark color was mainly due to its constant irradiation by the solar wind. Analyses performed in our laboratory by combined gas chromatography and mass spectrometry (GC-MS) confirmed the absence of organic matter in the lunar samples, with the exception of a few parts per million of carbon monoxide, methane and other simple compounds. The traces of amino acids found were presumably the product of reactions of HCN of cometary origin, or implanted as carbon, nitrogen, and hydrogen atoms from the solar wind.

13. Life on Mars and Europa ?

Exactly seven years later, on July 20, 1976, the first Viking spacecraft landed on the surface of Mars. The dust of Mars was red, in keeping with the popular name by which this planet is known. In this connection, I was reminded by D.M. Anderson, a colleague on the Viking molecular analysis team, that the redness of surface of Mars led me to forecast that we would not find life on this planet either. Molecular analysis for volatizable organic and inorganic compounds were carried out using an instrument (GC-MS) that was, in principle, similar to the one used in our laboratory for the lunar sample analysis. The new miniaturized apparatus was built according to the suggestions of K. Biemann, myself, and other members of the Viking project molecular team. The major difference between the GC-MS instrument in our laboratory and that on the Viking Mars Lander, was that the latter, instead of weighing two tons, it only weighed 20 kilograms. What were our findings ? We found no organic matter, not even in parts per billion, in either of the two landing sites on Mars, the plains of Chryse and Utopia, where the Viking landers actually landed.

On the other hand, one of the three biological experiments pointed to the rapid formation of the radioactive carbon dioxide. Levin and Straat (1976) suggested that this was the result of considerable microbial activity. I, however, countered that this was due to a relatively simple chemical reaction, in which the iron oxides and hydrogen peroxide in the analyzed Mars sample had oxidized to CO_2 the radioactive formic acid which Levin and Straat had included, among other metabolites, in the nutrient solution. This was 1976. In fact, 20 years earlier, while working on my doctoral thesis, I had studied the mechanism of formic acid oxidation in living beings. It is interesting to note that formic acid, when in the presence of hydrogen peroxide, is rapidly oxidized into carbon dioxide not only by enzymes such as catalase (Oro and Rappoport, 1959), which have an iron atom in their active center, but also by inorganic catalysts such as iron oxides (Fe^{2+}, Fe^{3+}), which are very abundant on Mars.

Furthermore, additional laboratory studies on the oxidation of organic substances by amounts of UV light which are comparable to those which fall on the surface of the red planet explained the total absence of organic matter on Mars. The half-life of residence of any meteorite or cometary organic matter exposed on the surface of Mars is barely a few months (Oró and Holzer), an instant in the geological time scale. We can thus say that there is no evidence for life at the two Viking landing sites on Mars, although it could be a worthwhile proposition for future space missions to revisit the planet in order to look for fossils which would show if life has existed there in the past. It is unlikely, although not totally impossible, that life may exist in some unique sites of Mars, such as a deep crevice or underground thermal spot where some of the ice of the permafrost may be temporarily converted into liquid water. However, the speculation that the oxygen on Mars is biologically produced does not pass the rigorous test of Occam's razor. In relation to Mars and Europa (Jupiter's satellite) there is excitement in the scientific community as well as in NASA centers and Headquarters concerning the possibility of microbial extraterrestrial life in the Solar System. The NASA spacecraft Galileo has obtained beautiful images of an array of linear canyons that have confirmed the frozen and cracked nature of its surface and the possibility of existence of life in the big ocean below. Indeed we reported this possibility sometime ago (Oró et al., 1992). Unfortunately the spectrophotometric finding of hydrogen peroxide and sulfuric acid in the surface of this satellite would lower the probability for the existence of microbial life in Europa. Concerning the possible presence of fossilized relics of past life in the Martian meteorite ALH 84001, nothing can be added to what we said before (Oró, and Cosmovici, 1997). We will have to wait until we receive returned samples from the red planet in order to determine the possible presence of relics of microbial life on Mars about 3,600 million years ago.

14. Life beyond the Solar System

It is reasonable to think that life exists in other planetary systems. An example of this may well be the orbital system around the star Beta-Pictoris, which is some 54 light years away from Earth. This star is intriguing because it is surrounded by a great ring of comet-like matter which emits intense infra-red radiation. It was first detected by the infrared telescope of the IRAS astronomical satellite, and photographed from the Las Campanas Observatory in Chile by American astronomers Smith and Terrile. Studies conducted by French astronomers over the past five years suggest that more than 100 comets fall on the central star each year. More recent estimates indicate that a total of 1,000 comets per year, each a kilometer in size, is necessary to explain the observations of the disappearance of dust cometary material into central area disk where planets my be present (Lagage and Pantin, 1994). If the orbital system had planets like Earth, life might well be emerging now on one of them around the star Beta-Pictoris. A recent report has been made on the finding a planetary system, with two Earth-like planets around a very distant neutron star, or pulsar, in the Virgo constellation and there is additional indirect evidence of the existence of planetary companions to nearby stars, and direct evidence for protoplanetary disks in the Orion Nebula. The emergence of life and intelligence on a planet like the Earth is possible, but may not be common in the Universe (Drake, 1963; Oro, 1995). NASA is encouraging the difficult astronomical search for extrasolar planets.

15. The discovery of extrasolar planets

During the last five years one the major discoveries of this century has taken place, that is the presence of planets beyond our Solar System at about 50 light years away. The first planet discovered was *51 Pegasi B*, by Michael Mayor and Didier Queloz of the Geneve Observatory, in Switzerland (Mayor et al., 1997). The planet has a mass equivalent to 0.6 Jupiter, and is very close to the star. There has been some debate about the nature of planets, so close to the central star. The following two planets were discovered by Geoff Marcy and Paul Butler, of the University of California at Berkeley, USA, from the Lick Observatory at Mount Hamilton (Butler and Marcy, 1997).

Among the planets discovered by Marcy and Butler were *70 Virginis B*, with a very large mass of 8.1 in relation to that of Jupiter and an orbit close to that of Mercury in our Solar System. *70 Virginis B* it is an extremely giant gaseous planet where the temperature in the atmosphere must be very hot, about 1000ºC. The following one was *47 Ursae Majoris*, with a mass also quite large, about 3.5 times that of Jupiter and located in an orbit a little further away than Mars in our Solar System. Therefore, it is possible that this giant planet has either frozen or liquid water, depending on the greenhouse effect of the gases of its atmosphere, the radiation received from the star and the heat developed in the interior of the planet by gravitational contraction.

More recently, in the past few years as many as 30 planetary systems have been discovered (Marcy and Butler, 1999) including the presence of multiple planet companions in the case of the star Upsilon Andromedae (Butler et al., 2000). However, we will have to wait until the large infrared interferometry telescope developed by NASA is deployed in order to determine the possible presence of extrasolar planets with orbits and masses comparable to that of the Earth in the Solar System.

The spectral detection of water vapor, free molecular oxygen and ozone would be indicators of the possible existence of life in these extrasolar planetary systems, If such data are obtained in the future, the statistics for the calculation of the possibilities of extraterrestrial intelligence would be significantly improved. It is difficult to guess but we will probably have to wait for at least two decades before we obtain more specific information bearing on theses extremely interesting and related discoveries.

16. Search for extraterrestrial intelligence (SETI)

It is still more difficult to say whether technologically advanced civilizations with the ability to transmit and receive intelligent signals exist in other planetary systems, Frank Drake (1963) developed an equation containing a series of variables from which he calculated that there is an advanced planetary civilization for every 10 million stars in our galaxy. How can we resolve the great enigma of the possible existence of extraterrestrial civilizations ? Several researchers, of whom Drake is a leading one, have suggested that it would be impossible using manned spaceships to explore our galaxy, since just one spacecraft would need all the energy produced by humans on this Earth to conduct an interstellar voyage to one of the nearer stars.

A group of American researchers, including Drake, Morrison and Tarter, as well as Oliver and Sagan – both of which passed away -, suggested tackling the problem passively, that is to say, by listening out for intelligent signals emitted by other civilizations. The radio-search for extraterrestrial intelligence, under the NASA SETI program started in 1992, at a fraction of cost of any other space project. The project was initiated by means of radiotelescopes using two different approaches. For the first, specific target approach, about 1,000 stars similar to the Sun have been selected, which are studied in detail by the radiotelescopic antennae in Goldstone (California) and Australia. The selected microwave range corresponds to that of the emission of hydrogen (H) as well as that of the hydroxyl group (OH). In common terms, this frequency range of the spectral region is referred to as the water window (H_2O). The second approach of detecting "intelligent" signals is by conducting a general sidereal sweep or scan using radiotelescopes. The projects were managed jointly by NASA's Jet propulsion Laboratory (JPL) in Pasadena, California and by NASA's Ames Research Center (ARC) in Moffett Field, California. There is an active complementation of the efforts of Dr. Frank Drake and the SETI Institute researchers by Guillermo Lemarchand in the Southern hemisphere in Argentina.

17. Epilogue: peace from cosmic evolution

As we have seen, the Universe is essentially made of H_2 and He, but is rich in organic compounds precursors of key biochemical molecules and therefore conducive to the emergence of life, given the right conditions. But even more surprising, especially for those who study the nervous system, the brain, memory and other mental processes, is the realization that some of the simple molecules involved in the transmission of nervous impulses in living beings (neurotransmitters), such as glycine, glutamic acid and gamma amino butyric acid (GABA), have been found in the Murchison meteorite. One could therefore say that the Universe is not only prepared for the emergence of life, but also for the appearance of intelligence ! This leads us to the intriguing corollary that the Universe might be populated by civilizations much more intelligent than those living on our small blue planet. Perhaps one day these advanced civilizations will be able to instruct us through interstellar communication as to the answers to human problems such as war, disease and old age. But until this utopian notion becomes a reality, we would do well to cherish our own small blue planet with all its varied and wonderful forms of life, the very ones that Darwin studied. After all, there is only one Earth in the Solar System.

The landing on the Moon by the Apollo astronauts, allowed them to see the Earth as a small, distant body lost in the immensity of space. They did not see the borders that separate people into different nations nor the color people's skin. The astronauts developed a global collective consciousness that we are all brother and sisters, and we would do well to share the limited resources of our little blue planet, and live in fraternal peace. After all, realistically, there is no other place better than the Earth. I believe this is the moral lesson of landing on the Moon, or as Neil A. Armstrong, said "That's one small step for man. One giant leap for mankind". We could extend this reflection into three ethical principles derived from a better knowledge of the cosmos and of the origin of life on Earth. Namely: **humility** (life comes from very simple molecules), **fraternity** (we have a common genetic origin, *Homo sapiens sapiens*) and **cooperation** (we need to share the limited resources of our planet)

18. References

Allamandola, L. J., Tielens, A. G.G.M. and Barker, J.R. (1989) Interstellar polycyclic aromatic hydrocarbons: the infrared emission bands, the excitation-emission mechanism and the a strophysical implications, *Astrophys. J. Suppl.* **71**, 733-755.

Basile, B., Middleditch, B.S. and Oro, J. (1984) Polycyclic hydrocarbons in the Murchison meteorite, *Organic Geochemistry* **5**, 211-216.

Butler, R. P. and Marcy G.W. (1997) The Lick observatory planet search, in: Cosmovici, C. B, Bowyer, S and Werthimer, D (eds.), *Astronomical and Biochemical Origins and The Search for Life in the Universe*, Editrice Compositori, Bologna, pp. 331-342.

Butler, R.P., Marcy, G.W., Fischer, D.A., Brown, T.W., Contos, A.R., Korzennik, S.G., Nisenson, P., Noyes, R.W. (2000) Evidence for Multiple Companions to Upsilon Andromedae (in press).

Cameron, A.G.W. and Benz, W. (1991) The origin of the Moon and the single impact hypothesis IV, *Icarus* **92**, 204-216.

Cronin, J.R. and Chang, S. (1993) Organic matter in meteorites: Molecular and isotopic analyses of the Murchison meteorite, in Greenberg, J.M. et al., (eds.), *The Chemistry of Life's Origins*, Kluwer Academic Publishers, Dordrecht pp. 209-258.

Delsemme, A. (1992) Cometary origin of carbon, nitrogen and water on the Earth, *Origins of life* **21**, 279-298.

Delsemme, A.H. (1999) Cometary origin of the Biosphere. 1999 Kuiper Prize Lecture. *Division of Planetary Sciences, American Astronomical Society.* Icarus (in press).

Drake, F. D. (1963) The radio search for intelligent extraterrestrial life, in: Mamikunian, G and Briggs, M.H. (eds.), *Current Aspects of Exobiology*, New York: Pergamon ress, pp.323-345.

Greenberg, J.M. and Hage J.I. (1990) From interstellar dust to comets: A unification of observational constrains, *Ap. J.* **361**, 260-274.

Hope, et al. (1997) Type II supernova matter in a silicon carbide grain from the Murchison meteorite, *Science* **272**, 1314-1317.

Lagage, O. O and Pantin, E. (1994) Dust depletion in the inner disk around Beta-Pictoris as a possible indicator of planets, *Nature* **369**, 628-630.

Lazcano, A. (1994) The RNA world, its predecessors and descendents, in: Bengston, S. (ed.), *Early Life on Earth*: Nobel Symposium **84**, Columbia University Press, pp. 70-80.

Levin, G. V. and Strat, P. A. (1976) Viking labeled release biology experiment: Interim results, *Science* **194**, 1322-1329.

Macià, E., Hernandez and Oro, J. (1997) Primary sources of phosphorus and phosphates in chemical evolution, *Origins of Life and Evolution of the Biosphere* **27**, 459-480.

Marcy, G. W. and Butler, R.P. (2000) Extrasolar planets: Techniques, Results, and The Future, in: C.J. Lada and N.D. Kylafis (eds.) *The Origin of Stars and Planetary Systems*, pp. 681-708.

Margulis, L. (1982) *Symbiosis in Cell Evolution*, Freeman, San Francisco, 419 pp.

Mayor, M. Queloz, D. Udry, S. and Halbachs, J.L. (1997) From brown dwarfs to planets,in: Cosmovici, C. B., Bowyer, S. and Werthimer, D. (eds.), *Astronomical and Biochemical Origins and The Search for Life in the Universe*. Editrice Compositori, Bologna, pp. 313-330.

Mumma, M.J. (1997) Organics in comets, in Cosmovici, C.B., Bowyer, S. and Werthimer, D. (eds.), *Astronomical and Biohcemical Origins and the Search for Life in the Universe*, Editrice Compositori, Bologna, pp. 121-142.

Oro, J. and Rappoport D.A. (1959) Formate metabolism by animal tissues II. The mechanism of formate oxidation, *J. Biol. Chem.* **243**. No. 7, 1661-1665.

Oro, J., Miller, S.L. and Lazcano, A. (1990) The origin and early evolution of life on Earth, *Annu. Rev. Earth Planet Sci.* **18**, 317-256.

Oro, J., Mills, T. and Lazcano, A. (1992) Comets and the formation of biochemical compounds on the primitive Earth – A review, *Origins of life* **21**, 267-277.

Oró, J., Squyres, S. W., Reynolds, R. T., and Mills, T. M. (1992) Chapter V Europa: Prospects for an ocean and Exobiological Implications. A: *Exobiology in Solar System Exploration*, NASA Special Publication 512, G. C. Carle, D. E. Schwartz, J. L. Huntington (eds.), pp. 103-125.

Oro, J., and Holzer, G. The photolytic degradation and oxidation or organic compounds under simulated Martian conditions, *J. Mol. Evol.* **14**, 153-160.

Oro, J. (1995) The chemical and biological basis of intelligent terrestrial life from an evolutionary perspective, in Shostak, G. Seth (ed.), *Progress in the Search for Extraterrestrial Life*, ASP Conference Series, Vol. 74, pp. 121-133.

Oro, J. (1996) Cosmic evolution, life and man, in: J. Chela-Flores and F. Raulin (eds.), *Chemical Evolution: Physis of the Origin and Evolution of Life*, Kluwer Academic Publishers, Dordecht, pp. 3-19.

Oro, J. and Cosmovici, C.B. (1997) Comets nad life on the primitive Earth, in: Cosmovici, C.B., Bowyer, S. and Werthimer, P. (eds.), in *Astronomical and Biochemical Origins and the Search for Life in the Universe*, Editrice Compositori, Bologna, pp. 97-120.

Oro, J., and Lazcano, A. (1997) Comets and the origin and evolution of life, in: Thomas, P.J. Chyba, C.F. and Mckay, C.P. (eds.), *Comets and the Origin and Evolution of Life*, Springer, New York, pp. 3-27.

Oró, J. (2000) Organic Matter and the Origin of Life in the Solar System, in BIOASTRONOMY '99: A new era in the Search for Life in the Universe (in press).

Schopf, J.W. (1983) *The Earth earliest biosphere: Its origin and evolution*, Princeton University Press, Princeton.

CHEMICAL EVOLUTION IN THE EARLY EARTH

A.NEGRON-MENDOZA AND S. RAMOS-BERNAL
Instituto de Ciencias Nucleares, U.N.A.M.
Circuito Exterior ,C. U. 04510, México D.F., México.

1. Introduction

How did life begin? It has been a mystery for humankind during the past and present. One of the few things that it is possible to affirm is that the complexity of the problem requires a mayor contribution from very many disciplines. However, this matter is subject of very many speculations that has been presented in all possible and different ways. As Bernal (1951) pointed out, a general solution to a problem of cosmic proportions demands a multidisciplinary approach.

Oparin and Haldane, in an independent way, proposed during the 1920's, the hypothesis that life was originated from its abiotic surroundings. It presupposes a long chemical evolution period in which the synthesis of bioorganic compounds was carried out from simple inorganic compounds under the influence of natural sources of energy. Therefore, a main objective in the chemical approach to this problem has been to disentangle the path by which organic compounds of biological importance could have been formed before the emergence of life.

Chemical evolution of organic matter is a part of an integral evolution of the planet and the universe. Thus, the chemistry involved on the prebiotic Earth must have been constrained by global physical chemistry of the planet. These constrains may be determined the temperature, pressure and chemical composition of the environment. These considerations need to be taken into account for possible pathways for prebiotic synthesis of organic compounds. The accumulation of organic matter on the primitive Earth and the generation of replicating molecules are two factors of prime importance in chemical evolution. This process may be considered to have taken place in three stages: the inorganic, the organic and the biological.

From the scientific point of view, the problem of the origin of life can be studied from two approaches: A) The analytical approach, in which information from astronomy, paleontology, biology, geology and other disciplines give us a panorama of the conditions that were probably present in the primitive Earth. B) By simulating in the laboratory the conditions proposed for the primitive scenario. These experiments are known as "prebiotic synthesis" or "simulated experiments" and their purpose is to synthesize compounds of biological significance.

In this review our aim will be to resume the major events that were the preamble for the emergence of life on Earth. Thus, it is important for us to learn from the proposed physical and chemical environments of the primitive Earth in order to proposed reactions that could conduce towards the so call event that is call life.

J. Chela-Flores et al. (eds.), Astrobiology, 71–84.

2. Analytical Approach

To account for the major event that led to the origin of life, an extensive study from several disciplines has been made by: examining sediments, looking for fossils that give information about the antiquity and evolution of life, from extraterrestrial sources as meteorites and lunar samples, by space missions, etc. In this part we will discuss briefly some environmental conditions about the early Earth.

Geological time encompasses the total history of the Earth. It is divided in two eons of markedly unequal duration. The earlier and longer is Precambrian Eon that extents from 4550 to 550 years ago, the later and shorter, is Phanerozoic Eon that expands from 550 million years to the present. These eons are divided in eras. Precambrian eon is divided in two eras: Achaean Era from 4550 to 2500 Ma ago and Proterozoic Era spanning from the end of the Achean to 550 Ma ago (Schopf, 1992).

When we study history at school, very often we study the development of human history, which is of about 10,000 years. But, in relation with the age of our planet about 4750 Ma, this period is incredible small. What happened before the event of life appeared? Was the early Earth similar to the present? How did life appear? How far back into the geological past can biology be traced? To answer these questions we can go back to the history of the Earth. Because it is on Earth where we have a great deal of information on the early stages still preserved in the rock record. We need to look for the fossilization process, dating techniques and a variety of other information to help understand this amassed, interesting field of the origin of life. Today, we can observe, that animals; plants and specially humans change drastically the environment. This supports the idea that the atmosphere/ lithosphere of our planet has been changed by biological activity. Many of these changes occurred in the Precambrian. The oldest rock is 3800 Ma from Isua, Greenland. Evidence of the first 600 million years of the Earth's history has not yet been found in the rock record.

2.1 EARLIEST EVIDENCE FOR LIFE

The earliest life forms can be traced by the study of fossil records. Fossils are remained or trace of prehistoric plants or animals, buried and preserved in sedimentary rock, or trapped in organic matter. Fossils represent most living groups we have discovered; also many fossils represent groups that are now extinct. Many factors can influence how fossils are preserved. Remains of an organism may be replaced by minerals, dissolved by an acidic solution to leave only their impression, or simply reduced to a more stable form. The fossilization of an organism depends on the chemistry of the environment and on the biochemical makeup of the organism. Fossils are most commonly found in limestone, sandstone, and shale (sedimentary rock). The rocks that contain the fossils are pressure-cooked, squeezed, heated (that is, metamorphosed) under conditions that can wipe out any evidence of life that the rocks originally contained (Schoft, 1992). Once the fossil is found, it is necessary to determine how old it is. For these purposes, there are dating techniques, which enable us to determine the absolute ages of rocks.

In a journey into the Precambrian Eon, the first evidence for life is found in the relatively well preserved rock 3500 to 3300 Ma in age. These are stromatolites (macroscopic structures formed by sediment-trapping algae), microfossils, and Rubuisco-type carbon isotopic values, dated about 3500 Ma found in the Warrawoona formation (Schopf, 1983, 1993). By this stage in the evolution of life, complex microbial stromatolitic ecosystems, possible oxygen-producing cyanobacteria, had already become established. Both microscopic fossils and fossil stromatolites require life to have originated on Earth prior to 3500 Ma ago. Controversial evidence for biologically mediated carbon isotope fractionation (Schidlowski, 1988) suggests that life may have existed by 3800 Ma ago. The origin of life coincided with the last stage of the heavy bombardment.

By 3000 Ma stromatolitic communities dominated by photosynthetic oxygen-producing cyanobacteria (Blue-green algae) were abundant and widespread. Between 2000 and 1000 Ma ago, the eucariotic cell with their nuclei, complex system of organelles and membranes developed and began to experiment with multicelled body structures. The evolution of the plants and animals most familiar to us occurred only in the last 570 million years. The principal Precambrian rocks that have been examined for their content of microfossils and organic compounds are 1) Bitter Spring Chert (900 Ma), Central Australia. 2) Gunflint chert (1900 Ma), Southern Ontario. 3) Sudan shale (2700 Ma), Northeastern Minnesota. 4) Bulawayan limestone (2700 Ma), Rhodesia. 5) Fig-tree chert (3100 Ma), Transvaal, South Africa. 6) Onverwacht (3200 Ma), Transvaal, South Africa.

2.1.1. Molecular Fossils.

Molecular or chemicals fossils are compounds found in ancient rocks. Analytical techniques allow us to distinguish if these chemical compounds have a non-biological origin, they are preserved products of prebiotic synthesis or they are conserved remains of living organisms. The biochemical organization of living systems provides chemical compounds like the enzyme ribulosa bisphosphate carboxylase/oxigenase, in shorthand Rubisco, which catalyzes the fixation of carbon dioxide. This Rubisco has unique properties, and it leaves a telltale isotopic signature in its products, a signature that can be decoded in organic matter having an age even as great as 3500 Ma (Schopf, 1992).

The most important information derived from these geological data is the time scale from the origin of life. The time available has been considerably compressed. The organisms in the Warrawoona Formation (3500 Ma ago) and the end of the heavy bombardment (4100 Ma) leave a short time for going from the primitive soup to the cyanobacteria.

2.2. PRIMITIVE SCENARIO

2.2.1. Time Scale

That life did evolve on the Earth, is a fact, but the question to answer is what was the planet like when the process began? Evidences suggest that the Earth was formed about 4.75 billions years ago (Tilton and Steiger, 1965). The evolutionary sequence presumable began with the origin of the universe and the formation of the elements from the primeval cloud of hydrogen gas.

Since the Earth was formed many changes were produced and establish the conditions for the emergence of life. The earliest forms of life are dated at 3500 Ma. Between the formation of the Earth and the development of life's forms was a very crucial period in that the evolution of the molecules arrived to a critical point were everything about life began. In this period many changes and unknown phenomena took place and originated the transformation from inorganic matter, to organic matter to life. This whole pattern introduces the idea that modern biological molecules may have had non-biological origins in the past. Therefore molecules that are important for the present organism, also were important at the time of the origin of life.

2.2. 2. Sources of Energy

It is obvious that a variety of energy sources must have been available for the formation of organic compounds upon the primitive Earth. Some of these may have been useful for particular aspects of this synthesis. The variety and intensity of the action may have caused different products. In the laboratory most of these forms of energy are used to simulate the primitive conditions.

To consider an energy source as a potential source for prebiotic synthesis, it needs to be available, abundant and efficient to induce chemical reactions. Different sources of energy have been proposed to contribute to the synthesis of organic matter. Ultraviolet light from the Sun, which produces a total energy of 260 000 cal cm^{-2} $year^{-1}$, is by far the most important. This is not surprising since life in our planet today is dependent on the Sun for its energy. The sunlight, which is most effective in the synthesis of organic compounds, is in the region that the atmospheric components could be excited. From the study of the photochemistry of these compounds it is showed that these regions are between 1200 and 2200 A.

Faithful simulation of geologically relevant prebiotic conditions would require the consideration of multiphase systems; chemistry occurring at atmospheric/ lithospheric, atmospheric/hydrospheric, and hydrospheric/lithospheric interfaces; and fluctuating climatic conditions. Under these conditions, natural energy sources not standardly considered in laboratory investigations, such as natural radioactive decay, and triboelectric energy (i.e., energy of mechanical stress) may have a great importance, because these natural scenarios, beside the requirements already mentioned, need a penetrating source of energy. To estimate the potential contribution from these energy sources it is necessary to evaluate the chemical effective energy dose, the energy transductor mechanism and the prebiotic dose rate.

Why ionizing radiation? Primordial radionuclides were made in the stars, by violent nuclear processes. Because of subsequent circulation of matter, they became incorporated into the Solar system. Since its formation our planet has been permanently exposed to ionizing radiation of both terrestrial and extraterrestrial origin. In this context, the application of ionizing radiation to prebiotic synthesis is based on its unique qualities: its omnipresence, its specific way of energy deposition, and its distribution. Reactions are through free radicals and the radiation furnishes pathways in the water in which chemical reactions occur. The use of this source of energy is substantiated by calculations of the energy available from the decay of radioactive elements like potassium-40, uranium-235, uranium-238 and thorium-232.

It is also possible to have other sources of energy like heat from volcanoes and hot springs, electric discharges from the atmosphere, cosmic rays, sonic energy generated from ocean waves; shock waves from thunderstorms or meteorite entering in the Earth's atmosphere. The problem of evaluating these energy sources is difficult, and in general entails non-equilibrium conditions, product quenching and protection.

2.2.3.Environmental Archean Conditions

Atmosphere. The environmental conditions in the primitive Earth are a matter of relevant importance. The starting point of the discussion must be around the nature of the Earth's primitive atmosphere, which supplies the raw material for the synthesis. While it is generally agreed that the primitive atmosphere did not contain more than a trace of free oxygen, widely different views have been expressed in relation to its constitution and redox character. When and how did the atmosphere start to accumulate molecular oxygen?

The Earth's atmosphere and oceans were formed along with the planet from impact degassing of in-falling planetesimals. The ocean was completely vaporized during at least part of the accretion period, creating a dense steam atmosphere that was hot enough to melt partially the Earth's surface. Evidence of the existence of this steam atmosphere is provided by neon isotopes (Kasting, 1983).

The steam atmosphere collapsed once main accretion process was over. Models of atmospheric evolution suggest that the Archean atmosphere was composed mainly of CO_2, H_2O and N_2 (Kasting, 1986, Walker, 1990). A higher atmospheric CO_2 level was particularly important. This is because carbon dioxide is an effective infrared absorber that could have warmed the Earth's surface by contributing to the greenhouse effect (Kasting, 1986). This provided surface temperatures of 30-50 Cat 3500-3200 Ma. (Lowe, 1992). Exactly how much CO_2 was present in the early atmosphere is uncertain. The estimate the abundance of carbon, in the crust, much of which stored in carbonate rocks, is about 10^{23}g. (Ronov and Yaroshevsky, 1967). Walker (1986) proposed that a 10-bar CO_2 atmosphere could have persisted for several hundred millions years if the early Earth was entirely covered by oceans (Kasting, 1983). This composition is consistent with the inventory and probable evolution of terrestrial volatiles. Also it is consistent with the inference that, because of the Archean Sun's lower luminosity, above the freezing temperatures were maintained by this CO_2-rich atmosphere on the early Earth by an enhanced greenhouse effect (Lowe, 1992; Kasting, 1992). The existence of a global greenhouse effect is also consistent with the absence of unambiguous glacial deposits until 2400-2200 Ma (Lowe, 1992).

The rise of atmospheric oxygen. Where was this oxygen coming from? Two sources of oxygen are considered plausible, one of these is abiotic: the photodissociation of water vapor by solar UV radiation in the upper layers of the atmosphere (Canuto et al., 1983). The second is a biotic source. Cyanobacteria, photosynthetic organisms that produced oxygen as byproduct, had first appeared 3500 Ma ago. They became common and widespread in the Proterozoic. The photosynthetic activity was primary responsible for the rise of atmospheric oxygen (Holland, 1994). Another source that recently has been proposed by Draganic et al. (1991) came from the decomposition of the ocean waters by the potassium-40 radiation.

Geological evidences, in particular, the distribution of pyritic conglomerates over the time suggests that oxygen production by cyanobacteria did not result immediately in a major increase of atmospheric oxygen. Instead three oxygen sinks consumed the molecular oxygen. 1) Reaction with volcanic gases, 2) Facultative aerobes microbes used oxygen and 3) The oxygen was burial in the iron-rich layers of banded iron formations (abbreviated BIF). Ultimately, about 2000-1800 Ma, after the oceans had been slowly, but irrevocably, swept free of dissolved iron; atmospheric oxygen levels began to rise. After this, a stable aerobic environment became established. This was the "first pollution crisis" that hit the Earth about 2200 Ma. It may seem strange to call this a "pollution crisis," since most of the organisms that we are familiar with not only tolerate but require oxygen to live. However, oxygen is a powerful degrader of organic compounds. Even today many bacteria are killed by oxygen. Organism had to evolve biochemical methods for rending oxygen harmless.

Beginning about 1700 Ma ago, banded iron formations disappeared from the geologic record and red beds (red-colored sediments) began to appear. Red beds are sedimentary rocks made of iron-bearing sand and mud eroded from rock deposited on land. The rock is red because it contains iron that has been oxidized (rusted). Its presence on land means that the eroding rock was exposed to free oxygen in the atmosphere. As it was mentioned earlier in this section this oxygen was produced during photosynthesis of the increasingly abundant microorganisms in the sea. Also, fossil soils have been found and they contain oxidized iron in their upper layers. This is another indicator of higher levels of oxygen in the atmosphere. While oxygen increased in the atmosphere, carbon dioxide decreased, as more photosynthesis took place. This is shown in the geologic record by the presence of limestone, which is made of calcium carbonate. From evidence found in some Precambrian rocks, it is fairly certain that several periods of widespread glaciation occurred on the continents, which suggests that the atmosphere was cooling during the Proterozoic Eon.

Lithosphere. There is considerable agreement that Archean Earth 3500-3200 Ma was dominated by oceanic lithosphere with micro-continents probably constituting less that 5% of the present continental area (Lowe, 1992). The composition of Archean seawater was controlled by its interaction with the oceanic crust and mantle (Lowe, 1992).

Hydrosphere. The chemical composition of the primitive ocean is not known. There is substantial evidence that the Archean oceans were strongly and permanently stratified, including a deep anoxic bottom layer and a thin, wind-mixed upper layer. This interpretation derives from the implications of climating modeling, the distribution and sedimentation of Archean banded iron formation (BIF), and the oxidation state of hallow-water Archean sediments. Although there is geological evidence of liquid water on the Earth after 3800 Ma, similar evidences do not exist from the period from accretion to 3800 Ma, the time of deposition of the oldest preserved sedimentary rocks. There is yet no geological evidences regarding atmospheric CO_2 and surface temperatures prior 3800 Ma (Lowe, 1992).

Geological evidences establish that the starting material needed for oxygenic photosynthesis water, and carbon dioxide, were present in the environment very early in the Earth history. For example the occurrence of liquid water is evidenced by ripple marks preserved in the upper units of Swaziland Supergroup, about 3300 Ma in ago. The presence of water is also well evidenced by the occurrence of pillow lavas (pillow-shaped masses of volcanic rock formed when lava flows into a lake or ocean and rapidly cools and solidifies) near the base of the Swaziland Supergroup, deposited 3500 Ma ago. Highly evidence of metamorphophosed Isua rocks indicates that both water and carbon dioxide were present in the environment as least as early as 3750 Ma ago.

2.2.4 Temperature of the Primitive Earth

Archean climate was controlled by a steady-state carbon cycle. Higher atmospheric carbon dioxide levels maintain a surface temperature of perhaps 30-50 C at 3500-3200 Ma. Prior 4000 Ma, surface temperatures may have been buffered at 90-100 C.

Impacts on Earth's surface. There are several evidences that show that the Earth must have been bombarded by large objects until about 3800 million years ago. Larger asteroids would supply enough energy to boil the ocean and kill all but the hardiest thermophilic bacteria. High temperatures can destroy prebiotic compounds dissolved in the ocean. This, any living system that may be originated very early in the history could not have persisted until the end of the major bombardment. This is one of the reasons to think that life must have arisen in ten million years or less (Miller, 1992). After 4.1-3.9 Ga, impacts declined.

3. Synthetic Experiments Approach

In addition to the analytical approach for the study of the origin of life, just briefly described, another approach is used to study chemical evolution by simulating in the laboratory the synthesis of compounds in conditions that are a recreation of the possible conditions of the early Earth.

How did life originate on the Earth? How were the complex molecules of life formed in the primeval environment? The exact process is not known, but it probably started by reactions among simple molecules brought to our planet by comets, and meteorites. Also by compounds formed by the action of different sources of energy on the Earth's primitive atmosphere. These simple compounds condensed either in water-or more likely on the surface of sediments. In particular the single units, which constitute the nucleic acids and the proteins, could have been synthesized by the action of various forms of energy in the primitive atmosphere. These small molecules may have been formed directly from the primordial atmosphere or may have resulted in a stepwise manner from reactive intermediates. In the synthesis of the polymers there is a removal of a molecule of water at every step of the condensation of two units. This process was to take place in an aqueous medium and is not energetically favorable. Such a sequence of events would have been feasible on the primordial Earth. Bernal (1951) was the first to point out that a lagoon when it dried up, would have been an ideal locale for the origin of large molecules. Organic material, adsorbed on the clay at the bottom of the lagoons, would have been exposed to solar radiation or other sources of energy, and dehydration would have taken place. Reactions of this type also may have occurred along the ocean shoreline. The lagoons that are strung along the waterfronts are occasionally found to be dry. At other times, at high tide for example, the water rushes in and fill these areas. This alternate drying and flooding would have been a very useful method for the synthesis of polymers and their removal into the prebiotic ocean.

Random chemical reactions could have formed a variety of organic compounds, many of which could not have led to life. Unproductive chemical diversity may have been limited by catalysis that favored the synthesis of specific molecules important for life. Those may have been organic compounds formed on Earth or by accretion from extraterrestrial sources.

3.1 CHEMICAL COMPONENTS IN CONTEMPORARY LIFE.

RNA Word. A fundamental concern in this field is what molecules were essential for the first life. form DNA (deoxyribonucleic acid) and proteins are the central biological polymers in present-day living organisms. DNA stores the genetic information. The DNA information is expressed in proteins. Proteins perform the structural and catalytic functions that are characteristic of a particular specie. RNA (ribonucleic acid) allows link between DNA and proteins. RNA carries the genetic information encoded in DNA to the molecular machinery for the synthesis of proteins.

A crucial problem for the studies about the origin of life is to establish whether the first organism contained proteins or RNA or both. The protein-RNA dilemma has been resolved, at least for the present, in favor of the RNA. It has been discovered the catalytic activity in certain RNA molecules, the ribozymes, (Cech and Bass, 1986). This indicates that RNA could carry out the function of both DNA and proteins, and if this is correct, this interpretation implies that the first organisms were composed of RNA. This hypothetical biosphere is called "RNA word" (Gilbert, 1986). Life eventually switched over from RNA to DNA and proteins, because these molecular specialists are able to perform the task of information storage and catalysis.

Other important molecules in present-day organism are the phospholipids, which composed the membrane. The energy that drives the biochemical reactions within a cell is stored in a molecule called adenosine triphosphate (ATP). Other molecules of interest are fatty acids; dicarboxylic acids related with metabolic processes; porphyrins-like pigments; nicotinamide and its derivatives; and sugars (which are also part of the nucleic acid backbone).

3.2. OUTLINE OF THE OPARIN-HALDANE HYPOTHESIS.

According to the Oparin-Haldane hypothesis living organisms arose naturally on the primitive Earth through a process of chemical evolution of organic matter. In this process the flux of energy through the prebiotic environment transformed simple molecules into complex bio-organic compounds. Eventually they became the precursors of proteins, nucleic acids and other biochemical compounds. One of the major contributions of Oparin was to propose that the first organisms were heterotrophic, that is, they used prebiologically produced organic compounds available in the environment. Haldane (1928) introduced the idea of primordial soup. He suggested that in bodies of water these reactions would be going on under the rather large concentrations of organic substances. In fact we need very concentrated primitive soup. Oparin, as well as Urey proposed that the early Earth had a reducing atmosphere composed of methane, ammonia, water vapor and molecular hydrogen. A basic pH and moderated temperatures were needed for the accumulation of organic compounds.

Before that Oparin was proposed that the origin and evolution of life involved a stepwise approach, other researchers had dealt with the problem from their own perspectives. Alfonso L. Herrera (1868-1942), was a Mexican scientist of great breath. In 1930s, had easily and quickly produced organized structures that had the look of a great variety of cells. Moreover, he started his experiments with substances such formaldehyde and ammonium thiocyanate. Decades after these experiments, new investigators found that the materials he used were organic substances abundant in the interstellar matter. In this he was far ahead of his time (Fox, 1988).

While many features of Oparin-Haldane scenario are untenable in the light of modern theory and knowledge they are still an important cornerstone in the study of the origin of life.

3.3. SIMULATED EXPERIMENTS

S. Miller, working with H. Urey at the University of Chicago carried out the first experiment in this field in 1953. The apparatus was designed to circulate methane, ammonia, hydrogen and water, though an electric spark. Electrodes were placed in a 5 l flask, and small tesla coil produced the spark (Figure 1). After a week of sparking, the water containing organic compounds was analyzed. The results were beyond belief. A number of life- related substances were synthesized. Among these substances were amino acids, simple fatty acids, and urea.

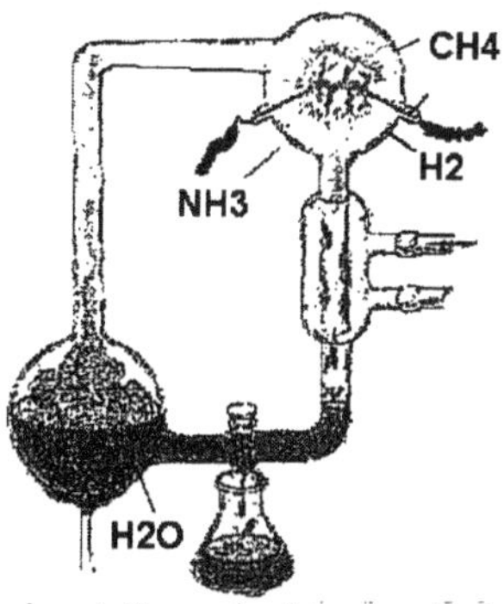

Figure 1. Diagram of the Miller experiment. Steam circulates from the boiling water at lower left, through the "atmosphere" of methane, ammonia and hydrogen.

After this experiment many others were performed using different sources of energy like heat, ionizing radiation, UV light, etc. These different types of experimentation have produced many building blocks of life such as amino acids, purines, pyrimidines, hydrocarbons, etc. Still, there are several compounds whose synthesis present several problems as low yield of formation, unrealistic experimental conditions, the optical activity problem, etc. Much effort should be given to this regard.

The modeling of the possible pathways for the formation of organic compounds on the primitive Earth is called "prebiotic synthesis" or "simulated experiment." In other words, chemical processes are the first steps in modeling the chemical events in the primitive Earth and other bodies in our solar system.

For a "simulated experiment" we need to take into account the following considerations: 1) to choose the raw material, and the energy source. 2) To allow the system to operate for some time, and 3) to perform a variety of analysis of the products.

There are several approaches to perform simulated experiments: To choose a mixture of gases that constituted the primitive atmosphere or gases detected in extraterrestrial bodies with an atmosphere. The mixture is exposed to an energy source, and the products detected are analyzed. In the case of the simulation of the atmospheres of extraterrestrial bodies, the results can be compared with the detection of compounds in the atmosphere that was simulated.

Since there are several ways to arrive to the same product is necessary to choose a pathway to visualize and simplify the experiment. In particular in synthetic chemistry the symmetry is of great importance for the simplification of such synthesis. Other two elements that can be very useful to visualize the synthesis are the followings: to observe structural relationships among reactants and products, to use chemical properties.

Lets examine the synthesis of succinic acid to illustrate the previous statements.

Succinic acid is an important compound in metabolic pathways and it is a precursor molecule in the synthesis of porphyrins. The structure of this acid is shown in Figure 2. It has a symmetry plane (axial symmetry) that divides in two identical parts this molecule. Firstly, this means that if we synthesize the fragment $\cdot CH_2COOH$ (since it has a non-share electron, it is a free radical), we have half of the succinic acid molecule. The problem now is to form the fragment $\cdot CH_2COOH$ and secondly to put together two of these fragments (this reaction is known as dimerization). The synthesis of succinic acid, a molecule of 4 carbon atoms has been reduced, just by symmetry considerations, to the synthesis of $\cdot CH_2COOH$ with 2 carbon atoms. How can CH_2COOH be synthesized? This fragment is very similar to CH_3COOH, acetic acid; the difference is an extra hydrogen atom in the acetic acid molecule. So, if we remove one H from acetic acid molecule, we will have the fragment that we were looking for. How can this be achieved? It is known in Radiation Chemistry, that the interaction of ionizing radiation with an aqueous solution of an organic compound (with H atoms next to a functional group, like the COOH group) will react abstracting an H atom. In these conditions the radical $\cdot CH_2COOH$ is formed, and later it will dimerize to form succinic acid

Since acetic acid was used as raw material for our synthesis, the next question would be may be acetic acid considered as a prebiotic compound? In order to answer this, it will be needed 1) to look for its prebiotic synthesis. The acetic acid molecule can be visualized by a rearrangement of its atoms. It is as formed by CH_4 and CO_2, compounds that we can find in the atmosphere or in other prebiotic synthesis. Would acetic acid be formed from the reaction of those gases? A search in the bibliography shows that this reaction was attempted and acetic acid was indeed formed with a high yield. Another way to synthesize acetic acid is from an electric discharge experiment starting with methane and water (Allen and Ponnamperuma, 1967). Now, for testing this scheme it is necessary to look for the synthesis of succinic acid from acetic acid. It has been reported that succinic acid is obtained in a very high yield from the gamma irradiation of dilute solutions of acetic acid (Negrón-Mendoza and Ponnamperuma, 1996).

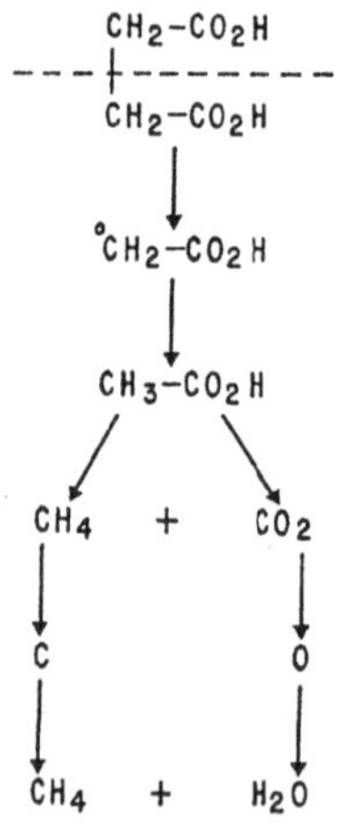

Figure 2. Synthesis of Succinic acid starting with acetic acid.

Another example of the simplification of a synthesis is showed in figure 3 for uracil. This molecule can be divided in two parts; this is shown in the Figure 3. It is needed to go backwards to try to get the backbone of those fragments and then related them with a known molecules. In this case, these are the urea and cyanoacetylene. In 1974, Ferris et al., reported the synthesis of cytosine and uracil starting with cyano-acetaldehyde, from the hydratation of cyanoacetilene and guanidine.

2) The second hint to simplify a synthesis is by structural relationships. For example, aspartic acid is an amino-

acid compound; and it is formed readily from many prebiotic experiments. Its structure is presented in Figure 4. This structure is very similar to the succinic acid (structure B) and malic acid (structure C). The difference resides in the NH_2, OH and H groups. So, it is very likely that if we have aspartic acid, the amino group can be removed by a simple chemical reaction and it will produce succinic or malic acids. In 1980 Aguilar et al., formed aspartic acid starting with succinic acid and one N source (ammonium ions), and malic acid from the irradiation of succinic acid. In the same manner irradiation of malic acid yields succinic acid (Castillo et. al., 1984). This is an example of structural relationships for the synthesis of compounds.

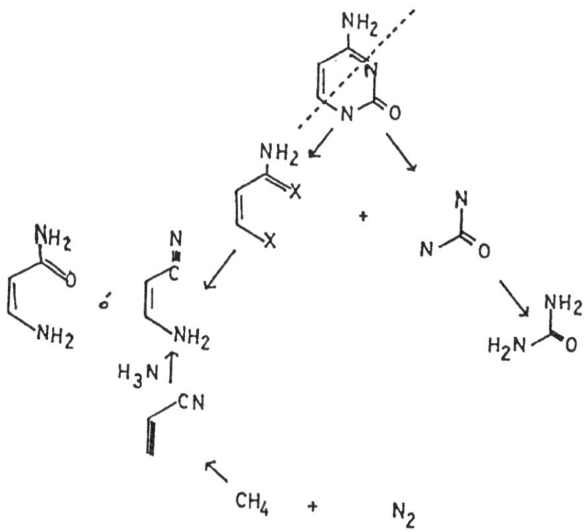

Figure 3. Synthesis of uracil from cyanoacetilene and guanidine..

(A)	H_2N-CH-CO_2H | CH_2-CO_2H	(B)	CH-CO_2H | CH_2-CO_2H	(C)	HO-CH-CO_2H | CH_2-CO_2H

Figure 4. A) aspartic acid, B) Succinic acid and C) malic acid

3) The use of chemical reactions. In most of the prebiotic experiments the raw material present chemical groups as C=O, CN or C=C. In an alkaline pH, which was believed to exist on the prebiotic oceans, and with the action of an energy source, condensation reactions are very likely to occur. For example the condensation reaction of formaldehyde in alkaline medium yielded carbohydrates (Gabel and Ponnamperuma, 1967). Another example of these condensations is the Strecker synthesis for amino acids from an aldehyde and ammonia. This is the mechanism that Miller proposed to explain the products of his first experiment, already discussed.

All these statements are just a guide for the planning of a prebiotic experiment in relation to the raw material. Another very important aspect to consider in planning a prebiotic synthesis, is to re-create in the laboratory geologically relevant scenarios for the primitive Earth, this is a very difficult task.

More realistically simulated environment that could be the resemblance, at the laboratory, of the primitive Earth would require the consideration of multiphase systems Chemical reactions may took place at hydrosphere-lithosphere, atmosphere-lithosphere interfaces. Thus it is relevant to consider the contribution due to the enhancement of chemical reactions by solid surfaces as it was proposed by Bernal (1951).

3.4. GROWTH OF MOLECULES TO BIOPOLYMERS

A whole set of small molecules upon which living organisms of today depend, have been formed, by a variety of energy inputs and using rather extensive raw materials. For a review of some of the synthesis of these small molecules see Miller (1992) and the references therein. However, there are many compounds that do not yet have adequate prebiotic synthesis (amino acids such as arginine, lysine and histidine).

An important step in these chemical sequences to yield compounds of biological relevance is a polymerization reaction. Fundamental molecules for life are biological polymers such as carbohydrates, nucleic acids and proteins. The synthesis of these compounds is a bottleneck in this synthetic approach. Not only because the condensation reaction is energetically unfavorable in an aqueous medium. A more important problem came with the information that these polymers carried out. For example, the word "evolution", it has a defined meaning of something that evolves. The word "voonetiu" has exactly the same letters that the word evolution, but it is meaningless. So, not only the number and kind of letters are important to build a word, but is essential for giving the information, the order in which they are put. The same problem happens with the proteins or nucleic acids. The protein molecule is made up of a string of amino acids. The individual amino acids are linked together by a peptide bond. Proteins perform the structural and catalytic properties, and the order of amino acids in which they are binding is important.

Although the problem of dehydration-condensation reactions to form polymers has not yet been solved, many approaches have been proposed. If the removal of water is necessary for the synthesis, perhaps a method of synthesis would involves the removal of that molecule of water by thermal methods. A typical experiment conducted by Fox, was to heat at 180 C a mixture of amino acids, with a large abundance of glutamic acid. This type of condensation yielded polymers of large molecular weight.

Another way to attack the problem is with the use of "condensing agents." These are compounds which can stored primary energy input to be used later in the hydration-condensation reaction. These compounds present multiple carbon-nitrogen bond as Cyanamid, dicyandiamide (see Miller, 1974 and the references therein).

Template polymerization. In present-day organism, by the process call template polymerization the genetic information in DNA is replicated by a complex process. Each of the two strands of DNA directs the synthesis of its complement. There are several very interesting results of potentially prebiotic template polymerization carried out by the group of Orgel. Their results showed that starting with purineribotides on a polypyrimidine template produced oligomers up to 50 units of nucleosides. However, the converse reaction of pyrimidine nucleosides on polypurine , does not work. This is a limitation of this model of polymerization.

3.5.MODELS FOR PRECELLULAR ORGANIZATION

At some stage in chemical evolution it would have been necessary to achieve a separation phase between an evolving organic system and the external environment. One of the factors in establishing an orderly system from a random pool of chemicals was probably the formation of a membrane. The set-up of monomers and biopolymers within such a membrane would have had the effect of both, concentrating and constraining their environment. This allows a faster synthesis of biopolymers within the protocellular structure (Oro, 1995). For this reason models of semipermable membranes are very important in the study of the origin of life.

Several theories exist today to explain how a biological entity could have emerged naturally out of the abundance of organic chemicals. Protocell arose as a result of natural physico-chemical interactions. However, there is not yet an answer on how the protocell was formed.

4. Concluding Remarks

The central question about life is how it was originated?. Early in the history some system of replication powered by external sources must have been formed. Chemical studies have provided to be fruitful avenues to our understanding for the transition from the non-living to the living. The discovery of amino acids and other compounds in extraterrestrial samples, such as meteorites, has given us some measure of conviction that many of the process we outlined for the infant Earth are commonplace in the universe. They constitute a part of the orderly sequence of cosmic evolution.

There is a remarkable progress in the understanding of the origin of life. We are still a long way from understanding the exact process to lead to the first living cell. Many answers to the questions that have been made are well supported by facts. Some of them are very likely models and others as plausible guesses that remain to be tested by future experiments.

Much has been learned over the past two decades about the antiquity and evolutionary history of Precambrian life; a great deal remains unknown. We now know that life existed at least as early as 3500 Ma ago. How much earlier did the origin of life occur?

Chemical evolution experiments have suggested that almost all bio-monomers can easily be produced by suitable simulation experiments. Only in the most general sense it is true. There are several difficulties. One crucial problem is the synthesis and stability of sugars. These molecules play a central role in our entire metabolism. However, prebiotic synthesis of them gave low yields. Also, they have very unstable nature.

Another difficulty with chemical evolution resides in the "selection problem" What chemical mechanism selects the molecules that constitute a specific polymer? Connected with this selection problem, still there are many bottlenecks such as the origin of the optical activity and the selection of one isomer for biological processes. How did the genetic apparatus evolve?

Of paramount importance, because it bridges the gap between chemical evolution and biological evolution, is the problem of precellular organization. The problem is to envision how the biologically important chemicals spontaneously organized themselves into a three dimensional matrix that would eventually acquire the characteristics of life form.

Finally, this attempt to draw from the scientific experience, some generalized concepts in chemical evolution has the intention to illustrate how by various pathways, quite diverse in their origin, gradually evolved into a common work and scientific interest. Each one of these independent discoveries has given rise to quite effective and profound changes in our environment and helps us to study the origin and evolution of life.

5. References

Aguilar, C. Ramírez, A. Negrón-Mendoza (1980), Estudio de la radiólisis del ácido succínico y del succinato de amonio en el contexto de la evolución Química. C.. In . A. Negrón-Mendoza y G. Albarrán (Ed.) Memorias del II *Simposio sobre Química Nuclear, Radioquímica y Química de Radiaciones.* México, D.F., p. 149.

Allan, W.V. and Ponnamperuma, C. (1967) Current trents in Mod. Biol. 1, **24**.

Bernal, J.D.: 1951, *The Physical Basis of Life*, Routledge and Kegan Paul,London

Calvin, M. (1969) *Chemica l Evolution*, Oxford University Press, Oxford.

Canuto, V.M., Levine, J.S., Augustsson, T.R., and Imhoff, C.L. (1982) UV radiation from the young sun and oxygen and ozone levels in the prebiological paleoatmosphere, *Nature*, **296**, 816-820.

Castillo, S, Negron, A, Draganic, Z.D., Draganic, I. The radiolysis of aqueous solutions of malic acid, *Rad. Phys. Chem.* **26**, 437-443.

Cech, T.R. and Bas, B.L. (1986) Biological catalysis by RNA, *Ann. Rev. Biochem.*, **55**, 599-629.

Draganic, I.G., Bjergbakke, Draganic, Z.D. and Sehested, K. (1991) *Precambrian Res.* **52**, 337-345.

Ferris, J.P. Sanchez, R.A., and Orgel, L.E. (1968) Studies in prebiotic synthesis. III. Synthesis of pyrimidines from cyanoacteline and cyanate, *J.Mol.Evol.* **11**, 293-311.

Fox, S. *The emergence of life. Darwinian evolution from inside.* (1988) Basic Books, Inc. Publishers, New York.

Gabel, N.W. and Ponnamperuma, C. (1967) Models for origin of monosaccharides. *Nature*, **216**, 453-455.

Gilbert, W. (1987) The exon theory of genes. Cold Spring Harbor *Sym. Quant. Biol.* **52**, 901-905.

Haldane, J.B.S. (1928) The origin of life,, *Racionalist Annual*, **148**, 142-154.

Holland, H.D. (1984) The *chemical evolution of the atmosphere and oceans*, Princenton University Press, N.J.

Kasting, J.F. (1983) Evolution of the Earth's atmosphere and hydrosphere, in Schopf, W. *Earth's earliest biosphere. It's origins and evolution*, Princenton University Press, p. 611-623.

Kasting, J. F. (1986) Climatic consequences of very high CO_2 levels in Earth's early atmosphere, *Origins of Life*, **16**, 186-187.

Lowe, D.R. (1992) Major events in the geological development of the Precambrian Earth in Schopf, W. And Klein, C. (Eds.) The Proterozoic Biosphere, Cambridge University Press.

Miller, S.L. and Orgel (1974) *The origins of life on Earth, Concepts in Biology*, Prentice Hall, New Jersey

Miller, S.L. (1992) *The Prebiotic Synthesis of Organic Compounds as a Step Toward the Origin of Life in*

Negron-Mendoza, A. And Ponnamperuma, C. (1976) Formation of biologically relevant carboxylic acids during the gamma irradiation of acetic acid, *Origin of Life*, **7**, 191-196.

Ponnamperuma C., (1972), *The Origins of Life*, Thames and Hudson, London.

Oparin A. I., (1972), *Exobiology*. Ponnamperuma (Ed.) North-Holland, Amsterdam 1-15.

Oro, J. (1995) Chemical Synthesis of lipids and the origin of life. *J. Biol. Phys.* **20**, 135-147.

Ronov, A.B. and Yaroshevsky, A.A.(1967), Chemical structure of the Earth's crust, *Geochemistry*, **11**, 1041-1066.

Schidlowski, M.(1988) A 3,800-million-year isotopic record of life from carbon in sedimentary rocks, *Nature*, 313-318.

Schopf. W. Ed. *Major events in the history of life* , Jones and Barlett Publishers, Boston.

Schopf. J.W (1992) *The oldestFossils and What They Mean*, inSchopf. W. Ed. *Major events in the history of life* , Jones and Barlett Publishers, Boston.

Tilton G.R. y Steiger R.H., (1965), *Science*, **150**,1805-1808.

Schopf, J. W. and Parker,B.M.(1983) Early archean (3.3-billion to 3.5-billion-year-old) microfossiles from Warrawoona Group, Australia, *Science* ,**237**, *70-73*

Schopf J. W. y Walter M. R., (1983), *Earth's Earliest Biosphere*, Princenton University Press, New Jersey 214-239.

Schopf, J.W (1993) Microfossils of the early Archean apex chert: new evidence of the antiquity of life, *Science*, **260**, 640-646.

Walker, J.C.G. (1990) Precambrian evolution of the climate system. *Paleography, Paleclimatology, Paloecology* **82**, 261-289.

NITROGEN FIXATION IN PLANETARY ENVIRONMENTS: A COMPARISON BETWEEN MILDLY REDUCING AND NEUTRAL ATMOSPHERES

RAFAEL NAVARRO-GONZÁLEZ
Laboratorio de Química de Plasmas y Estudios Planetarios
Instituto de Ciencias Nucleares
Universidad Nacional Autónoma de México
Circuito Exterior, Ciudad Universitaria
Apartado Postal 70-543, México D.F. 04510 México.

Abstract. The chemical composition of the atmosphere is of paramount importance in the type and efficiency of nitrogen fixing molecules form from it. The availability of such molecules is considered essential for the emergence of life on Earth and elsewhere. Here I review the production of HCN and NO under two different planetary environments that are relevant for Titan and early Earth and Mars.

1. Introduction

Nitrogen is a basic constituent of nearly all biomolecules essential to life (*e.g.*, nucleic acids, proteins, energy transfer molecules (ATP) and coenzymes). As a result living organisms incorporate usable forms of nitrogen for their growth and survival. Therefore, the availability of nitrogenated species must be a prerequisite for the emergence of life, as we know it. The principal repository of nitrogen in Titan, the early Earth and early Mars is atmosphere where it has been present in molecular form. Dinitrogen is however chemically inert under normal atmospheric conditions. Its conversion into reactive forms, a process referred to as nitrogen fixation, is thermodynamically favored but kinetically restricted (Howard and Rees, 1996). In this paper I review the production of nitrogen species by different energy sources under two different planetary environments: mildly reducing (CH_4 and N_2) and neutral (CO_2 and N_2) atmospheres. Emphasis is given to the mechanisms and chemical yields of production of nitrogenated species in order to get a quantitative view of the importance of the different energy sources in bringing about chemical reactions under two different planetary environments.

2. Mildly reducing atmosphere

Titan, the largest moon of Saturn, is the only satellite in the Solar System with a dense atmosphere (~1.5 bar) composed mainly of nitrogen, a small fraction of methane (0.5-3.4%) and traces of several organics (C_2H_2, C_2H_6, C_2H_4, C_3H_8, C_3H_4, C_4H_2) (Hunten *et al.*, 1984). Hydrogen cyanide is the main N-organic species detected in the stratosphere. Its mixing ratio was measured to vary from 1.6×10^{-7} to 1.4×10^{-6} in the equator and North

J. Chela-Flores et al. (eds.), Astrobiology, 85–96.

Pole, respectively. The second species in abundance is cyanoacetylene with a mixing ratio spanning from $\leq 1.5\times10^{-9}$ in the North Pole to 1.7×10^{-7} in the equator. Cyanogen is present with lower mixing ratios: 1.1×10^{-8} (equator) and $\leq 1.5\times10^{-9}$ (North Pole). Acetonitrile is a trace species detected in the stratosphere but its mixing ratio was not estimated.

A variety of energy sources contribute to the formation of N-containing species in Titan's stratosphere and troposphere. These include Solar UV light, cosmic rays, electrons from Saturn's magnetosphere, lightning and corona discharges and shock waves from meteor impacts and lightning. A detail account on the mechanism of formation of HCN by different energy sources is given below.

2.1 PHOTOCHEMICAL

Nitrogen is sufficiently stable that its photolysis is negligible, and generally only participates in photochemical reactions as a third body (Canuto *et al.*, 1983; Yung *et al.*, 1984). Methane is photochemically reactive in the vacuum UV. Its absorption spectrum is a continuum beginning at about 145 nm, with two maxima in its absorption curve centered in the region between 100 and 80 nm; there is no minimum at any wavelength down to 0.25 nm (Sun and Weissler, 1955; Ditchburn, 1955). The photochemical decomposition of methane has been studied in detail and its quantum yields at 121.6 nm are indicated in reactions 1a through 1e (Gordon and Ausloos, 1967; Rebbert and Ausloos, 1972; Slager and Black, 1982):

$$CH_4 \longrightarrow CH_3 + H \qquad (1a)$$
$$CH_4 \longrightarrow CH_2 + H_2 \qquad \phi = 0.580 \qquad (1b)$$
$$CH_4 \longrightarrow CH_2 + 2H \qquad \phi = 0.510 \qquad (1c)$$
$$CH_4 \longrightarrow CH + H_2 + H \qquad \phi = 0.060 \qquad (1d)$$
$$CH_4 \longrightarrow C + 2H_2 \qquad \phi = 0.004 \qquad (1e)$$

Reactions 1b and 1c are the principal photodissociative channels; reactions 1d and 1e are minor pathways, and reaction 1a is the least important (Salger and Black, 1982). The major products from the photolysis of methane are H_2, C_2H_6, C_3H_8, n-C_4H_{10}; other hydrocarbons such as C_2H_4, C_2H_2 are also formed but in low yield (Mahan and Mandal, 1962; Magee, 1963; Braun *et al.*, 1966).

The report on the photochemical synthesis of HCN from CH_4-N_2 mixtures by Dodonova (1966) was highly controversial. The author detected the formation of HCN in very low yield (20 nmoles) after irradiating (125-170 nm) for 10 hours a one to one mixture of CH_4-N_2 at 4-6 mbar. The mechanism of formation was not studied but it was suggested that photoactivated N_2 and its interaction with CH radicals could account for its formation (Dodonova, 1966). Chang *et al.* (1979) repeated this experiment using a monochromatic hydrogen lamp emitting at 123.6 nm. HCN or other N-containing organics were not detected. Experiments performed at higher wavelengths did not produce any detectable amount of N-containing compounds (Bossard *et al.*, 1981).

Contrary to the unsuccessful efforts of Chang *et al.* (1979) to duplicate the experiments of Dodonova (1966), the formation of HCN in photolytic experiments has also been demonstrated by Scatterwood *et al.* (1986) using the UV light emitted from an arc discharge in nitrogen. The energy yield of production (P) of HCN was determined to be $(3\text{-}7)\times10^{13}$ molecule J^{-1} in mixtures containing 3 to 10% CH_4 in N_2. The synthesis of HCN in these experiments can be explained from the reaction of CH_2 radicals with N_2 as demonstrated by Laufer and Bass (1978). These authors studied the kinetics of the reaction (2) using flash photolysis on CH_2CO-N_2 and CH_2N_2-N_2 gas mixtures, and determined that its rate of reaction is small ($\leq6\times10^4$ dm^3 $mole^{-1}$ s^{-1}).

$$CH_2 + N_2 \longrightarrow HCN + NH \tag{2}$$

Another possible channel for the formation of HCN is the reaction of CH radicals with N_2. Braun *et al.* (1967) have presented kinetic evidence for the reaction (3) using flash photolysis (136 nm) on CH_4-N_2 mixtures. The rate constant was estimated to be 4.3×10^7 dm^3 $mole^{-1}$ s^{-1}, but the products were not identified (Braun *et al.*, 1967).

$$CH + N_2 \longrightarrow \text{Products} \tag{3}$$

The Solar flux incident on Titan's atmosphere is about 1.1% the Earth's value. About 10^{-10}-10^{-9} erg cm^{-3} s^{-1} are deposited in the upper atmosphere by short-wavelength (<155 nm) ultraviolet light (Sagan *et al.*, 1992).

2.2 RADIOCHEMICAL

Irradiation of N_2 leads to its dissociation, reaction 4 (Harteck and Dondes, 1957, 1958a,b). The effect of ionizing radiation on methane has been extensively studied, and this system is reasonably well understood (Lind and Bardwell, 1926; Hoing and Sheppard, 1946; Lampe, 1957; Maurin, 1962; Hauser, 1964; Rebbert and Ausloos, 1973; Arai *et al.*, 1981a,b). The mechanism of decomposition of methane involves ionic and radical processes. The principal ion formed at atmospheric pressure is CH_5^+ (Field *et al.*, 1963); its origin and fate during radiolysis are (Maurin, 1962):

$$N_2 \longrightarrow 2N \tag{4}$$

$$CH_4 \longrightarrow CH_4^* \tag{5a}$$

$$CH_4 \longrightarrow CH_4^+ + e \tag{5b}$$

$$CH_4^* \longrightarrow CH_3 + H \tag{6}$$

$$CH_4^+ + CH_4 \longrightarrow CH_5^+ + CH_3 \tag{7}$$

The major products from the radiolysis of methane are H_2, C_2H_6, C_3H_8 and n-C_4H_{10} (Lind and Bardwell, 1926; Hoing and Sheppard, 1946; Lampe, 1957; Maurin, 1962; Hauser, 1964; Rebbert and Ausloos, 1973; Arai *et al.*, 1981a,b). The energy yield of production of HCN in a CH_4-N_2 mixture as a function of composition, temperature, pressure and dose rate was studied by Zhdamirov *et al.* (1971). The P(HCN) value has a well-marked dependence on the composition of the mixture, reaching its maximum value of $\sim6\times10^{16}$ at 5-7% methane content. The temperature dependence of P(HCN) over the range 30-400°C is not of the usual Arrhenius type; showing two activation energies: 2.9 kJ $mole^{-1}$ and 16.2 kJ $mole^{-1}$ at low (30-100°C) and high (200-400°C) temperature, respectively. P(HCN) can be increased up to 4×10^{17} at 400°C and atmospheric pressure (Zhdamirov *et al.*, 1971). Although a mechanism was not proposed by the authors, the strong dependence of P(HCN) on dose rate (increasing 4-5 times by changing the dose rate from 10^{13} to 10^{16} eV cm^{-3} s^{-1}) suggests that HCN is formed by radical-radical reactions. A likely channel for its formation is:

$$CH_3 + N \longrightarrow HCN + H_2 \quad (8)$$

Armstrong and Winkler (1955) studied such reaction, and determined that its rate coefficient is about 1.3×10^{-6} mole s^{-1}. Another source of HCN at low dose rates could be the reaction of active nitrogen with CH_4 (reaction 9). Froben (1974) detected the radical H_2CN from this reaction by electron-spin resonance at 77 K.

$$CH_4 + N \longrightarrow H_2CN + H_2 \quad (9a)$$
$$H_2CN \longrightarrow HCN + H \quad (9b)$$

2.3 ELECTRICAL

The mechanism of decomposition of methane is not well understood (Bossard *et al.*, 1982). The major products are H_2, C_2H_2, C_2H_4, C_2H_2, and higher hydrocarbons (Ponnamperuma and Woeller, 1964; Ponnamperuma and Pering, 1966; Ponnamperuma *at al.*, 1969). Their relative yields depend on the pressure of the system and current of the electric discharge. At low pressure ethane is the most important hydrocarbon (glow discharge), but as the pressure is raised to atmospheric levels, acetylene becomes the principal product (spark discharge) (Fujio, 1963; Weiner and Burton, 1953; Sieck and Johnsen, 1963; Borisova and Eremin, 1968). At lower currents (silent and corona discharge), ethane is the most abundant hydrocarbon even at atmospheric pressures (Lind and Glockler, 1929, 1930; Lind and Schltze, 1931; Ponnamperuma and Woeller, 1964; Ponnamperuma and Pering, 1966; Ponnamperuma *et al.*, 1969).

The first report of electrosynthesis of HCN from a simulated atmosphere of CH_4-N_2 is that of Sanchez *et al.* (1966). Toupance *et al.* (1975) carried out a series of investigations on the yield of formation of HCN as a function of N_2 percentage in CH_4-N_2 mixtures. They found that maximum production of HCN occurs in mixtures composed of 70% N_2 (Toupance *et al.*, 1975). Molecular hydrogen formed *in situ* (Raulin *et al.*, 1982) or added prior to electrolysis (Stribling and Miller, 1987) decreases the rate of formation of HCN. The chemical yield of production of HCN has been determined to be in the range from $\sim5\times10^{16}$ to $\sim1.5\times10^{17}$ molecule J^{-1} for sparks and laser-induced plasmas, respectively in mixtures containing 10% CH_4 (Scattergood *et al.*, 1989). This value decreases by a factor of 2 in mixtures containing 3% CH_4. The most likely channels for the formation of HCN in the hot channel of lightning sparks are reactions 8 and 10 (Navarro-González, 1989).

$$CH_2 + N \longrightarrow HCN + H \quad (10)$$

The corona discharge synthesis of HCN by corona discharges has been studied by Navarro-González and Ramírez (1997). The P(HCN) value has been estimated to vary from $\sim 2\times10^{14}$ to $\sim 6\times10^{15}$ molecule J^{-1} for negative and positive corona discharges (Ramírez and Navarro-González, 2000). Negative corona discharges behave similarly to ionizing radiation in the mode of propagation of electrons through the medium while positive corona discharges have in addition ultraviolet photons that excite the surrounding molecules. Therefore the mechanisms previously discussed for photochemical and radiochemical synthesis of HCN applied to corona discharges.

2.4 PYROLYTICAL

The thermal decomposition of methane has been extensively studied under a variety of conditions (Skinner and Ruehrwein, 1959; Kevorkian *et al.*, 1960; Kozlov and Knorre, 1963; Palmer and Hirt, 1963; Kondratiev, 1965; Harting *et al.*, 1971). The rate of decomposition of methane is independent of pressure, reaction conditions (flow, static, or shock wave pyrolysis), and presence of an inert gas; the energy of activation calculated under various conditions is about 423 kJ $mole^{-1}$ in the temperature range of 900-2,000 K (Skinner and Ruehrwein, 1959; Kevorkian *et al.*, 1960, Kozlov and Knorre, 1963; Palmer and Hirt, 1963; Kondratiev, 1965; Harting *et al.*, 1971; Chen *et al.*, 1975).

The dissociation reaction leading to the decomposition of methane is shown in reaction 1a. Dimerization of CH_3 or H leads to the formation of the initial products C_2H_6 or H_2, respectively. Ethane is however, rapidly consumed during the pyrolysis leading to the formation of ethylene, the secondary product. The latter further reacts to produce acetylene, the tertiary product, as the extent of the reaction is increased (Chen *et al.*, 1975; Roscoe and Thompson, 1985).

The shock wave pyrolysis of a CH_4-N_2 atmosphere was studied by Rao *et al.* (1967). Mixtures of 5%% CH_4 and 5% N_2 in argon were shock-heated to temperatures between 1500 and 6000 K. Hydrogen cyanide is formed, starting from 2500 K and increasing in yield up to about 20% at 5000 K (Rao *et al.*, 1967). The second order rate of formation for HCN was determined as $k\ (CH_4 + N_2) = 9.3\times10^{-3} - 5.0T^{-1}$, in $dm^3\ mole^{-1}\ s^{-1}$. The mechanism of formation was not studied but a possible scheme consistent with the mechanism of methane pyrolysis was suggested (reactions 11 to 16, followed by reactions 8 to 10). The energy yield of production of HCN was calculated to be $\sim 2\times10^{17}$molecule J^{-1} (Bar-Nun and Shaviv, 1975).

$$N_2 \longrightarrow N_2^* \quad (11)$$
$$N_2^* + CH_3 \longrightarrow CH_2 + N_2 + H \quad (12)$$
$$N_2^* + CH_2 \longrightarrow CH + N_2 + H \quad (13)$$
$$N_2^* + CH \longrightarrow CHN_2^* \quad (14)$$
$$CHN_2^* + CH \longrightarrow 2HCN \quad (15)$$
$$CHN_2^* \longrightarrow HCN + N \quad (16)$$

3. Neutral atmosphere

At the end of the heavy bombardment process, conditions on Earth and Mars may have been favorable for the emergence of life. According to the orthodox theory advanced by Oparin (1924, 1936) and Haldane (1928), and examined experimentally by Miller (1953, 1955), the atmosphere played an important role in the formation of raw materials necessary for the synthesis of the building-blocks of life (*e.g.*, amino acids, purines, pyrimidines and sugars). The early terrestrial (Kasting, 1993) and Martian (McKay and Stober, 1989) atmospheres are believed to have been dominated by carbon dioxide, molecular nitrogen, water vapor and traces of methane and molecular hydrogen. The production of nitrogenated species under this environment is described below.

3.1 PHOTOCHEMICAL

No nitrogen fixation occurs over most of the spectrum (Raulin *et al.*, 1982). Dissociation of molecular nitrogen occurs only at very short wavelengths (< 90 nm) according to reaction 4. Zahnle (1986) has estimated the contemporaneous production of ionospheric nitrogen atoms to be about (1-6)$\times10^{13}$ g yr^{-1}. During the T-Tauris phase of the early Sun, this value is expected to be greater at least an order of magnitude. These nitrogen atoms must then be transported to the stratosphere where they can be converted into HCN molecules by their reaction (8) with methelene radicals from methane photolysis (section 2.1). The downward flow of active nitrogen species from the thermosphere to the stratosphere requires a source of CH_4 which may not be plausible. In the absence of CH_4, atomic nitrogen reacts with carbon dioxide producing nitric oxide (reaction 17). The latter is then effectively converted into molecular nitrogen before reaching the stratosphere according to reaction 18.

$$N + CO_2 \longrightarrow NO + CO \quad (17)$$
$$NO + N \rightarrow N_2 + O \quad (18)$$

3.2 RADIOCHEMICAL

Carbon dioxide is practically inert to ionizing radiation (Lind and Bardwell, 1925). This great stability has been attributed to its efficient back reaction (19) (Hirschfelder and Taylor, 1938).

$$CO_2 \leftrightarrows CO + O \quad (19)$$

Atomic oxygen readily reacts with atomic nitrogen (reaction 4) to yield nitric oxide:

$$O + N + N_2 \longrightarrow NO + N_2 \quad (20)$$

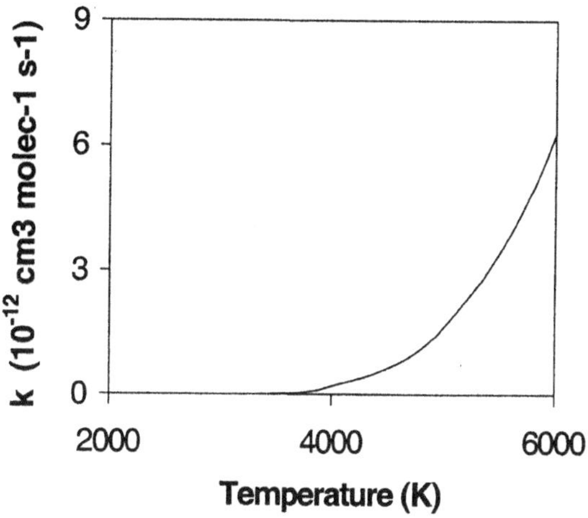

Figure 1. Variation of the rate of reaction of atomic oxygen with molecular nitrogen as a function of temperature.

The rate of this reaction has been measured to be 5.5×10^{-33} (T/298 K) exp(155 (K) /T), in cm^6 $molecule^{-1}$ s^{-1} (Campbell and Gray, 1973). There are no reports however on the chemical yield of production of nitric oxide by ionizing radiation in CO_2-N_2 mixtures. An order of magnitude estimate of P(NO)~10^{14}-10^{15} molecule J^{-1} has been recently reported by Navarro-Gonzalez et al. (1998).

3.3 ELECTRICAL

Our group is currently studying the corona discharge chemistry of a CO_2-N_2 atmosphere. Carbon monoxide and oxygen are the main products. Their chemical yields of production are P(CO) ~ 5.1×10^{15} molecule J^{-1} and P(O_2) ~ 2.2×10^{15} molecule J^{-1} for negative corona discharges (Navarro-González and Ramírez, unpublished results). The production of carbon monoxide and molecular oxygen is explained by reactions 21 and 22.

$$CO_2 \longrightarrow CO + O \qquad (21)$$

$$O + O \longrightarrow O_2 \qquad (22)$$

There was no production of nitrogenated products in these experiments. The following products were specifically searched for: HCN, NO, NO_2 and N_2O. Nitric oxide would be expected to arise from reaction 23:

$$O + N_2 \longrightarrow NO + N \qquad (23)$$

Nevertheless, the rate constant of this reaction is temperature dependent according to $k=5.5\times10^{-10}$ cm^3 $molecule^{-1}$ s^{-1} exp(- 40669 (K)/T) (Monat *et al.*, 1979). Figure 1 shows the variation of this rate coefficient with temperature. It can be seen that for NO to be produced, the temperature of the system must exceed 3,500 K. Since corona discharges are cold plasmas (*e.g.*, the temperature of electrons is in excess of 10,000 K but the ions and neutral species are at room temperature), nitrogen fixation cannot occur in neutral atmospheres.

For the case of lightning, the only published experimental work on the production of NO is that of Levine *et al.* (1982), in which lightning was simulated by a 1-cm electrical spark in an atmosphere composed of 95.9% CO_2, 3.97% N_2, and other minor trace species. P_{NO} was found to be $(3.7 \pm 0.7) \times 10^{15}$ molecule J^{-1}. Theoretical calculations of P_{NO} vary within an order of magnitude from $\sim 4 \times 10^{15}$ to $\sim 3 \times 10^{16}$ molecule J^{-1} in CO_2-N_2 dominated atmospheres (Yung and McElroy, 1979; Chameides *et al.*, 1979; Chameides and Walker, 1981; Mancinelli and McKay, 1988). These variations are basically due to differences in the mechanism thought to fix nitrogen in lightning discharges, namely around or inside the hot lightning channel which results in different freeze-out temperatures for NO. The calculations performed by Mancinelli and McKay (1988) are based on a more realistic model of the lightning chemistry inside the hot cannel, and suggest that P_{NO} is $\sim(1\text{-}3)\times10^{16}$ molecule J^{-1}. Since lightning is a hot plasma (*e.g.*, the temperature of electrons, ions and neutrals is in equilibrium and exceeds 10,000 K), formation of nitric oxide proceeds according to reaction 23.

3.4 PYROLYTICAL

Figure 2 shows the equilibrium mixing ratios of the various species produced by heating an atmosphere of CO_2-N_2 in the presence of traces of methane, and hydrogen. Nitric oxide is the sole nitrogen fixing product arising from this mixture. The chemical reaction leading to its formation is given by equation 23. It can be seen that NO becomes an important product only at temperatures above 2,000 K. The chemical yield of production of NO by shock heating produced by bolide impacts has been estimated by Kasting (1990). The mass of some impactors was so large that their plumes arose far above the atmosphere and considerable erosion of volatiles occurred. The shock processing yield per unit of impactor energy decreases as the energy of the impactor increases. Kasting (1990) has calculated the average P_{NO} value of $\sim 3 \times10^{14}$ molecule J^{-1} by integrating over an impactor distribution extending from 10^{13} to 10^{27} J.

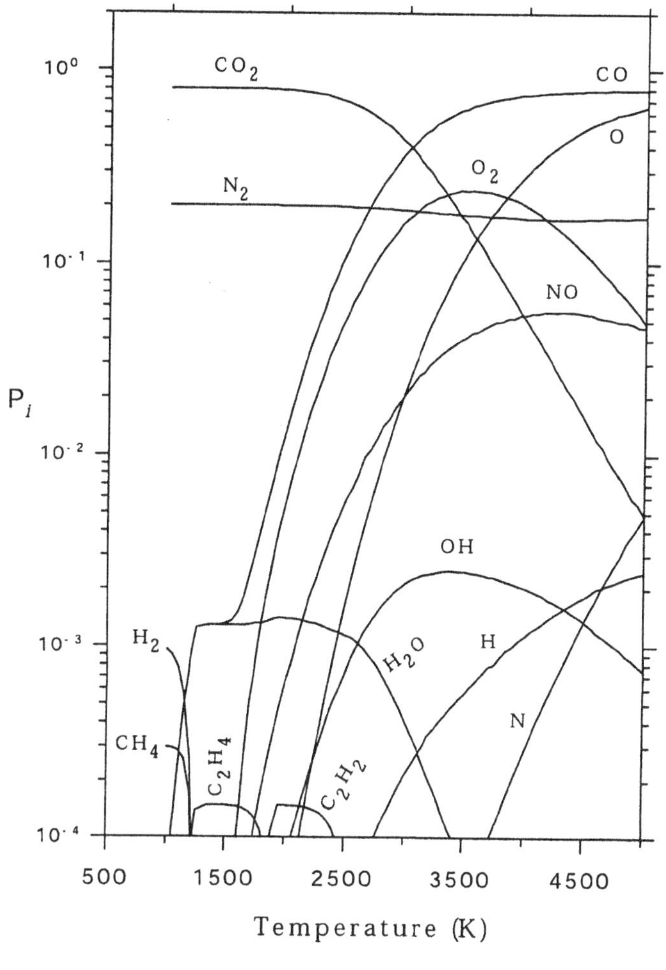

Figure 2. Thermochemical equilibrium partial pressures, P_i, of chemical species derived from a canonical primitive Earth's atmosphere (79.90% CO_2, 19.97% N_2, 0.10% H_2 and 0.03% CH_4 at 1 bar) at various temperatures. The chemical species considered in the computations were: H, H_2, O, O_2, O_3, OH, H_2O, H_2O_2, C_g, C, CO, CO_2, CH, CH_2, CH_3, CH_4, CHO, HCHO, HCO_2H, CH_3OH, C_2, C_2H, C_2H_2, C_2H_3, C_2H_4, C_2H_5, C_2H_6, CH_3CHO, C_2H_5CHO, C_3H_8, C_3H_6, C_3H_4, C_2H_5OH, N, N_2, N_2O, NO, NO_2, CN, HCN, C_2N_2, CH_3CN, C_2H_5CN, CH_2CHCN, HC_3N, HOCN, NH, NH_2, NH_3, N_2H_4, CH_3NH_2, $C_2H_5NH_2$.

Table 1. Order of magnitude estimates of the chemical yields of production (P given in molecule J^{-1}) of nitrogen fixation in mildly reducing and neutral atmospheres

Energy source	CO_2-N_2 mixtures	N_2-CH_4 mixtures
Photochemical	No net production	10^{13}-10^{14}
Radiochemical	10^{14}-10^{15}	10^{17}
Electrical:		
Coronal	No N-fixation	10^{14}-10^{16}
Lightning	10^{15}-10^{16}	10^{16}-10^{17}
Pyrolytical	10^{14}	10^{17}

4. Concluding remarks

An inventory of the chemical yields of production of N-fixation in mildly reducing and neutral atmospheres is given in Table 1. HCN was the main product arising in mildly reducing environments while NO was the sole molecule formed in neutral atmospheres. Nitrogen fixation occurs with all the energy sources examined in mildly reducing atmospheres and is much more efficient than in neutral atmospheres. Another relevant parameter is the abundance of a given energy source in a planetary environment in order to determine their relative roles in the production of nitrogenated molecules. For the case of the early Earth this comparison has already been done by Chyba and Sagan (1992) and Navarro-González *et al.* (1998) for organics and nitrogenated species, respectively.

Acknowledgements

The research reported here was supported by a grant from the National Autonomous University of Mexico (DGAPA-IN102796). The author is grateful to Professor Julian Chela-Flores for the excellent organization of the first Iberoamerican School on astrobiology.

References

Arai, H., Nagai, S., Matsuda, K. And Hatada, M. (1981a), *Radiat. Phys. Chem.* **17,** 151.
Arai, H., Nagai, S., Matsuda, K. And Hatada, M. (1981b), *Radiat. Phys. Chem.* **17,** 217.
Bar-Nun, A. and Shaviv, A. (1975), *Icarus* **24**, 197.
Borisova, E.N. and Eremin, E.N. (1968), in Pechuco, N.S. (ed.), *Organic Reactions in Electric discharges*, Consultants Bureau, New York, 52.

Bossard, A., Raulin, F., Mourey, D. and Toupance, G. (1981), in Wolman, Y. (ed.), *Origin of Life*, D. Reidel Publ. Co., Holland, 83.
Bossard, A., Raulin, F., Mourey, D. and Toupance, G. (1982), *J. Mol. Evol.* **18**, 173.
Braun, W., Welge, K.H. and McNesby, J.R. (1966), *J. Chem. Phys.* **45**, 2650.
Braun, W., McNesby, J.R. and Ross, A.M. (1967), *J. Chem. Phys.* **46**, 2071.
Campbell, I.M. and Gray, C.N. (1973), *Chem. Phys. Lett.* **18**, 607.
Canuto, V.M., Levine, J.S., Augustsson, T.R., Imhoff, C.L. and Giampapa, M.S. (1983), *Nature* **305**, 281.
Chameides, W.L., Walker, J.C.G. and Nagy, A.F. (1979), *Nature* **280**, 820.
Chameides W.L. and Walker, J.C.G. (1981), *Origins Life Evol. Biosph.* **11**, 291.
Chang, S., Scattergood, T., Arowitz, S. and Flores, J. (1979), *Rev. Geophys. Space Phys.* **17**, 1923.
Chen, C.J., Back, M.H. and Back, R.A. (1975), *Can. J. Chem.* **53**, 3580.
Chyba, C. and Sagan, C. (1992), *Nature* **355**, 125.
Ditchburn, R.W. (1955), *Proc. Roy. Soc. (London)* **A229**, 44.
Fujio, C. (1963), *Bull. Chem. Soc. (Japan)* **5**, 249.
Gordon, R. and Ausloos, P. (1967), *J. Phys. Chem.* **46**, 4823.
Haldane, J. B. S. (1928), *Ration. Ann.* **148**, 3.
Harteck, P. and Dondes, S. (1957), *J. Chem. Phys.* **27**, 546.
Harteck, P. and Dondes, S. (1958a), *J. Chem. Phys.* 28, **975.**
Harteck, P. and Dondes, S. (1958b), *J. Phys. Chem.* 63, **956.**
Harting, R., Troe, J. and Wagner, H.G.G. (1971), *13th Symposium on Combustion*, Combustion Institute, Pittsburgh, 147.
Hauser, W.P (1964), *J. Phys. Chem.* **68**, 1576.
Hirschfelder, J.O and Taylor, H.S. (1938), *J. Chem. Phys.* **6**, 783.
Hoing, E.R. and Sheppard, C.W. (1946), *J. Phys. Chem.* **50**, 119.
Howard, J.B. and Rees, D.C. (1996), *Chem. Reviews* **96**, 2965.
Hunten, D.M., Tomasko, M.G. Flaser, F.M. Samuelson, R.E. Strobel , D.F. and Stevenson, D.J. (1984), in *Saturn*, Gehrels T. and Matthews M.S. (eds.), University of Arizona Press, Tucson, Arizona, 671.
Kasting, J. F. (1993), *Science* **259**, 920.
Kevorkian, V., Heath, C.E. and Boudart, M. (1960), *J. Phys. Chem.* **64**, 964.
Kondratiev, V.N. (1965), *10th Symposium on Combustion*, Combustion Institute, Pittsburgh, 319.
Kozlov, G.I. and Knorre, V.G. (1963), *Russ. J. Phys. Chem.* **9**, 1128.
Lampe, F.W. (1957), *J. Am. Chem. Soc.* **79**, 1055.
Laufer, A. and Bass, A. (1978), *Combustion and Flame* **32**, 215.
Levine, J.S., Gregory, G.L., Harvey, G.A,, Howell, W.E., Borucki, W.J. and Orville, R.E. (1982) *Geophys. Res. Lett.* **9**, 893.
Lind, S.C. and Bardwell, D.C. (1925), *J. Am. Chem. Soc.***47**, 2675.
Lind, S.C. and Bardwell, D.C. (1926), *J. Am. Chem. Soc.***48**, 2335.
Lind, S.C. and Glockler, G. (1929), *J. Am. Chem. Soc.* **51**, 2811.
Lind, S.C. and Glockler, G. (1930), *J. Am. Chem. Soc.* **52**, 4450.
Lind, S.C. and Schltze, G.R. (1931), *J. Am. Chem. Soc.* **53**, 3355.
Magee, E.M. (1963), *J. Chem. Phys.* **39**, 855.
Mahan, B.H. and Mandal, R. (1962), *J. Chem. Phys.* **37**, 207.

Mancinelli, R. L. and McKay, C. P. (1988), *Origins Life Evol. Biosph.* **18**, 311.
Maurin, J. (1962), *J. Chim. Phys.* **59**, 15.
McKay, C. P., and Stoker, C. R. (1989), *Rev. Geophys.* **27**, 189.
Miller, S. L. (1953), *Science* **117**, 528.
Miller, S. L. (1955), *J. Am.. Chem.. Soc.* **77**, 2351.
Monat, J.P., Hansen, R.K. and Kruger, C.H. (1979), *Symp. Int. Combust. Proc.* **17**, 543.
Navarro-González, R. (1989), The role of hydrogen cyanide in chemical evolution, Ph. D. Thesis, University of Maryland, College Park.
Navarro-González, R. and Ramírez, S.I. (1997), *Adv. Space Res.* **19(7)**, 1121.
Navarro-González, R., Molina, M.J. and Molina, L.T. (1998), Geophys. Res. Lett. **25**, 3123.
Oparin, A. I. 1924. *Proiskhozhdenie Zhizni.* Isd Moskovskii Rabotchii, Moscow.
Oparin, A. I. 1936. *The Origin of Life.* Macmillan, New York.
Palmer, H.B. and Hirt, T.J. (1963), *J. Phys. Chem.* **67**, 709.
Ponnamperuma, C, and Pering, K. (1966), *Nature* **209**, 979.
Ponnamperuma, C. and Woeller, (1964), *Nature* **203**, 272.
Ponnamperuma, C., Woeller, F., Flores, J., Romiez, M. and Allen, W. (1969), *Adv. Chem. Ser.* **80**, 280.
Ramírez, S.I. and Navarro-González, R. (2000), this volume.
Rao, V.V., MacKay, D. And Trass, O. (1967), *Can. J. Chem. Eng.* **45**, 61.
Raulin, F., Mourey, D. and Toupance, G. (1982), *Origins Life,* **12**, 267.
Rebbert, R.E. and Ausloos, P. (1972), *J. Photochem.* **1**, 171.
Rebbert, R.E. and Ausloos, P. (1973), *J. Res. N.B.S.* **77A**, 109.
Roscoe, J.M. and Thompson, M.J. (1985), *Int. J. Chem. Kinetics* **17**, 967.
Sagan, C., Thompson, W.R. and Khare, B.N. (1992), *Acc. Chem. Res.* **25**, 286.
Sánchez, R.A., Ferris, J.P. and Orgel, L.E. (1966), *Science* **154**, 784.
Scattergood, T.W., McKay, C.P., Borucki, W.J., Giver, L.P., Van Ghyseghem, H., Parris, J.E. and Miller, S.L. (1989), *Icarus* **81**, 413-428.
Sieck, L.W. and Johnsen, R.H. (1963), *J. Phys. Chem.* **67**, 2281.
Skinner, G.B. and Ruehrwein, R.A. (1959), *J. Phys. Chem.* **63**, 1736.
Slager, T.G. and Black, G. (1982), *J. Chem. Phys.* **77**, 2432
Stribling, R. and Miller, S.L. (1987), *Origins Life* **17**, 261.
Sun, H. and Weissler, G.L. (1955), *J. Chem. Phys.* **23**, 1160.
Toupance, G., Raulin, F. and Buvet, R. (1975), *Origins Life* **6**, 83.
Weiner, H. and Burton, M. (1953), *J. Am. Chem. Soc.***75**, 5815.
Yung, Y.L., Allen, M. and Pinto, J.P. (1984), *Astrophys. J. Suppl. Ser.* **55**, 465.
Yung, Y. L. and McElroy, M. B. (1979), *Science* **203**, 1002.
Zahnle, K. J. (1986), *J. Geophys. Res.* **91**, 2819.
Zhdamirov, G.G., Kornienko, V.N., Borosov, E.A., Rumyantsev, Y.M. and Dzantiev, B.G. (1971), *High Energy Chem.* **5**, 475.

Section 3:
Biological bases for the Study of the Evolution of Life in the Universe

DARWINIAN DYNAMICS AND BIOGENESIS

JESUS ALBERTO LEON
Grupo de Biología Teórica
Instituto de Zoologìa Tropical
Facultad de Ciencias U.C.V.
Caracas, Venezuela

Adaptation is perhaps the most conspicuous attribute of living beings, as compared with inanimate natural objects or processes. Adaptation consists of two kinds of correspondence: one between the organisms structure (and activities) *and* the environment in which it lives (which includes other living beings); another between the parts of the organism themselves, each of which seems to contribute specifically to the overall performance, within a hierarchical arrangement of "functions" (Brandon 1990, Rose & Lauder 1996). This, of course, reminds artefacts, whose structure is articulated so as to allow specific performances. Aristotle (e.g. Edel 1982) recognised four kinds of factors (*aitia*) that would entirely explain an object; since they would give answers to the four manners of asking why: what is it made of? (material factor); how is it produced, how it works? (efficient factor); what is it, how is it arranged? (formal factor); and what is it for (final factor). These "causes" can obviously be applied to artefacts, but not to natural physical things or processes. Living things, however, are natural objects which also seem to have design and directionality, are "adapted" and endowed with "adaptations" which perform functions (León 1992). The solution to this conflict comes from the mechanism of natural selection, which Darwin adumbrated. Natural selection can work on physical systems that have the properties of reproduction (multiplication plus heredity) and variation. These entities will always exist as collections (populations), thanks to reproduction. Variation is due to occasional failures in heredity (genetic errors), But variation implies (usually) variation in reproductive success, so that the population composition will change. Those variants endowed with characteristics which help to survive and reproduce better will come to predominate in the population, since they will transmit this superiority to their descendants, which will do the same and increase their number in the same environment. Accidental modification of existing heritable features will just occur, for better or worst; they are not produced *for* improving an organism. But if they serve in that environment, they will contribute to an increased reproductive success (fitness) and so will spread. Thus, if one is to account for the

J. Chela-Flores et al. (eds.), Astrobiology, 99–107.

evolution of adaptive complexity by natural selection in the context of biogenesis, it is crucial to consider the emergence of the simplest form of reproduction, i.e. molecular replication of biopolymers.

1. Darwinian dynamics of replicators

Suppose we entertain the notion that the first protagonists of natural selection were naked polymers capable of self-replication. This capability would depend of the physical property of "complementarity", which allows a molecule to serve as a template for the production of its copy. Some oligonucleotides displaying this property without enzymatic help have been synthesised (von Kiedrowski, 1986; Rebek 1994). Also, by introducing molecules of polycitidine (poly-C) in solutions of activated guanosine, Inoue & Orgel (1983) were able to obtain complementary molecules of polyguanosine (poly-G) 30 or more monomers long. Based on this property, one can postulate what Michod and his colleagues (Bernstein *et al*, 1983; Michod, 1999) have called a Darwinian Dynamic as a kinetic model of natural selection:

$$\frac{dn_i}{dt} = n_i r_i \qquad r_i = b_i - d_i \tag{1}$$

for the change in numbers n_i of each replicator i, where b_i is the *per capita* rate of (template mediated) synthesis and d_i the breakdown rate. Or, if the availability of activated monomers (resources R, presumably provided by pre-biotic synthesis) is taken into account:

$$\frac{dn_i}{dt} = n_i\left(\frac{b_i R}{K_i + R} - d_i\right) \tag{2}$$

where, besides the birth and death parameters, a Michaelis constant K_i is introduced as a measure of the replicator capacity at utilising resources. b_i, d_i, and K_i are heritable capacities that combine in an appropiate index of fitness (success) to determine which replicator will predominate under natural (molecular) selection, without (eqn. 1) of with (eqn. 2) resource limitation. Such fitness indexes are:

$$\begin{array}{ll} b_i - d_i & \text{for eqn (1)} \\ (b_i - d_i)/d_i K_i & \text{for eqn (2)} \end{array} \tag{3}$$

They can be used to identify the fittest replicator (maximal fitness) in the appropiate circumstances, so that the Darwinian (originally Spencerian) "survival of the fittest" holds.

2. Limitations

We must face at once two difficulties. One is due to sub-exponential growth, the other to the inaccuracy of copying unaided by replicases (imperfect heredity).

The complementary binding of template and copy can partially inhibit growth, so that it will be slower than the exponential growth implied by equation 1. For instance, if

$$\frac{dn_i}{dt} = rn_i^p. \tag{4}$$

then

$$n_i(t) = \left\{r_i t - r_i pt + n_i(0)^{1-p}\right\}^{1/(1-p)} \tag{5}$$

If, due to partial inhibition, $0<p<1$, we will have the sub-exponential growth mentioned above. In case $p=½$ (a kinetic obtained by von Kiedrowski, 1986, for oligonucleotides) this is called parabolic growth. The relative concentrations of two parabolically growing replicators will be

$$\frac{n_1(t)}{n_2(t)} = \frac{\left\{\sqrt{n_1(0)} + r_1 t/2\right\}^2}{\left\{\sqrt{n_2(0)} + r_2 t/2\right\}^2} \tag{6}$$

which leads to the "survival of everybody" (Szathmáry 1991), i.e. no predominance of any selective type, since in the limit $n_1(t)/n_2(t)$ tends to r_1^2/r_2^2. So here the Darwinian "survival of fittest" fails, although this implies a varied population of coexistent replicators, available for association in further stages of evolution.

If replication is not enzymatically aided, there is a higher probability of incorporating the mistaken monomer at each place, when copying a template. Since these errors are independent, the probability of a right copy decreases multiplicatively with the length of the template. Eigen (1971) called attention to this fact, establishing an "error threshold of replication". So, early genomes would have consisted of small (less than one hundred nucleotides) independently replicating entities. But they compete with each other and the fittest would win. Hence the paradox: no large genome without enzymes, and no enzymes without a large genome. This was called by Maynard Smith (1983) the Catch-22 of molecular evolution, using the title of the well-known novel of Joseph Heller.

3. Hypercycles

Eigen (1971) proposed as a solution what he called the hypercycle. This is a cycle of replicators, with each member catalysing the replication of the next one. The name points out that each member undergoes a self-replicative cycle within the superimposed cycle of heterocatalysis (Eigen & Schuster 1979).

Although the heterocatalytic aid was originally attributed to simple polypeptides, the discovery of catalytic activity in RNA by Altmann and Cech strengthened the plausibility of schemes like hypercycles, and made single-stranded RNA (ssRNA) the main candidate to be the original replicator, leading to proposals (Alberts, 1986; Gilbert, 1986; Lazcano, 1986) summarised in Gilbert's concept of the "RNA world" (1986).

But hypercycles suffer two difficulties. One concerns invasibility, the other, permanence.

Can a population of self-replicators be invaded by what could be called an auto-hypercycle (AHC), i.e. a polymer which besides direct self-replication can also help itself catalytically? One would think yes at first sight, yet the answer depends on the costs of self-help. Following Michod (1999), we write the kinetics

$$\frac{dn_1}{dt} = n_1(r_1 - \Psi)$$
$$\frac{dn_2}{dt} = n_2(r_2 + Bn_2 - \Psi) \qquad (7)$$

where $\Psi = r_1 n_1 + r_2 n_2 + Bn_2^2/N$ is the *per capita* production of the whole system, and is substracted in the equation to provide regulation (this idea comes from Eigen, 1971).

The additional term in the second equation Bn_2, includes a "benefit" coefficient and the concentration n_2 of the same "self-helpers", since these molecules have to meet in order to self-catalysis to be accomplished. $N=n_1+n_2$ is total concentration. The condition for type 2 (the AHC) to invade is $r_2>r_1$, and this will be difficult, since r_2 is depressed by the cost of helping (when a type 2 molecule is helping another, it cannot be self-replicating).

Also, if some pure auto-hypercycle is established, it cannot be invaded by another, even if the second has improved capacities. This can be seen if in Eqn. 4 we make $p=2$, so obtaining a case of hyperbolic growth. Such difficulty of invasion is due the requirement of bi-molecular encounters, hard to obtain at very low concentrations. This is a problem shared by many models of the evolution of cooperation in other evolutionary settings. It has been called, within the present framework, the "cost of rarity", the "survival of the first" (Michod 1983) and the "survival of the common" (Szathmáry & Maynard Smith 1997).

The permanence of established hypercycles is endangered by "selfish" mutants which turn any of its members into a better target for hetero-help and/or a worst ribozymic "helper". Such a "parasitic" mutant will be favoured by natural selection, and so will disrupt the hypercycle.

4. Fostering cooperation

Several ways out of these difficulties have been suggested to understand the evolution of "cooperation" between replicators. The first is to regard the members of the hypercycle as recurrent mutants of a selfish sequence, which gives them a hike out of rarity, and is then replaced (Michod 1999). For instance, if we substract to the first eqn. 7, and add to the second, a term μn_1 (where μ is a mutation rate), we convert it into another system which can have a unique stable equilibrium with n_1 excluded, and $n_2=N$, provided $BN>r_1-r_2-\mu$, which is relatively easy to hold (Michod 1999).

Other ways of favouring cooperation (in the form of hypercycles or in other arrangements) involve surfaces with local heterogeneities, allowing some population structure of the replicators or, even further, some degree of encapsulation by membranes.

Suppose that groups of replicators adhere to some crevices in the rock, forming what Wilson (1980) called "trait groups", before being washed back to the sea. The composition of the groups will vary. If there is one "altruistic" replicator which helps others (included its own type) to replicate, it would be selectively excluded in a homogeneous solution. However, in a structured model things can go the other way around. Groups with a high proportion of the cooperator will be more productive, and will contribute more to the population pool when washed away. Precise conditions for this to work have been worked out (Michod 1983, Szathmáry & Maynard Smith 1997, Michod 1999) and resemble Hamilton's rule of kin selection.

The importance of membranes in the origin of life is widely recognized today (Lazcano *et al* 1992). A prebiotic model of membranes was suggested by Deamer and Oró in 1980. Some studies with model systems like liposomes, small vesicles composed of fluid lipid bilayers, have produced encapsulation of large molecules. For example, Deamer and Barchfeld (1982) did it by cycles of drying and wetting. Whatever the method, encapsulation is a way of protecting cooperative systems of replicators. Thus, encapsulated hypercycles are not vulnerable to cheaters and interlopers. Rather "altruists" which favour the permanence and growth of the hypercycle are selectively preferred. The reason is the following: assuming that cells in which the total number of replicators increases most rapidly also divides most rapidly, then vesicles containing the most rapidly growing hypercycle would increase in frequency, as compared with other types. As Szathmáry & Maynard Smith (1997) say "good" hypercycles (with efficient heterocatalysis) can be favoured over "bad" ones. As Michod (1999) say, by putting everybody in the same boat, everybody's self-interest becomes more closely aligned with the interest of the group. So, dynamic encapsulation introduces a new level of selection: between-vesicles selection, as well as between-replicators selection.

Indeed, the replicators included within the vesicles do not have to be linked in an ordered kinetic scheme like the hypercycle. Szathmáry & Demeter (1987) Proposed a model in which cooperation is looser. The replicators simply contribute each a certain function, which makes better for the proto-cell a composition including several types of them. For example, if there are two types, a composition with equal numbers is "optimal", in the sense of including faster growth, and so faster division rate. Within the vesicles, one type replicates more quickly, and so tends to outgrowth the other, but there are chance variations as well, analogous to genetic drift, due to the small number of replicators. The presence of chance gives the model its name: stochastic corrector. When the total number of molecules within a proto-cell rises to some critical value, the cell divides, with molecules being distributed randomly to the two daughter cells. Thus, we have "group selection" favouring proto-cells with both kinds of template, and "individual selection" plus drift within each proto-cell. The mathematical result is that the population of "cells" reaches an equilibrium with a constant fraction of "optimal" proto-cells.

5. The beginning at the end

Say that the end precedes the beginning
T.S. Elliot

We have explored the consequences of replication, without worrying about the origin of replicators. Such beginning should be briefly discussed now, at the end.

The traditional view of biogenesis started with the so called "prebiotic soup" of Oparin-Haldane. This conceived the abiotic establishment of a chemically diverse environment in which the first polymers capable of replication (and others) could have arisen. Research has shown that a wide variety of compounds can be produced indeed. However, as Wächtershäuser (1997) has pointed out, some of the supposed prebiotic reactions demand conditions incompatible with those of others. This would require "several separate cauldrons with prebiotic broth". There are also difficulties in the formation of ribose and pyrimidines, and in polymerisation in aqueous solution. All this has led to the idea of surface metabolism, linked to the exergonic formation of pyrite, posed by Wächtershäuser (1988a, 1988b, 1990, 1992, 1993). He has suggested detailed schemes which could make understandable an "autotrophic metabolism first" notion of biogenesis, in an Iron-Sulphur World. Since all this would happen on surfaces von Kiedrowski (1996) has spoken of the "primordial crêpe" and Maynard Smith & Szathmáry (1995) of the "primitive pizza".

The new view, then, moves away from the heterotrophic, diffusion based approach, to envisage proto-cells growing autotrophically from a progression of simpler precursor structures beginning with an ancestral, elementary molecular array, anchored to molecular surfaces and always integrated. These views have received inspiration from the idea of metabolite channelling (Edwards, 1996). On the pyrite surfaces, lipids, peptides and nucleic acids would have formed sequentially, with the products modifying the primary reactions, in a way akin in some respects to embryonic development, as Edwards (1998) has pointed out. The formation, growth and positioning of the molecular components of these primordial structures would have continuously transformed the whole. Thus, semi-cells and proto-cells would have appeared on these surfaces, already containing the replicators, and some primordial form of metabolism providing for growth.

Acknowledgements

Most of the ideas put forward in this essay come -almost to the point of plagiarism- from the work of three friends: Rick Michod, Eörs Szathmáry and John Maynard Smith, one of which (JMS) allowed me the privilege of being his D. Phil. student twenty five years ago. Another friend, Tomás Revilla, former student of mine, shared ideas and discussions, and helped practical aspects of the writing. Thanks to them all.

References

Alberts, B.M. (1986) The function of the heredity materials: Biological catalyses reflect the cell's evolutionary history. *Amer. Zoologist* **26**, 781-796.

Bernstein, H., Byerly, H.C., Hopf, F.A., Michod, R.E., & Vemulapalli, G.K. (1983) The Darwinian dynamic. *Quart. Rev. Biol.* **58**, 185-207.

Brandon, R.N. (1990) *Adaptation and Environment*. Princeton University Press. Princeton, New Jersey.

Deamer, D.W. & Barchfeld, G.L. (1982) Encapsulation of macromolecules by lipid vesicles under simulated prebiotic conditions. *J. Mol. Evol.* **18**, 203-206.

Deamer, D.W. & Oró, J. (1980) Role of lipids in prebiotic structures. *ByoSystems* **12**, 167-175.

Edel, A. (1982) *Aristotle and his Philosophy*. Univ. Of North Carolina Press. Chapel Hill.

Edwards, M.R. (1996) Metabolite channeling in the origin of life. *J. Theor. Biol.* **179**, 313-322.

Edwards, M.R. (1998) From a soup or a seed? Pyritic metabolic complexes in the origin of life. *Trends Ecol. Evol.* **13**, 178-181.

Eigen, M. (1971) Self-organisation of matter and the evolution of biological macromolecules. *Naturwissenschaften* **58**, 465-523.

Eigen, M. & Schuster, P. (1979) *The Hypercycle*. Springer-Verlag. Berlin.

Gilbert, W. (1986) The RNA world. *Nature* **319**, 618.

Inoue, T. & Orgel, L.E. (1983) A nonenzymatic RNA polymerase model. *Science* **219**, 859-862.

Lazcano, A. (1986) Prebiotic evolution and the origin of cells. *Treb. Soc. Cat. Biol.* **39**, 73-103.

Lazcano, A., Fox, G.E. & Oró, J. (1992) Life before DNA: The origin and evolution of early archaean cells. In: R.P. Mortlock (ed.) *The evolution of metabolic function*. CRC Press. Boca Ratón, Florida.

León, J.A. (1992) Bioteleología. *Cuadernos de Episteme* **5**, 31-66.

Maynard Smith, J. (1983) Models of evolution. *Proc. Roy. Soc. Lond.* **B219**, 315-325.

Maynard Smith, J. & Szathmáry, E. (1995) *The Major Transitions in Evolution*. Freeman & Co. Oxford.

Michod, R.E. (1983) Population biology of the first replicators: on the origin of the genotype, phenotype and organism. *Amer. Zoologist* **23**, 5-14.

Michod, R.E. (1999) *Darwinian Dynamics*. Princeton Univ. Press. Princeton, New Jersey.

Rebek, J. (1994) Synthetic self-replicating molecules. *Scientific American* **271**, 48-55.

Rose, M.R. & Lauder, G.V. (eds.) (1996) *Adaptation*. Academic Press. San Diego, California.

Szathmáry, E. (1991) Simple growth laws and selection consequences. *Trends Ecol. Evol.* **6**, 366-370.

Szathmáry, E. & Demeter, L. (1987) Group selection of early replicators and the origin of life. *J. Theor. Biol.* **128**, 463-486.

Szathmáry, E. & Maynard Smith, J. (1997) From replicators to reproducers: the first major transitions leading to life. *J. Theor. Biol.* **187**, 555-571.

von Kiedrowski, G. (1986) A Self-replicating hexadeoxynucleotide. *Angew. Chem. Int. Ed. Engl.* **25**, 932-935.

von Kiedrowski, G. (1996) Primordial soup or crêpes? *Nature* **381**, 20-21.

Wächtershäuser, G. (1988a) Before enzymes and templates: theory of surface metabolism. *Microbiol. Rev.* **52**, 452-484.

Wächtershäuser, G. (1988b) Pyrite formation, the first energy source of life: a hypothesis. *Syst. Appl. Microbiol.* **10**, 207-210.

Wächtershäuser, G. (1990) Evolution of the first metabolic cycles. *Proc. Natl. Acad. Sci. U.S.A.* **87**, 200-204.

Wächtershäuser, G. (1992) Groundworks for an evolutionary biochemistry: the iron-sulphur world. *Prog. Biophys. Mol. Biol.* **58**, 85-201.

Wächtershäuser, G. (1993) The cradle chemistry of life. *Pure Appl. Chem.* **65**, 1343-1348.

Wächtershäuser, G. (1997) The origin of life and its methodological challenge. *J. Thoer. Biol.* **187**, 483-494.

Wilson, D.S. (1980) *The Natural Selection of Populations and Communities.* Benjamin Cummings. Menlo Park, California.

EVOLUTION OF ADAPTIVE SYSTEMS

HERNÁN J. DOPAZO
Laboratorio de Biología del Comportamiento.
Instituto de Biología y Medicina Experimental. (IBYME-CONICET).
Vuelta de Obligado, 2490 (1428).
Buenos Aires. Argentina.
hdopazo@dna.uba.ar

1. Abstract

Adaptive systems are those biological and non-biological assemblages able to evolve by natural selection. They are formed by entities with the property to replicate with errors in such a way that these variants bias directly, or indirectly, their own frequency in later generations. Neo-Darwinism focused on the study of a single adaptive system, i.e.: populations of multicellular organisms with sexual reproduction and early somatic and germ line differentiation[1]. Evolutionary biologists however recognize that entities like genes, chromosomes, cells, organisms, kin, animal societies, cultural characters and computer programs are indeed able to evolve by differential reproductive success of heritable variants. The objective of this paper is to overview the main ideas and concepts dealing with the origin and evolution of individuality at different hierarchical levels of biological organization[2]. First, I'll describe the evolutionary dynamics in three abstract spaces. This will led us to distinguish replicator versus interactor concepts and codical versus material domains. This will allow us too to introduce the problem of the units of selection and the concept of organism from an evolutionary perspective. After, I'll follow sketch the major transitions in evolution, many of them originate the components of alternative adaptive systems. Cooperation and conflicts were recurrent processes guiding the construction of alliances between units of selection. Cooperation by kin selection, trait- group selection, reciprocal altruism and byproduct mutualism will be differentiated, at the same time fraternal and egalitarian alliances will be distinguished. Exceptions to cooperation and mechanisms to avoid selfish interest of free riders are also discussed. I'll conclude with an outline of the alternative ways by which hereditary information was stored and transmitted during the course of biological evolution from one generation to the next.

---

[1] *Some chordates and artrophods show this kind of development. Most phyla of animals, plants, fungi and protoctista don't. The inheritance of acquired features is possible in these last taxa [6].*
[2] *Life can be divided into a series of increasing levels of organization. Eldredge [21] differentiates genealogical hierarchy (genes, chromosomes, genomes, organisms, kin group, demes, species and monophyletic taxa), from ecological hierarchy (molecules, cells, tissues, organs, organisms, populations, communities and regional biota). This has been an important attempt to organize evolutionary theory in a rather comprehensive framework.*

J. Chela-Flores et al. (eds.), Astrobiology, 109–120.

2. Darwinian Dynamics

Schuster [65] has recently pointed out that evolutionary dynamics take place in three different abstract spaces: the concentration, sequence and shape space. Fig. 1.

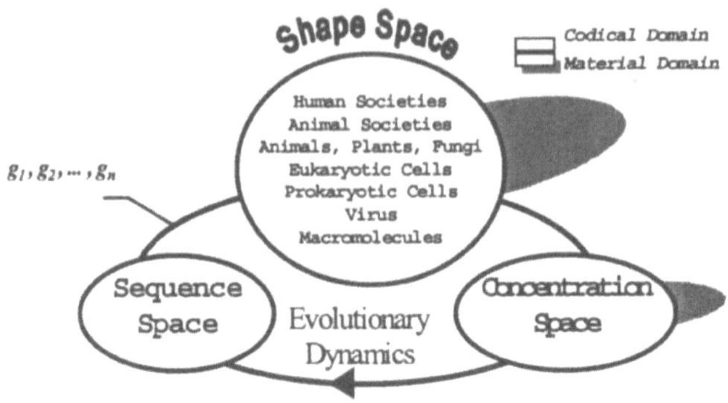

The *concentration space* is familiar to evolutionary ecologists and chemists. Processes occurring within it are measured by changes over time in the proportion of different interacting material entities –alleles, virus, cells-, or organisms-. A great diversity of growth models, ranging from single species to interacting populations describes the change in frequency [55, 64]. The Darwinian process of struggle for existence takes place in this ecological arena, where entities compete and cooperate with the blind purpose to survive and reproduce.

The *sequence space* represents all possible combinations of a particular carrier of information [19]. It is a high dimensional space in which one sequence is surrounded by all possible combinations of "one-mutant" neighbors. The dimension of this space is huge[3]. In the case that RNA or DNA codifies information, the sequence space is proportional to 4^N. A convenient unit to measure distances in this space is the Hamming distance. A DNA string of length N= 10 represents a single point in a space greater than a million sequences of which 30 differ only in one nucleotide or a Hamming distance[4]. As a consequence of the astronomically large universe of possible genotypes, biological evolution only has visited an irrelevant subspace of this informational universe since the origin of life [65]. A population can be represented as a cloud of points surrounding one that identifies the average sequence of a population. Adaptive evolution in this space can be viewed as the gradual movement of this point with an associated reduction of its dispersion. The reduction of variation is a consequence of natural selection acting on genetic variance, and as Fisher [23] deduced, the process definitively stops when sequences are identical (genetic variance reduces to zero) or when their material effects do not bias their own representation in the next generation. This may be so because se-

[3] *A short sequence of 10 amino acids represents a single point in a space of $20^{10}=10^{13}$ sequences! Where 20 corresponds to the number of different amino acids.*
[4] *A sequence of RNA or DNA of length N has 3N neighbor sequences in this space.*

quences unfold distinct phenotypic structures or different structures have identical reproductive success. In both cases evolution by natural selection stops and the adaptive system reaches to a maximum of adaptation.

The *shape space* unfolds the expression of all the information (genotypes) in a given environment. If we think how a particular information, biological or non-biological, is expressed we find secondary and tertiary structures of macromolecules, prokaryotic and eukaryotic forms, multicellular organisms, animal and human societies. Several qualitative attempts have been done to deal with the properties of the entities inhabiting this. Alberch [2] has emphasized that phenotypes are well-buffered systems, resilient to environmental and genetic perturbations. Phenotypes are distributed in a finite set of bounded domains (steady states) of continuous variation (tissue types, taxonomic units, etc.) separated by discontinuities because a wide variety of morphologies are never produced. Transitions from one steady state into another is non-random, some morphologies seem to be recurrent and more likely to occur than others are. In an exhaustive study of the secondary structures of RNA sequences, some of these properties were recently found [36]. These studies show a striking attribute: the shape space is cover by a small section of the sequence space. This means that, in small spaces around any arbitrary chosen sequence all the secondary structures (common phenotypes) of the RNA are found. As a corollary, evolution could run faster than previously thought. It is the unnecessary to search for all the huge sequence space to find alternative phenotypes. Although a useful metric seems to be difficult to define, fitness values between alternative forms have an important evolutionary significance in this space.

Evolutionary dynamics run through g_i generations in a continuous cycling between spaces (Fig.1). The genotype-phenotype mapping defines the relevant phenotype properties in the shape space that makes an entity able to cooperate and compete in the concentration space. According to their reproductive success, alternative arrays of information proliferate in the sequence space to begin a new (g_{i+1}) round of evolution.

3. The Problem of the Units of Selection

The shape and the concentration space support material entities that interact with the environment. Are defined within the *material domain* and display properties like charge, density, volume, weight, color, etc. Alternatively, the sequence space just contains units of information (codex) described by bits, redundancy, fidelity and meaning in the *codical domain* (Fig.1). The replication of the codex depends on the imposition of a pattern by any kind of entity that functions as interactor in the material domain. Selection in the codical domain is not based on individual success of each codex; rather it is based on the average phenotypic effects imposed by a complete codex of interactors in the concentration space. The distinction between the material and the codical domain aids to conceive life as information and to differentiate levels of biological organization on which natural selection acts directly and on which levels it affects incidentally [75]. During years, biologists and philosophers of sciences have discussed the problem of the *units of selection* [6, 34, 40, 43, 44]. The answer is not evident when several levels share the benefits of adaptations; i.e.: what benefits an organism seems to be good for its group and its species [74]. Furthermore, conflicts between levels are frequent, i.e.: what benefits a gene may not benefit its organism [11], and what benefits an organism is often in contest with its group [33]. This issue has a more comprehensive interest indeed; a general theory of evolution should include a precise and accurate account of why adaptations evolve.

The concepts of *replicator*[5] and *interactor*[6] [34], turn the units of selection controversy in the resolution of two different questions: at what levels does replication occur, and at what levels does interaction occur. These questions can be solved if we look for entities whose frequency is adjusted by natural selection and those that display adaptations respectively. There is a delicate agreement by which lower levels of the genealogical hierarchy like genes, non-recombinant chromosomes and genomes of asexual individuals are considered replicators. This is so because they have the sufficient permanence in the material domain for natural selection to change their frequency. Other entities do not qualify. It is impossible to adjust the frequency of any entity between successive g_i generations if it has ceased to exist. Sexual genomes are broken down by meiosis each generation. Likewise, organisms and groups, are ephemeral entities. While the first ones becomes dust by death the second disperse diminishing their heritability[7]. An alternative is to consider the entire gene pool of a species. When a species splits in two, the daughter species carries most of the ancestral genes, so that, the species heredity is ensured. Therefore, genes[8] and gene pools have the potential to be immortal in the material domain and hence, their information perpetuates ageless in the codical domain[9].

Replicators may code for adaptations but do not necessarily exhibit adaptations. They must produce interactors which elaborate adaptations (structures and strategies) able to cope with the material domain[10]. Biologists have recognized adaptations that benefit genes [11, 35, 46], organelles [18], cells [6, 52, 53], organisms [12], kin [39, 56, 78] and groups of unrelated individuals [28, 73] in struggle with other units of selection.

3.1. THE CONCEPT OF ORGANISM

Natural selection produces adaptations at different levels of organization because different interactors display heritable variations in fitness. Within this framework an important question needs to be solved: why do adaptations appear mainly at the organismic level? Evolutionary biologists have a simple answer. The *organism* is the main receptor of adaptations because it is the unit that unmistakably shows heritability. A new mutation expressed by any kind of interactor in the material domain (whether a gene, organelle, cell, organism, or group) must be passed on to the offspring of that unit in the next generation to increase its frequency in the codical domain. Such mutation is easily expressed in the organism, but not so easily in other units. The organism is a consoli-

---

[5] *Replicator is "an entity that pass on its structure directly in replication".*

[6] *Interactor is "an entity that interacts directly as a cohesive whole with its environment in such a way that replication is differential".*

[7] *Natural selection can not work if there is not heritability.*

[8] *Williams [74] defined genes as those entities "which reliably survive the process of meiosis intact".*

[9] *This scheme, nicknamed the "eye-gene view" arises from the Fisherian tradition of population genetics.*

[10] *Vehicle was the first term suggested by Dawkins [14]. Szathmáry [69] just recently proposed reproducers, because neither vehicles nor interactors suggest that they may be unit of selection.*

dated unit of design because it has evolved efficient adaptations to solve intra-organismic conflicts that atomize it. Entities that successfully resolve internal contests are candidates to classify as organisms in spite of the ability to develop physical contours. The old idea suggesting that human societies [13] and colonies of social insects [77] are superorganisms is more than an insubstantial analogy; it has justification within the current *multilevel selection theory* [5, 66]. This is the evolutionary framework dealing with the origin of the novel components of adaptive systems emerged during the history of life.

4. The Major Transitions in Evolution

Buss [6] emphasized that the history of life is a history of transitions between units of selection. Such a history does not list important phenotypic changes like the vertebrate's conquest of land, the origin of flowering plants and reptile's embryonic membranes. Major evolutionary transitions involve changes in the way biological information is stored and transmitted from one generation to the next during evolution [51]. These transitions indeed do create new interactors of further adaptive systems.

TABLE 1. Major evolutionary transitions

Lower level		*Higher level*
1. Replicating molecules	...>	Populations of molecules in compartments
2. Independent replicators	...>	Chromosomes
3. RNA as gene and enzyme	...>	DNA + Protein *(genetic code)*
4. Prokaryotes	...>	Eukaryotes
5. Asexual clones	...>	Sexual populations
6. Protists	...>	Animal, plants, fungi *(cell differentiation)*
7. Solitary individuals	...>	Colonies *(nonreproductive castes)*
8. Primate societies	...>	Human societies *(language)*

Table 1 urges two questions. How groups of formerly independent entities coalesce into integral wholes and, how natural selection governs changes of the inheritance rules that indeed guide the ways natural selection forms its products. These questions involve two major topics of the current evolutionary theory, the origin of individuality and the evolution of heredity respectively.

4.1. THE ORIGIN OF INDIVIDUALITY

A common feature of transitions 1, 2, 4 - 7, is that entities able to replicate as a single unit before the transition only can do it as part of a larger whole after it. Reproductive restrain of lower units in favor of a higher unit needs an explanation in terms of its immediate selective advantage. Why natural selection favoring selfish behavior does not disrupt integration after each transition? Evolutionary biologists answer this question taking into account two different problems: the emergence of cooperation and the regulation of conflicts between units of selection.

4.1.1. Cooperation Among Units of Selection
Conventional expositions of Darwinism suggest that natural selection evolves traits that cause individuals to have more offspring than their competitors, not fewer. Under this scheme, there is an apparent paradox to explain biological altruism[11]. This paradox was first completely resolved by Hamilton [31] who noticed that when groups of inbred individuals as in a hive of social insects, help their relatives to breed, leave more gene copies than when they breed themselves. A gene may increase its frequency either by enhancing the reproduction of itself or by increasing the reproduction of the same gene in a different body. All sorts of behavior that seem puzzling when seen through the lens of the individual become clear when seen through the eye-gene view. Natural selection working on groups of relatives is known as *kin selection* and constitutes, along with *trait group selection*[12] [76], *reciprocal altruism*[13] [71] and *byproduct mutualism*[14] [8], the main hypothesis that biologists have to explain the evolution of unselfish behaviors.

Fruitful attempts to join all categories of cooperation under a single theoretical umbrella are based on trait-group selection model [67], game theory [17] and the formalism created by Price [58, 59]. *Price's equation* differentiates within and between-group components of selection and is helpful to predict the evolution of integrative wholes in each transition. In Price's model, cooperation leads to a cohesive whole when fitness covariance at the level of the higher unit increases enough to overcome within-unit mutants defecting cooperation [32, 27, 53, 67]. Cooperative units of selection build up two kinds of alliances during transitions, the egalitarian and the fraternal. The libertarian route is the option for most units of selection that behave as free riders [60].

Egalitarian Alliances. Protocells, chromosomes, eukaryotic cells and sexual reproductive form egalitarian alliances. They develop by symbiosis of different, nonfungible units attending diverse functions. Division of labor is the main initial advantage. Be-

---

[11] *Altruistic traits reduce the fecundity or increase mortality of an interactor for benefit of others. Consider a population composed of altruist and nonaltruist individuals. Everyone has a certain number of offspring in absence of altruism. In addition, each altruist behaves in a way that decreases its own number of offspring and increases the number of offspring of a single recipient in the population. Altruists can benefit from other altruists, but they also experience the cost of their own self-sacrificial behavior. Selfish types do not experience any cost and can benefit from all altruists in the population. It is obvious that selfish types always have more offspring than altruist ones and will be favored by natural selection. Despite the fact that altruists increase their own fitness, they still get extinct because they increase the fitness of others even more. Altruism can not evolved by natural selection.*

[12] *Trait-group selection relies on the superior group-level production of traits in a structured population. Altruist genes can increase its frequency if the within-group cost is offset by some between-group benefit. Cooperation can evolve when groups are able to export altruist genes to other groups, otherwise selfish genes replace altruist ones within each group [67]. The power of group-level selection overcoming within-group interest is clearly exemplified by the evolution of intermediate degrees of parasite's virulence [22, 43, 48] and in the female-biased sex ratio evolved in insects [10].*

[13] *Reciprocity involves the exchange of beneficence between two or more interactors. There is an implication of at least temporary expense associated with each beneficent act. In the long run, however, each participant experiences a net benefit, provided that neither individual cheat (fail to reciprocate). Reciprocity may be expected to evolve only when sufficient safeguards against cheating exist [3]. Reciprocity has been studied in insects, fishes, non-primate mammals and non-human primates [16].*

[14] *Byproduct mutualism or pseudo-reciprocity applies when cheating is usual. It holds when the best one can do for oneself also happens to be the best for the other. In these, the return benefit for a beneficent act is a by product or incidental effect of a selfish behavior. Cooperative pairs of hunting lions [56] and dwarf mongoose-hornbill mutualism [61] are intraspecific and interespecific examples respectively.*

cause each unit has independent reproduction ability, the main problem at its origin is to solve different conflicts among replicators.

The *hypercycle* model consists in an array of unlike cooperative replicators [20]. If the hypercycle is enclosed in compartments, defecting mutants can not parasite protocells[15]. The next transition focuses on how unrelated genes can tie to form chromosomes at a cost of larger replication times. Cooperation emerges if positive synergistic effects arise between genes [50]. This effect coupled with a synchronized replication time avoids the temptation to defect of lower units. The origin of eukaryotic cells brought together genetic replicators form different prokaryotic lineages [47]. The evolution of cooperation by reciprocity is possible if selfish interests of organelles are overcome [26]. The origin of sex, is one of the greatest puzzles in evolutionary theory [49]. Müller's [54] hypothesis considers that each partner supplies some good genes and some bad genes, with sufficient asymmetry of functions to form a progeny without deleterious mutations. This can be a sufficient explanation of sex if the rate at which asexual populations suffer mutational decay is enough to overcome asexuality's short-term advantage within populations. Here we have, once again, the problem of two competing levels of selection.

Fraternal Alliances. Multicellular organisms and colonies with nonreproductive castes form fraternal alliances. They evolved by kin selection of fungible units differentiated by epigenesis to attend dissimilar functions. The initial advantage of these alliances resides in the economy of scale those units develop. Because units are genetically identical, a single unit can in theory employ reproductive labors without contests among units. Soma sacrifice reproduction as a natural altruistic act in favor of their genetically identical germ-line neighbors. This harmony is destroyed however, if a selfish mutant arises. The early germ line segregation process probably evolved in animals as a fair mechanism to decrease cell lineages competition among mutant-cells to become gametes within organism [6, 52]. Some animals, notably ants, bees, wasp and termites live in colonies in which only a few individuals reproduce. The origin of insect societies probably was predisposed by haplodiploid genetic systems. Queens fertilized by a single male produce diploid females and haploid sons. Below this system the coefficient of relationship among sisters is ¾, between mother and daughter is ½., and between son and daughter is ¼. A female ant or bee would gain nothing by leaving the colony. Their genes replicate faster raising female eggs. Workers, increase fitness by kin selection distorting the colony sex ratio in favor of them [72], or producing sons by parthenogenesis. Animal societies display possibilities within the group, between individual selection processes.

4.1.2. Avoiding Conflicts Between Units of Selection

The achievement of individuality at the higher levels requires the suppression of disruptive effects of selection at the lower levels. *Policy* and *contingent irreversibility* are common adaptations that benefit higher units against the selfish advantage of the lower units of selection. The concept of a "parliament of genes" is a useful image of the first

[15] *Szathmáry and Demeter [70], proposed the stochastic corrector model in which additional variance among compartments increases when the number of molecules within them is small. In such a case more effective compartment selection is feasible.*

mechanism. *Genetic conflicts* are solved by mutations that compensate selfish interest of mutant genes [41, 42, 45]. In organisms with separate sexes, cytoplasmic factors favor transforming sons into daughters, this strategy however, is restored by mutations of nuclear genes [9]. Similarly, a fair insect society makes use of honeybee workers to policy and preferentially destroys worker-laid eggs rather then queen-laid eggs [62]. Likewise, the queens of the eusocial naked mole rats control the pace of work and the reproduction of competitive females through physical dominance [63]. By chance, independent reproduction of the lower units may be lost once the transition has been established. Irreversibility limits the selfish behavior of lower units of selection. They have not future as free living units. Some examples will make the issue clear. Mitochondria have lost genes most of which were translated into the nucleus, consequently they can not live like free prokaryotes. The same logic compromises somatic cells of animals and individuals of social insects. Although cancer cells may overrate the pace of standard cell replication, they have no future as unicellular eukaryotes. Eggs produced by honeybee workers develop into sons who have not prospect to set up a new colony. Sexual reproduction can be though as a contract between independent reproductive units. Exceptions favoring selfish strategies of asexual reproduction have evolved many times in different taxa. However mammals never produce asexual offspring probably because DNA methylation patterns[16] selectively mark male and female genes to become functional in different tissues [4].

4.2. THE EVOLUTION OF HEREDITY

Inheritance systems may be limited if the number of possible codex is similar to the number of realized sequences, or unlimited when the formers are higher. In the first case, natural selection quickly comes to a halt and the optimal sequence invades the population. Otherwise, if an informational system is unlimited, adaptive systems can evolve indefinitely[17]. Additionally, information may be analogue or digital depending on if it is stored on diffuse systems or template forms. Only the last ones are able to change digit by digit. Several *hereditary systems* emerged during evolution that competed and alternated in a way that a general trend from limited-analogues to unlimited-digitals can be traced [38].

The primary replication system is represented by any autocatalyc cycle in which a chemical compound A undergoes a series of transformations to produce a new A (e.g.: A-B-C-D-2A). *Autocatalysis* is the simplest form of self-maintenance and it probably produced a rich chemical environment before the origin of life[18]. Such cycles however fail to show heredity. Inheritance systems assure that if a variant A_1 appears at the end of the cycle, two A_1 must emerge after a new round (e.g.: A_1-B_1-C_1-D_1-2A_1). Autocatylic replication codes analogue information. There is no possibility to change parts without changing the whole replication system. This is the reason why analogue systems can replicate few states. As soon as alternative cycles evolved, they could compete

[16] *Methylation is a reversible mechanism used to mark particular sequences of DNA in gametogenesis. Genomic imprinting occurs when parents differentially express these. This epigenetic inheritance system evolved from bacteria as part of the restriction-modification system that protects prokaryotes against parasites.*
[17] *Adaptive systems can explore by mutation alternative codex inside the huge sequence space.*
[18] *Growth requires a fair round of autocatalysis.*

for resources and intermediates, and so were naturally selected as units of selection [69].

The origin of large templates invites to solve Eigen's paradox[19]. The *hypercycle* model elucidates the dilemma and once ribonucleotide replication evolved (RNA-like system) digital information are stored and transmitted. Ribozymes-like entities controlled metabolism and inheritance functions of primitive cells in the *RNA world.* This scenario changed with the origin of the *genetic code.* According to the coding coenzyme handle hypothesis [68], some ribozymes were able to bond trinucleotides with amino acids specifically that made catalysis of other ribozymes more efficient. Evolution replaced the first ones by different protein aminoacyl-tRNA synthetases and the last ones by mRNA. Further labor division arose when DNA replaces RNA, in this case an unlimited-digital inheritance system evolved by competence and cooperation.

Cells divide and frequently show some kind of memory. Maintenance of differentiated states usually does not involve DNA changes. These states derived from three different epigenetic mechanisms of inheritance [37, 38]. The simplest *epigenetic inheritance system* (EIS's) is based on auto-regulated gene products like those controlling maternal early development in *Drosophila.* Structural inheritance system is based on three-dimensional structures which act as templates for transmitting cell form, cytoskeletal and cortical organization in mitotic and meiotic cell descendants in ciliates. The chromatin–marking system is based on the inheritance of chromatin marks such as DNA methylation patterns or patterns of proteins bound to DNA. EIS's probable evolved in prokaryotes to deal with fluctuating periodical environments changing too fast to make a genetic mutation benefit and too slow to make non-heritable gene expression useful [37]. Both, the steady-state and the structural inheritance system are codified by a limited set of analogue instructions. Alternatively, the chromatin marking system controls an unlimited set of possible orders digitally adjusted making the origin of multicellular organisms possible.

The large gap between animal and human communication systems turns complex an adaptive explanation for the origin of language. *Human language* probably evolved from a simpler representation plan. Selective pressures favoring cohesion of large social groups probably forced the origin of a poor grammar and syntax [1]. Language of babies younger than two years, those of the adults deprived of language exposure during the critical developmental stage and the pidgin forms emerging from communities of adults deprived of a common language, are candidates to qualify as *protolanguages* [57]. For natural selection acting on grammar abilities, genetic variation is necessary and this evidence has been found [29]. Although protolanguage can store large amounts of information, it is limited when compared with mature language able to create massive unambiguous sentences and ideas. Human language abilities developed by using templates derived from universal grammar rules [7]. Universality hints to a common genetic program probably acquired by Baldwinian selection. This Lamarckian simile Darwinian mechanism leads that behavioral skills proved during life can be genetically assimilated in future generations [15].

---

[19] *This states that originally, there was an error threshold for the accuracy of replication bellow which useful enzymes that ensure accurate replication can not be produced. The paradox is no enzyme without large genome, and no large genome without enzymes.*

5. Conclusions

1- The hierarchical structure of life does not result from an inherent law of cumulative progress imposed to any adaptive systems. It follows from major events of natural selection favoring cooperation rather than competition between units of selection.
2- Replicators and interactors are useful mutually nonexclusive concepts applied to different units of selection. Replicators have the ability to perpetuate ageless in the codical domain, while interactors generally are fungible entities carrying most adaptations to confront with the material domain.
3- Transitions of individuality are recurrent solutions essayed by adaptive systems facing with the informational ruin of replicators and the environmental challenges [25].
4- Consolidate wholes are recognized once natural selection overrides selfish interest of the lower units. Only in this case, their adaptive traits can increase in frequency.
5- Since cooperation is synergism with the selection's stamp of approval, self-maintaining entities probably would coalesce into higher units if "the tape of life reruns again" [30, 24].

6. Acknowledgements

I am grateful to Guillermo Lemarchand who invited me to participate in the Caracas's School of Astrobiology. I also want to thank to Julian Chela Flores, ICTP (Italy) and SETI Institute (USA) that made it possible. I want to acknowledge the CONICET fellowship support.

7. References

1. Aielo, L. and Dunbar, R. (1993) Neocortex size, group size and the evolution of language. *Current Antrophology* **34**, 184-93.
2. Alberch, P. (1980) Ontogenesis and morphological diversification. *Amer. Zool.* **20**, 653-67.
3. Axelrod, R. and Hamilton, W. D. (1981) The evolution of cooperation. *Science* **211**, 1390-6
4. Barlow, D. (1993) Methylation and imprinting: from host defense to gene regulation. *Science* **260**, 310.
5. Boehm, C. (1997) Impact of human egalitarian syndrom on Darwinian selection mechanics, *Am. Nat.* **150**, S100-S121.
6. Buss, L. W. (1987) *The evolution of individuality*, Princeton, N.J., Princeton Univ. Press.
7. Chomsky, N. (1965) *Aspects of the theory of syntax*, MIT Press.
8. Connor, R. C. (1986) Pseudo-Reciprocity: investing in mutualism. *Anim. Behav.* **34**, 1562-6.
9. Cosmides, L. and Tooby, J. (1981) Cytoplasmic inheritance and intragenomic conflict, *J Theo. Biol.* **89**, 83-129.
10. Colwell, R. K. (1981) Group selection is implicated in the evolution of female-biased sex ratios, *Nature* **263**, 401-4.
11. Crow, J. F. (1979) Genes that violates Mendel's laws, *Sci. Am.* **240**, 134-46.
12. Darwin, C. R. (1859) *On the origin of species*, J. Murray, London.
13. Darwin, C. R. (1871) *The descent of man and selection in relation to sex*, J. Murray, London.
14. Dawkins, R. (1976) *The selfish gene*, Oxford Univ. Press, Oxford.
15. Deacon, T. (1997) *The symbolic species. The co-evolution of language and brain.* Norton & Co. NY.
16. Dugatkin, L. A. (1997) *Cooperation among animals. An evolutionary perspective,* Oxford Series in Ecology and Evolution, Oxford Univ. Press, Oxford.
17. Dugatkin, L. A. and Mesterton-Gibbons, M. (1995) Cooperation among unrelated individuals: reciprocal altruism, byproduct mutualism and group selection in fishes, *Biosystems* **37**, 19-30.

18. Eberhard, W. G. (1980) Evolutionary consequences of intracellualar organelle competition, *Q. Rev. Biol.* **55**, 231-49.
19. Eigen, M. (1992) *Steps towards life*, Oxford Univ. Press, Oxford.
20. Eigen, M. and Schuster, P. (1979) *The hypercycle, a principle of natural self-organization*, Springer-Verlag, Berlin.
21. Eldredge, N. (1985) *Unfinished Synthesis. Biological hierarchies and modern evolutionary thought*, Oxford Univ. Press, Oxford.
22. Ewald, P. W. (1993) *Adaptation and disease*, Oxford Univ. Press, Oxford.
23. Fisher, R. (1930) *The genetical theory of natural selection*, Oxford Univ. Press, Oxford.
24. Fontana, W. and Buss, L. (1994) What would be conserved if "the tape were played twice", *PNAS* **91**, 757-61.
25. Frank, S. A. (1996) The design of natural and artificial adaptive systems, in M. Rose and G. Lauder (eds.), *Adaptation*, Academic Press, N.Y. pp. 451-505.
26. Frank, S. A. (1997) Models of symbiosis, *Am. Nat.* **150**, S80-S99.
27. Frank, S. A. (1998) *Foundations of social evolution*, Princeton Univ. Press, Princeton, N. J.
28. Goodnight, C. J. and Stevens, L. (1997) Experimental studies of group selection: what do they tell us about group selection in nature?, *Am. Nat.* **150**, S59-S79.
29. Gopnik, M. (1990) Feature-blind grammar and dysphasia, *Nature* **344**, 71.
30. Gould, S. J. (1989) *Wonderful life. The burgess shale and the nature of life*, Norton & Co. NY.
31. Hamilton, W. D. (1964) The genetical evolution of social behavior. I. *J Theo. Biol.* **7**, 1-16.
32. Hamilton, W. D. (1975) Innate social aptitudes of man: an approach from evolutionary genetics, in R. Fox (ed.), *Biosocial Anthropology*, Wiley, NY, pp. 133-155.
33. Hardin, G. (1968) The tragedy of commons, *Science* **162**:1243-8.
34. Hull, D. L. (1980) Individuality and selection, *Ann Rev. Ecol. Syst.* **11**: 311-32.
35. Hurst, L. (1996) Adaptation and selection of genomic parasites, in M. Rose and G. Lauder (eds.), *Adaptation*, Academic Press, N.Y. pp. 407-49.
36. Huynen, M. A., Stadler, P. F. and W. Fontana (1996) Smoothness within ruggedness. The role of neutrality in adaptation, *PNAS* **93**, 397-401.
37. Jablonka E. and Lamb M. J. (1995) *Epigenetic inheritance and evolution, the Lamarckian dimmension*, Oxford Univ. Press. Oxford.
38. Jablonka, E. and Szathmáry, E. (1995) The evolution of information storage and heredity, *TREE* **10**, 206-11.
39. Jarvis J. U., O'Rian M. J., Bennett N C. and Sherman P. (1994). Mammalian eusociality. A family affair, *TREE* **9**, 47-51.
40. Keller, L. (1999) *Levels of selection in evolution*, Princeton Univ. Press, Princeton, N. J.
41. Leigh, E. G. (1977) How does selection reconcile individual advantage with the good of the group?, *PNAS* **74**, 4524-46.
42. Leigh E. G. (1991) Genes, bees and ecosystems: the evolution of the common interest among individuals, *TREE* **6**, 257-262.
43. Lewontin, R. C. (1970) The units of selection, *Ann Rev. Ecol. Syst.* **1**: 1-18.
44. Lloyd, E. A. (1988) *The structure and confirmation of evolutionary theory*, Princeton Univ. Press, Princeton, N. J.
45. Lyttle, T. W. (1979) Experimental population genetics of meiotic drive systems. II. Accumulation of genetic modifiers of segregation distorter in laboratory populations. *Genetics* **91**, 339-357.
46. Lyttle, T. W. (1991) Segregation disorders. *Ann Rev. Gen.* **25**, 511-57.
47. Margulis, L. (1981) *Symbiosis in cell evolution*, W. H. Freeman, San Francisco.
48. May, R. M. and Anderson, R. M. (1983) Epidemiology and genetics in the coevolutionof parasites and hosts, *Proc. R. Soc. Lond. B* **219**, 281-313.
49. Maynard Smith, J. (1978) *The evolution of sex*, Cambridge Univ. Press, London.
50. Maynard Smith, J. and and Szathmáry, E. (1993) The origin of chromosomes I. Selection for linkage, *J Theo. Biol.* **164**, 437-46.
51. Maynard Smith , J. and Szathmáry, E.(1995) *The major transitions in evolution*, W. H. Freeman, San Francisco.
52. Michod, R. E. (1997) Evolution of the individual, *Am. Nat.* **150**, S5-S21.
53. Michod, R. E. (1999) *Darwinian dynamics. Evolutionary transitions in fitness and individualitty*, Princeton Univ. Press, Princeton, N. J.
54. Müller, H. J. (1964) The relation of recombination to mutational advance, *Mutat. Res.* **1**, 2-9.
55. Murray, J. D. (1993) *Mathematical Biology*, 2nd ed, Springer, Berlin.

56. Packer C., Gilbert D. A., Pusey A. E. and O'Brian S. J. (1991) A molecular genetic analysis of kinship and cooperation in African lions, *Nature* **351**, 562-5.
57. Pinker, S. (1994) *The language instinc*, Morrow, N. Y.
58. Price, G. R. (1970) Selection and covariance, *Nature* **227**, 529-31.
59. Price, G. R. (1972) Extension of selection covariance mathematics, *Ann. Hum. Genet.* **35**, 485-90.
60. Queller, D. C. (1997) Cooperation since life began, *Q. Rev. Biol.* **72**, 184-8.
61. Rasa, O. A. (1983) Dwarf mongoose and hornbill mutualism in the Taru Desert, Kenya. *Behav. Ecol. Sociob.* **12**, 181-190.
62. Ratnieks, F. L. and Visscher, P. K. (1989) Worker policing in the honeybee, *Nature* **342**, 796-7.
63. Reeve, H. K. (1992). Queen activation of lazy workers in colonies of the eusocial naked mole-rat, *Nature* **358**: 147-149.
64. Roughgarden, J. (1979). *Theory of population genetics and evolutionary ecology: an introduction* Macmillan, NewYork.
65. Schuster, P. (1996). How does complexity arise in evolution. *Complexity*. Pp. 22-30.
66. Seeley, T. D. (1997) Homey bee colonies are group level adaptive units, *Am. Nat.* **150**, S22-S41.
67. Sober, E. and Wilson, D. S. (1998) *Unto others. The evolution and psychology of unselfish behavior*, Harvard Univ. Press, Cambridge.
68. Szathmáry, E. (1993) Coding coenzyme handles: a hypothesis for the origin of the genetic code, *PNAS* **90**, 9916-20
69. Szathmáry, E. (1999) The first replicators, in L. Keller (ed.), *Levels of selection in evolution*, Princeton Univ. Press, Princeton, N. J., pp. 31-52.
70. Szathmáry E. and Demeter, L. (1987) Group selection of early replicators and the origin of life, *J. Theor. Biol.* **128**, 463-86.
71. Trivers, R. L. (1971) The evolution of reciprocal altruism, *Q. Rev. Biol.* **46**, 35-57.
72. Trivers, R. L. and Hare, H. (1976) Haplodiploidy and the evolution of the social insects, *Science* **191**, 249-63.
73. Wade, M. J. (1976) Group selection among laboratory populations of *Tribolium*, *PNAS* **73**, 4604-7.
74. Williams, G. C. (1966) *Adaptation and natural selection*, Princeton Univ. Press, Princeton, N. J
75. Williams, G. C. (1992) *Natural selection: domains, levels and challenges*, Oxford Univ. Press, Oxford.
76. Wilson, D.S. (1975) A theory of group selection, *PNAS* **72**, 143-6.
77. Wilson, E. O. (1975) *Sociobiology: the new synthesis*, Belknap Press of Harvard Univ. Press, Cambridge, Mass.
78. Woolfenden, G. E. and FitzPatrick, J. W. (1984) *The Florida scrub jay: demography of a cooperative-breeding bird*, Princeton Univ. Press, Princeton, N. J.

CONTEMPORARY CONTROVERSIES WITHIN THE FRAMEWORK OF THE EVOLUTIONARY THEORY

ALICIA MASSARINI
Consejo Nacional de Investigaciones Científicas y Técnicas (CONICET)
Grupo de Investigaciones en Biología Evolutiva (GIBE)
Facultad de Ciencias Exactas y Naturales. Depto. de Cs. Biológicas, Universidad. de Buenos Aires. Ciudad Universitaria, Pab. II, 4to. Piso.
1428 Nuñez, Buenos Aires, Argentina.
alicia@bg.fcen.uba.ar

1. Introduction

The Evolutionary Theory is, without a doubt, the paradigm that structures and makes up modern biology. Ever since Darwin proposed in 1859 the process of natural selection as the causal mechanism of the biological evolution, the theory has been enriched by several contributions of different branches of biology. Nevertheless, the development of this theory has not been unidirectional, because in the last 150 years the original Darwinism has gone through moments of intense crisis. The objective of this work is to present a review of the main discussions raised in the last three decades between different trends of evolutionary biology.

One of the most controversial aspects can be synthesised in the following question: can the process of natural selection explain of all the evolutionary change? On the matter, Darwin (1859) himself showed a cautious attitude, when considering the scope of the selective process, in the prologue of "The origin of species" he affirmed: *"...I am convinced that the natural selection has been the most important process of modification, although not the only one".*

Nevertheless, the population geneticists that in the decade of 1940 contributed to consolidate the Synthetic Theory, toughened their position around this argument. Within the framework of the evolutionary synthesis, every characteristic of an organism was comprehended like an adaptation and therefore, like the result of the process of natural selection. According to J. Huxley (1953): *"For what we know up until now, the natural selection is not only inevitable and an effective evolution factor, but the only effective evolution factor."* This was the dominant conception from 1940 until the end of the 1950-decade; but in the decade of 1960, this position, that soon was named panselectionism, began to face different objections.

2. The neutralism and the "ticking" of the evolution

The geneticists of the beginning of the XX century proposed that in each environment exist a "normal or wild type" allele for each gene, represented in most of the individuals of the population. The mutations were considered an "oddity of nature", so that the populations would constitute genetically homogeneous groups. Then, the evolution

J. Chela-Flores et al. (eds.), Astrobiology, 121–133.

would consist in the replacement of the wild type by a successful mutant that, due to the action of natural selection, would constitute itself in the new normal type.

Later, the investigations of the population geneticists that worked with larger samples from natural populations, showed the existence of a great amount of variability hidden until then. These evidences make a rupture with the idea that the populations are genetically uniform. In accordance to these geneticists; the variability was a consequence of the adaptation to different sub-environments, in which different selective pressures operate.

During the 60's new questions arose, since the use of biochemical and molecular techniques revealed that variability at the molecular level was even much more extensive than was previously known. These new evidences made some evolutionary biologists to ask themselves: Which is the real meaning of all this variability? Will the different alleles have differential adaptive values? Is all the genetic variability explainable in terms of natural selection?

In answer to this challenge, some Japanese and North American geneticists proposed that a good part of the variation at molecular level could not be affecting the fitness of the organisms, and represents only a neutral "noise" of the system. For these geneticists, this did not mean to deny neither the process of natural selection nor the existence of the adaptations as a result of this process. The proposal only underlies that the power of natural selection may be much weaker than it was believed up to then.

According to Motoo Kimura, Japanese biologist creator of the school known as "neutralism", most of the genetic variants at the molecular level do not confer advantage nor disadvantage to the carrier. For that reason, they are able "to drift" in the populations without being detected by natural selection, so that they are lost or they are randomly fixed by genetic drift (Kimura, 1965).

One of the evidences that sustain the proposal of the neutralism is the hypothesis of the "molecular clock" (Zuckerland and Pauling, 1965). If the molecular structure of some proteins is analysed, it can be observed that this molecule shows changes between different groups of organisms. These are attributed to the genic mutations that were accumulated since the lineages separated. If large groups of organisms are analysed, for example different groups of vertebrates, it can be seen that a correlation exists between the time passed from the divergence and the amount of accumulated changes. This correlation suggest that exists a relatively constant rate of change. The mutations appear and they are regularly incorporated to the populations, while natural selection does not eliminate nor favours them, because most of them are neutral. (Kimura, 1968).

Nevertheless, since all the proteins do not seem to evolve at the same rate, some objections to this method were raised. The different mating systems, life cycles, population structures and other biological characteristics, make the rates of change not constant when different groups of organisms are compared (Britten, 1986; Wu and Li, 1985). Although it has been shown that the molecular clock does not have a universal ticking, these critics do not invalidate its utility, but allow to restrict the criteria with which it must be used (Avise, 1994; Hillis *et al.*, 1996).

In this way, it is considered that the rate of change must be calculated as an average of a great number of protein systems, so that the clock must be "calibrated" for each lineage on the basis of "accurate" data coming from the fossil record. Whith these considerations, the molecular clock is an important tool to the phylogenetic studies.

The discussion between neutralists and the biologists who explain all the variability as a result of the action of the natural selection, so called panselectionists, remains open (Kreitman, 1996; Otha, 1996). According to the neutralists, genes are drived in the populations by random processes, and they are generally of little utility for the carrier. According to the unconditional neodarwinists this idea is untenable, because as Mayr (1963) says: "*It is very improbable that a gene remains selectively neutral for indefinite time.*"

Numerous population studies take as a frame of reference either of these interpretations. Which of them is more representative? Which one allows to give account more suitably of the processes that explain the patterns of the variability in natural populations? The last word is not said. The investigations continue, although a conclusive answer seems to be still distant, because the problem presents numerous difficulties. One of the most important ones is that, in general, it is very difficult to test the adaptive value of the different forms of a molecule. In many cases its functionality is not known and it is ignored what consequences could have these variants for the organism as a whole.With regard to this last problem, the fragmented or integrated study of the characteristics of the organisms is another topic to be discussed. In relation to this problem, we will consider another important source of criticism to the neodarwinism.

3. The critic to the "adaptationist program"

Towards the end of the 1970 decade, the palaeontologist Stephen J. Gould and the population geneticist Richard Lewontin presented what it is known as the critic to the "adaptationist program" (Gould and Lewontin, 1979). These two important evolutionary biologists proposed a new way to analyse the adaptations and to evaluate the role of the natural selection.

In order to present their position, they propose an architectonic analogy: if the vault of the ceiling of a gothic building is observed, structures in form of triangles that are located in the highest part can be appreciated, generally beautifully ornamented. If one thinks about the structure and the function of those triangular structures, is easy to conclude that they were designed specially, with the intention of being carriers of the beautiful paintings that adorn them. Actually, those structures are the inevitable result of the juxtaposition of two gothic arcs: whereas the arc has a function, the triangular structures are only a collateral result. Only when the constructive process and the limitations of the architectonic design are known, it is possible to recognise the triangular structures as what they are: inevitable consequences.

Supported in this image, Gould and Lewontin ask themselves: Why did we insist on watching every characteristic of an organism as if it has a function, a purpose? Between the characteristics of the living organisms, some or many of them could be emergent or consequences of the restrictions that the development or the organisation of the organism as a whole imposes in itself, so that these ones should not be interpreted like adaptations.

Other problem that appears is how to limit the characters to be analysed. For example, the leg of the animals: is an evolutionary unit, so that its adaptive function can be discussed. If this is true, what can we say about a part of the leg, for example the foot, or only a toe or the bone of the toe?

The evolution of the human chin is a good example to illustrate this problem. This structure is highly developed in the humans, whereas neither the young or adult anthropoid primates have chin. All the attempts to explain the chin of the man as a specific adaptation failed. Finally, analysing the developmental patterns, it was stated that there are two growth zones in the inferior jaw: the basal zone of the jaw and the alveolar zone, located in the superior part. In the human evolutionary lineage, both zones became smaller in respect to their ancestors, but it is in the alveolar zone where the growth rate has been reduced the most (Stark and Kummer, 1962). In this way, the chin appears as a result of the different relative growth rates of the distinct regions of the jaw. If this process is identified, the chin appears more as a mental representation than an evolutionary unit. Together with its recognition as an emergent, the problem of its adaptive interpretation disappears.

3.1. ALTERNATIVE INTERPRETATIONS

Recognising the many factors other than adaptation that may account for an organism's features, Williams (1966) wrote that "*adaptation is often recognised in purely fortuitous effects, and natural selection is invoked to resolve problems that do not exist*". In the same way, Gould and Lewontin, in their critic to the adaptacionist program do not deny the existence of the adaptation, but they extend the interest to a wider view, that allows to consider the whole organism. This view, also proposes to incorporate alternative explanations to natural selection in order to explain the characteristics of the organisms, such as the following:

Characters fixed by genetic drift. It is probable that many changes in the evolutionary process are due to randomness. An example of this are the evidences that the neutralist school presents, that shows that a large part of the substitutions of amino acids in the evolution came from the random fixation of mutations in small populations. By effect of genetic drift, non-neutral characters can also be fixed in small populations, independently of its adaptive value.

Adjustment to the environment without genetic base. The adjustment of an organism to its environment does not always have a genetic base and therefore it is not always the result of a process of natural selection. The form of the corals and the sponges are generally, well adjusted to the directions of the currents that these organisms must support, but these forms are not hereditary, they are directly produced by the environment. These types of features are not subject to natural selection.

Differential growth. Many changes occur as a result of alometries, the differential growth of parts of an organism, which result in the modification of the proportions between them. That is the case of the human chin, previously presented.

Changes in the development times. Important evolutionary changes are consequences of heterochronic processes, this means a change in the developmental rates that can produce the acceleration, the retardation, or the interruption of that process. An example is the neoteny that involves the acquisition of the sexual maturity in an immature or young stage. Among acari, there are species that have their life cycle so accelerated that the sexual maturity settles down in the larval stage. Thus, the youthful characteristics of these organisms cannot be interpreted like adaptive, but as a consequence of the heterochronic process.

Neutral or disadvantageous characters that are carried by others with adaptive value. Some selectively neutral characters, or even disadvantageous ones, could be expressed only by being associated to others, which have a selective advantage. The changes in a gene can have multiple effects in the physiology or the development of an organism. This is known as a pleiotropic effect. Natural selection can favour the presence of a gene by some of the effects that it shows, but other effects will be only carried along.

Different solutions for the same problem. There can be different answers for the same requirement, but the different alternatives not always constitute specific adaptations. For example: the African rhino has a horn and the Asian two. In both cases the horns are protective structures against depredation, but it is not justified to affirm that one horn is especially adaptive for the African savannah and two horns for the conditions in India.

Architectonic emergents. As it was previously discussed, selective processes did not construct the structures that are architectonic emergents. These characteristics can be neutral but also, once established, they can be incorporated to play some functional role. For example in certain bivalve molluscs the shells displayed ornamentations that constitute an emergent of the pattern of deposition of minerals during their growth. For some species the ornamentations became adaptive since they favour mobility, while in other sessile species these structures did not have any utility. In a third group, some of the lines of the ornamentation are transparent so that they allow the passage of light that, in this case, is used by a symbiotic alga that lives in the interior of the mollusc (Seilacher, 1972). The example shows how a basic pattern, that was fixed for being an emergent of the design, can remain like a neutral characteristic or be incorporated selectively to carry out a function.

3.2. APTATION, ADAPTATION, EXAPTATION

With regard to the example of the shell of the bivalve molluscs, a new question can be raised. If an adaptation is a characteristic that establishes itself gradually, by means of natural selection, is it valid to consider as an adaptation a structure that is an architectonic emergent and later on is incorporated to perform a function? The question is extensive to the cases in which an adaptive structure establishes through a selective process in order to play a role, but in another stage of the evolution, the same structure, is selected to carry out a new role. To understand better this problem let us analyse some examples.

It has been proposed that feathers in the ancestor of birds, was selectively favoured because they provided a greater efficiency for the thermoregulatory function. Later, when the feathers already existed, they began to play its present role, as well as flight efficiency. Another similar case is the swimming bladder, an organ that have all the finishes, whose function is to regulate the depth of flotation. Many anatomists think that the swimming bladder was constituted on the base of the previously existing structure of the lung of the primitive lunged fishes. Both examples show a structure that was modelled by natural selection to play a rol that in another stage can be selectively incorporated to perform a completely different function.

Let us go back now to the question raised previously: is it valid to affirm that these second or third functions are adaptations? The question aims to distinguish between the result and the process. Although this discussion can seem irrelevant, it is meaningful if we understand that we are trying to identify the mechanisms that explain the genesis and the maintenance of new characteristics, and to define the raw material on which the evolutionary process works.

The biologists who pointed out this problem were Stephen Gould and Elizabeth Vrba (1982), that proposed a series of new concepts that allow to distinguish the different cases. For the secondary adaptive functions, they introduce the exaptation concept. Thus, an exaptation is a selectively incorporated characteristic from a previously existing one, either neutral, or modelled by natural selection for a different function. In the nomenclature introduced by these authors, the adaptations as well as the exaptations conform aptations, that is, characteristics that increase the aptitude of the organisms. In the same way, the neutral characters or those detrimental ones that they had fixed by chance, are called non-aptations.

Thus, natural selection has a great amount of raw material to incorporate exaptations, since the sources of the exaptations are all the characteristics fixed by chance, the architectonic emergents and all the previous adaptations. When a characteristic is incorporated as exaptation, is said to be coopted. The exaptations are at random respect to their cooptation, as well as the mutations are at random respect to their selective value. The relevance of this approach is that it allows to distinguish the features of the organisms from the processes by means they were established.

This way to conceive the aptations also implies to leave aside the preadaptation concept. This term, frequently used in evolution texts, dwells on the existence of incipient stages of complex structures that, although they do not have any function, might be the base of future adaptations. For example: a proto-wing would not be functional to fly, so it cannot be considered an adaptation. However, as the proto-wing would constitute a necessary step for the gradual construction of a wing, it would represent a preadaptation. This type of explanation entails a clearly finalist conception. How can a structure be pre-adapted for a function for which it was still not selected? Only if one thinks that a preestablished plan exists, an *a priori* purpose, is possible to justify such concept.

The concepts of aptation, exaptation and non-aptation, in addition to adaptation, lets us analyze the way in which the evolutionary process operates in a broader sense, and makes unnecessary the finalist concept of preadaptation. On the other hand it is shown that the variability on which the natural selection operates, is not only the mutational change. It involves all the adaptive characteristics and the non-aptations, that have the potentiality of being coopted for different functions in relation to new requirements of the environment.

[T]he constraints of the evolutionary change

[An]other objection that receive the classic position is expressed in this question: can [nat]ural selection, acting in a constant way, perfecting increasingly the characteristics of [the] species until reaching optimal adaptations? The answer of the adaptacionist [pro]gram's critics is an absolute no. Natural selection does not do "what it wants" but [wh]at it can". The raw material is not an amorphous mass, but real, imperfect, and [stru]ctured organisms. Then, the limiting factors or constraints appear. On one hand, [the]re are intrinsic conditioners of the organism, such as the restrictions imposed by the [hist]ory, the patterns of development and the design of each group. On the other hand, [the]re are extrinsic or ecological factors. As far as this aspect, it has been claimed the [con]venience of reconsidering the definition of niche, and the idea that natural selection is [abl]e to produce an increasing adaptation to a certain niche. In order to discuss the [con]straints of the selective process, the analyses of the intrinsic and extrinsic factors are [com]plementary.

THE INTRINSIC FACTORS

[Th]e population geneticists present the evolutionary process like the change in the genic [fre]quencies of a population through the time. However, it was shown that not all the [al]lic combinations are possible.

With regard to this problem, Gould and Lewontin (1979) propose that the [org]anisms do not behave like billiard balls, that roll freely in a plane, impelled from [her]e to there by different selective pressures. Since the organisms are integrated systems, [and] only few combinations are apt, the evolutionary change would rather look like the [mo]vement of polyhedrons that can only fall on some of their faces. Once a group [bec]omes stabilised on a face, inertia would establish, so that it would not be as easy to [cha]nge to a new position. The consequence of this idea is that the most outstanding [evo]lutionary change, represented by the change of face of the polyhedron, would not be [gra]dual and permanent, but rather discontinuous.

If natural selection, operating through the change in the genic frequencies, was [the] only process able to explain the diversity of life, the possibilities would be [lim]itless. We know, nevertheless that this is not true. The possible variants are [rig]orously limited by different factors. First, the history of the organism that is [com]prised in the genetic program; implying that for a certain set of genes it is only [pos]sible to change within certain limits. Secondly, the pattern of embryonic [dev]elopment, which comprises the complex sequence of changes, that takes place by [inte]grated "blocks", cannot be disassembled piece by piece. Finally, the possibilities of [cha]nge are restricted by the structure, the architecture of the organism, and by the [cha]racteristics of the material that the organism is constructed.

A mistaken idea about evolution presents the natural selection like a perfecting [prin]ciple, so free of ties in its work that the organisms end up playing an engineering [gam]e that necessarily leads to an optimal design. But the proof that an evolutionary [pro]cess has occurred and not a rational supernatural agent that has constructed the [org]anisms, is the limitation that show a history of descendants and refute the creation [fro]m nothing. The animals cannot develop a series of advantageous forms because the [inh]erited structural schemes prevent it.

On the matter, it does turn out interesting to emphasise a reference, in which Darwin raises this aspect, in the "*Origin of species*": "*... We must neither be marvelled because all the dispositions of the nature are not – up to where we can judge - absolutely perfect, as in the case of the same human eye, nor that some of them are far away from our idea of the suitable thing. We do not have to be astonished of the sting of the bee, that when is used against an enemy, causes the death of the own bee. Not because such a great number of drones are produced for a single act, and afterwards they are killed by his sterile sisters. Neither because of the amazing waste of pollen of our firs, nor of the instinctive hatred of the queen of the bees towards its own fecund daughters; or of the ichneumonids that fed in the interior of the body of live caterpillars. The rare thing, within the theory of the natural selection, is that more cases of absolute lack of perfection have not been discovered.*"

4.2. TO FOLLOW THE ENVIRONMENT

The present point of view on the adaptation is that the environment creates certain "problems" that the organisms need "to solve" and that the evolution by natural selection, constitutes the mechanism to reach these solutions. Natural selection is the process by means of which the organism tries a better solution to the problem that the environment proposes, being the adaptation the final result. However, if the evolution is the process of increasing adaptation of the organisms to its niche, the niches must pre-exist to the species that will adapt to them. Thus, empty niches awaiting to be occupied by new species would exist. Here a true problem appears, related with the definition of the niche concept.

The ecological niche could be defined as the result of the pluridimensional interaction between the environment as a whole, and the mode of life of an organism. These complex interactions involve many factors: physical, biological extrinsic (such as resources and predators), and biological intrinsic to the organism (among others: ethological, activity cycles, mobility rules, etc.).

In fact, in absence of organisms, the world could be divided in an infinity of arbitrary niches. However, like an *a priori* entity, the niche idea loses all explanatory and predictive power. For example: we can define that there exists a terrestrial niche, for a herbivorous animal, of nocturnal habits, that moves slithering. Therefore we anticipate the existence of snakes that feed on vegetables. Nevertheless, there are no grazing snakes, although many of them move on the grass.

The idea of a niche that pre-exists to the real species neglects the role of the own organisms in the creation of the niche. The species do not undergo the environment in a passive way but create and define the place they inhabit. The trees reconstruct the ground in which they grow, fertilising it with their leaves and airing it when sinking their roots. The herbivorous animals change the composition of species of the grass on which they feed, harvesting them in a differential form, fertilising the ground with its manure and altering the physical structure of the land. There is a constant interaction between the organism and the environment. Thus, although the organism can be in a process of adaptation to the surroundings by means of natural selection, the evolution of the own organism is changing these circumstances.

On the other hand, if the niche is conceived in a static form and all the characteristics of the organisms are interpreted like adaptations to those niches, we get a paradox: "if all the characteristics of the organisms are adaptations, the evolution no ıger has no effect, because the organisms are already adapted."

An alternative model that offers an answer to the outlined problems is the Red ıeen hypothesis proposed by Leigh Van Valen (1973). A scene of the book of Lewis ıroll "*Through the Looking-Glass*" inspired the name of this hypothesis. Alice, allenged by the Queen of hearts, tries to advance in a landscape that is reflected in a irror to her side, but as the landscape moves very fast as if it was a film, she must run thout stopping to be always in the same place. Van Valen proposes that something ;e this happens in nature. Small changes in the life conditions, which are given not ly by abiotic changes but also by changes in other related species, cause the niche to modified. The niche is continuously changing, and the species are following the anges by means of the selective process. Thus, natural selection allows the organisms maintain their level of adaptation, but does not improve it.

The proposal of following the environment provides a satisfactory answer to the oblem of the niche definition. According to this conception, in a world without life, it impossible to define the different niches because the organisms themselves are eating them. From this approach, the adaptation concept is modified remarkably. It is longer a process of increasing improvement, that leads to an ideal design for optimal age of a certain niche. Now it changed to be a slow movement of the niche through ne and space, followed by a change in the species that are always one step back, ghtly not so well adapted.

The alterations of the niche will be generally small and gradual so that the new ches to which the organisms must adapt are probably closely together from the old es, in the multidimensional space of the niches. Yet, if the deterioration of the niche abrupt, the species has a great probability of being extinguished, for not being able to ıy coupled to the requirements imposed by the change.

The most outstanding evidences for the Red Queen hypothesis come from the alysis of the rates of extinction of numerous species in the fossil record. These data ow that the probability of a species of being extinguished seems to be constant within group of organisms, independently of its time of existence. In other words: species at have existed during long time have the same probability of being extinguished than hers that are more recent (Flessa, 1986, Hoffman, 1991). If natural selection was creasingly improving the adjustment of populations to the environments, then one ould expect that the probability that a species is extinguished in the next period to be ıaller for those that have existed from long time ago. The data of the fossil record monstrate that the probability of extinction of a species is characteristic of the group which it belongs, but is independent of the time of persistence, so that all are equally ılnerable to the extinction and have an equivalent adaptive condition. These evidences rrespond with a model of multidimensional and dynamic niche, followed by the ecies.

On the other hand, it is necessary not to lose sight, that adaptation is a relative ndition and that the relation between natural selection and adaptation is not ciprocal. Whereas a greater relative adaptation will always be selected, the natural lection does not always lead to a greater adaptation. Even though a species survives r a long time, always exists the possibility that a new form would be originated that ıs a greater reproductive rate, or a greater capacity in the employment of the same sources, so that it is able to cause the extinction of the ancestral form.

5. More critics: patterns and processes in macroevolution

In addition to the discussion about the power of natural selection to explain the evolutionary change, other two basic postulates of the modern synthesis have been objected: the gradualist argument and the reductionist argument. The first argument, was already present in the view of Darwin, proposes that all the evolutionary processes respond to the slow and gradual accumulation of genetic changes throughout the generations. The second argument is that the macroevolutionary changes, such as the appearance of the great groups of organisms or the mass extinctions, can be understood in terms of the same processes that explain the adaptation of the populations in the microevolutionary level.

5.1. THE PATTERNS OF MACROEVOLUTIVE CHANGE

Darwin saw the evolutionary process like a series of gradual and continuous changes. Nevertheless, the fossil record showed serious discontinuities and, in most of the cases, the intermediate forms were absent. This was one of Darwin's greatest preoccupations, since this problem could make stagger his theory. Although this difficulty was not solved while he lived, Darwin trusted that, with time, new fossils would be filling the incomplete sequences. However, this did not happen.

The main objection to the gradualist vision comes from the palaeontologists Gould and Eldredge (1977) who observed, from the data analysis of the fossil record, that new species generally appear by abrupt processes, in very brief periods in the geologic time scale. These authors also remark that, once established, species stay without greater changes during long periods, that is, they remain in "evolutionary stasis", until a new species is eventually originated.

From this interpretation, the fossil record turns out to be a faithful reflection of the history of life, since the absence of transitional forms is due to that, in many cases, these did not exist. In this approach, known as "punctuated equilibria model", the evolution is an essentially discontinuous and non-gradual process (Gould and Eldredge, 1977, 1993; Stanley, 1979). The models of quantum speciation, that explain the origin of new species in short times from the fixation at random of drastic genetic changes (Mayr, 1954; Templeton, 1980; Carson, 1982), contributed to lay the foundations of the discontinuist proposal.

Later paleontological investigations showed the analysis of fossil sequences that support the punctuated equilibria hypothesis (Malmgren *et al.*, 1983; Bell *et al.*, 1985). Nevertheless, records that put in evidence the gradual change within certain lineages have also been reported (Sheldon, 1987). The discussion referred to the way and the rate in which the evolution operates remains still open (Gould and Eldredge, 1993; Levinton, 1988). Nowadays it is not considered that the gradualist and discontinuist models are mutually exclusive, but it is discussed which of these two models is more representative of the preponderant patterns in the evolution of life.

2. THE MACROVOLUTIONARY PROCESSES

nother polemic topic of the discussion is if macroevolution can be understood in terms ꞉ the microevolutionary processes. According to evolutionary biologists in favour of e synthesis, the diversity of genera, families, orders and even of taxonomic groups of gher rank, would be explained as a result of the adaptive process. The action of natural ·lection would extend at the macroevolutionary level, extrapolating the processes that ɪppen within populations to explain the patterns of large-scale evolution.

In this point, critics of synthesis propose that the macroevolutionary ɪenomena respond to patterns and processes that belong to them, which cannot be ·duced to the microevolutionary processes. They claim that gradual evolutionary ɪange by natural selection, operates so slowly within established species that it cannot ꞉count for the major features of evolution, because evolutionary change tends to be ɔncentrated within speciation events. In this sense, the patterns of transpecific ʼolution would be determined by the process of species selection, which is analogous ı natural selection, but acts upon species within higher taxa rather than upon dividuals within populations (Stanley, 1975).

At the genetic level, this view corresponds with the knowledge that in the ɜnetic program exists as well a hierarchical structuring. This means that all genes do ɔt have the same evolutionary importance. Most of them are structural genes, that ɔdify information for a structural protein, but some genes control the expression of ꞉her genes. Generally, a change in a regulator gene would have undesirable ɔnsequences, but occasionally a successful mutation could produce surprises that ɪange the course of evolution.

The "evolutionary novelties", such as the appearance of new organs, new ·ganization plans or new groups of organisms, would not arise by the slow ꞉cumulation of changes in the structural genes. Genetic changes related with acroevolutive processes are probably at the level of master genes, able to modify the ꞉pression of other genes, the development times, the proportions between different ructures, and other changes with remarkable consequences in the structure of the ·ganism. Thus, more than the selective processes, they would be the species selection, ɪe alometric changes, the heterochronies, and other alternative processes, the echanisms able to explain the macro-evolutionary change.

The evolution of evolution

ne of the most outstanding aspects of Darwin's Evolutionary Theory is that it was not ·ystallised in its original postulates, but constituted the starting point of a complex ɪeoretical scheme that was enriched by numerous later contributions. The present ɔntroversies about the protagonism of the natural selection, the mode, the rates, and ɪe hierarchical structuring of the evolutionary change, are nourished by nobel views. eeping the interpretation frame that still provides the classic Darwinism, the ʼolutionary theory recreates and grows.

As Osvaldo A. Reig (1984), outstanding Argentine evolutionary biologist, ɪid:

"We possibly are, in the dawn of a new synthesis or an expansive development of the modern synthesis, that is outlined as a hierarchic theory of the evolution able to surpass the reductionists limitations of the original Darwinism and its version in the modern synthesis, admitting the existence of different ambits and levels of manifestation of the evolutionary processes".

References

Avise, J.C. (1994) *molecular markers, natural history and evolution*, Chapman & Hall, New York.

Bell, M.A., Baumgartner, J.V., and Olson, E.C. (1985) Patterns of temporal change in single morphological characters of a Miocene stickleback fish, *Paleobiology*, **11**, 258-271.

Britten, R.J. (1986) Rates of DNA sequence evolution differ between taxonomic groups, *Science* **231**, 1393-1398.

Carson, H.L. (1982) Speciation as a major reorganization of polygenic balances, in C. Barigozzi (ed.), *Mechanisms of speciation*, Alan R. Liss, New York, pp. 411-433.

Darwin, C. (1859) *On the Origin of Species*, John Murray, London.

Flessa, K.W. et al. (1986), Causes and consequences of extiction, in D.M. Raup and D. Jablonsky (eds.), *Patterns and Processes in the History of Life*, Springer-Verlag, Berlin, pp. 235-257.

Gould, S.J. and Eldredge, N. (1977) Punctuated equilibria: The tempo and mode of evolution reconsidered, *Paleobiology*, **3**, 115-151.

Gould, S.J. and R.C. Lewontin (1979), The spandrels of San Marco and the Panglossian paradigm, *Proc. R. Soc. Lond. B.*, **205**, 581-598.

Gould, S.J. and Vbra, E.S. (1982), Exaptation –a missing term in the science of form-, *Paleobiology*, **8**, 4-15.

Gould, S.J. and Eldredge, N. (1993) Punctuated equilibrium comes of age, *Nature*, **366**, 223-227.

Hillis, D.M., Mable, B.K. and Moritz C. (1996) Applications of molecular systematics: the state of the field and a look to the future, in D.M. Hillis, C.Moritz, and B.K. Mable (eds.), *Molecular systematics*, second edition, Sinauer Associates, Sunderland MA., pp. 515-543.

Hoffman, A. (1991) Testing the Red Queen Hypothesis, *J. evol. Biol.*, **4**: 1-7.

Huxley, J. (1953) *Evolution in action*, Harper & Row, New York.

Kimura, M. (1965) A stochastic model concerning the maintenance of genetic variability in quantitative characters, *Proc. Natl. Acad. Scie. USA* **54**, 731-736.

Kimura, M. (1968) Evolutionary rate at the molecular level, *Nature* **217**, 624-626.

Kreitman, M. (1996) The neutral theory is dead. Long live the neutral the neutral theory *BioEssays*, **18**(8), 678-683.

Levinton, J.S. (1988) *Genetics, Paleontology, and Macroevolution*, Cambridge University Press, Cambridge.

Malmgren, B.A., Berggren, W.A, and Lohmann, G.P. (1983) Evidence for punctuated gradualism in the late Neocene *Globorotalia tumida* lineage of planctonik Foraminifera, *Paleobiology*, **9**, 377-389.

Mayr, E. (1954) Change of genetic environment and evolution, in J. Huxley, A.C. Hardy, and E.B. Ford (eds.), *Evolution as a Process*, Allen and Unwin, London, pp. 157-180.

Mayr, E. (1963) *Animal species and evolution*, Harvard University Press, Cambridge, MA.

Ohta, T. (1996) The neutralist–selectionist debate, *BioEssays*, **18**(8), 673-677.

Reig, O.A. (1984) La Teoría de la Evolución a los 125 años de la aparición de "El origen de las especies", *Academia Nacional de Medicina de Buenos Aires*, Separata Vol. 62.

Seilacher, A. (1972) Divaricate patterns in pelecypod shells, *Letaia*, **5**: 325-343.

Sheldon, P.R. (1987) Parallel gradaulistic evolution of Ordovician trilobites, *Nature*, **330**, 561-563.

Stanley, S.M. (1975) A Theory of Evolution Above the Species Level, *Proc. Nat. Acad. USA.*, **72**(2), 646-650.

Stanley, S.M. (1979) *Macroevolution: Pattern and Process*, W.H. Freeman, San Francisco, CA.

Starck, D. and Kummer, B. (1962), Zur Ontogeneses des Schimpancenschaedels, *Anthrop. Anz.*, **25**, 204.

Templeton, A.R. (1980) The theory of speciation via the founder principle, *Genetics*, **94**, 1011-1038.

Van Valen, L. (1973) A new evolutionary law, *Evol. Theory*, **1**: 1-30.

Williams, G.C. (1966) *Adaptation and Natural Selection*, Princeton University Press, Princeton, NJ.

Wu, C.-I. and Li, W.H. (1985) Evidence for higher rates of nucleotide substitution in rodents than in man, *Proc. Natl. Acad. Sci. USA*, **82**, 1741-1745.

Zuckerland, E. and Pauling, L. (1965) Evolutionary divergence and convergence of proteins, in V. Bryson and H.J. Vogel (eds.), *Evolving Genes and Proteins*, Academic Press, New York, pp. 97-166.

MOLECULAR BIOLOGY AND THE RECONSTRUCTION OF MICROBIAL PHYLOGENIES: Des Liaisons Dangereuses?

A. BECERRA, E. SILVA, L. LLORET, S. ISLAS,
A. M. VELASCO, and A. LAZCANO

Facultad de Ciencias, UNAM
Apdo. Postal 70-407
Cd. Universitaria, 04510 México, D.F., MEXICO

1. Introduction

Only half-a-century after the DNA double chain model was first suggested, molecular biology has become one of the most provocative, rapidly developing fields of of scientific research, that has led not only to tantalizing new findings on processes and mechanisms at the molecular level, but also to major conceptual revolutions in life sciences. Is there any hope of developing methodological approaches and theoretical frameworks not only to make sense of the overwhelming growing body of data that this relatively new field is producing, but also to use them to develop a more integrative, truly multidisciplinary understanding of biological phenomena? As Peter Bowler wrote a few years ago, Charles Darwin and his followers were accutely aware that "evolutionism's strength as a theory came fom its ability to make sense out of a vast range of otherwise meaningless facts" (Bowler, 1990). This situation has not changed. Evolutionary biology may be in a state of major turmoil, but its unifying powers have not diminished at all. In fact, they probably represent one of the most promising possibilities of overcoming the perils of reductionism that have plagued molecular biology since its inception.

Molecular approaches to evolutionary issues are a century old. The possibility of developing a sucessful blending between them may have been first suggested by the American-born British biologist and physician George H. F. Nuttall, who in 1904 published a book summarizing the results of the detailed comparison of blood proteins that he had used to reconstruct the evolutionary relationships of animals. "In the absence of palaentological evidence", wrote Nuttall (1904), "the question of of the interrelation-ship amongst animals is based upon similarities of structure in existing forms. In judging of these similarities, the subjective element may largely enter, in evidence of which we need but look at the history of the classification of the Primates" Such subjective element, Nuttall believed, could be succesfully overcomed by

J. Chela-Flores et al. (eds.), Astrobiology, 135–150.

constructing a phylogeny based not on form but on the inmunological reactions of blood-related proteins.

Although the comparative analysis of biochemical properties, metabolic pathways and, in few cases, morphological characteristics, had provided some useful insights on the evolutionary relationships among certain microorganisms, until a few years ago the reconstruction of bacterial phylogenies and the understanding of microbial taxonomy were both viewed with considerable skepticism. This situation has undergone dramatic changes with the recognition that proteins and nucleic acid sequences are historical documents of unsurpassed evolutionary significance (Zuckerkandl and Pauling, 1965), and has led to a radical renovation of the phylogeny, classification, and systematics of prokaryotic and eukaryotic microbes (Woese, 1987).

But these changes have also sparked new debates, and have led to an increased appreciation that the scope and limits of molecular cladistic methodologies require clarification. As shown by the current controversies on the characteristics of the first organisms, the origin of the different components of the eukaryotic cell, and the soundness of traditional taxonomic systems, the development of the full potential of molecular cladistics will depend not only on methodological refinements to improve the algorithms used for reconstructing evolutionary history from molecular data, but also on the critical reexamination of its theoretical framework, which includes a number of central concepts, most of which were grafted from classical evolutionary theory into molecular biology. Here we discuss some of these issues, and review briefly some of the major contributions that they have promoted in our understanding of previously uncharacterized early periods of biological evolution.

2. On the nature of eukaryotic cells

The awareness that genomes are extraordinarily rich historical documents from which a wealth of evolutionary information can be retrieved has widened the range of phylogenetic studies to previously unsuspected heights. The development of efficient nucleic acid sequencing techniques, which now allows the rapid sequencing of complete cellular genomes, combined with the simultaneus and independent blossoming of computer science, has led not only to an explosive growth of databases and new sophisticated tools for their exploitation, but also to the recognition that different macromolecules may be uniquely suited as molecular chronometers in the construction of nearly universal phylogenies.

A major achievement of this approach has been the evolutionary comparison of small subunit ribosomal RNA (rRNA) sequences, which has allowed the construction of a trifurcated, unrooted tree in which all known organisms can be grouped in one of three major (apparently) monophyletic cell lineages: the eubacteria, the archaebacteria, and the eukaryotic nucleocytoplasm, now referred to as new taxonomic categories, i.e., the domains *Bacteria*, *Archaea*, and *Eucarya*, respectively (Woese et al., 1990). There is strong evidence that the identification of these lineages is not an artifact based solely upon the reductionist extrapolation of information derived from one single molecule. While trees based on whole genome information have confirmed at a broad level rRNA-based phylogenies (Snel et al., 1999; Tekaia et al., 1999), it is also true that the congruence between rRNA genes and other molecules is not always ideal, and anomalous phylogenies have been reported (Rivera and Lake, 1992; Gupta and Golding, 1993). At the time being there is no general explanation to account for these peculiar topologies, and the possibility that we may have to restrict ourselves to empirical characterizations of such cases should be kept in mind. However, a large variety of phylogenetic trees constructed from DNA and RNA polymerases, elongation factors, F-type ATPase subunits, heat-shock and ribosomal proteins, and an increasingly large set of genes encoding enzymes involved in biosynthetic pathways, have confirmed the existence of the three primary cellular lines of evolutionary descent (Doolittle and Brown, 1994), between which extensive horizontal transfer events have taken place (Doolittle, 1999).

The ensuing tripartite taxonomic description of the living world fostered by Woese and his followers has been disputed by a number of workers, who contend that both eubacteria and archaebacteria are *bona fide* prokaryotes, regardless of the pecularities that separate that separate them at the molecular level, both are prokaryotes (Mayr, 1990; Margulis and Guerrero, 1991; Cavalier-Smith, 1992). Furthermore, because of their very nature, molecular dichotomous phylogenetic trees cannot be drawn which include anastomozing branches corresponding to the lineages which gave rise to the different components of eukayotic cells. Accordingly, Margulis and Guerrero (1991) have argued that although molecular cladistics is now a prime force in systematics, phylogenetically accurate taxonomic classifications should be based not only on the evolutionary comparison of macromolecules, but also on metabolic pathways, chromosomal cytology, ultrastructural morphology, biochemical data, life cycles, and, when available, paleontological and geochemical evidence.

While molecular phylogenies have confirmed the endosymbiotic origin of plastids and mitochondria, a number of trees also suggest that a major portion of the eukaryotic nucleocytoplasm originated from an archaebacteria-like cell whose descendants form the monophyletic eucaryal branch (Gogarten-Boekels and Gogarten, 1994). As asserted by Woese and his collaborators, although the presence of endosymbionts is of critical importance to the eukaryotes, it is undeniable that the latter "have a unique, meaningful phylogeny" (Wheelis et al., 1992). While such view assumes an absolute continuity between the nucleocytoplasm and its direct ancestor, the holistic arguments advocated by Margulis and Guerrero (1991), Cavalier-Smith (1992), and others, emphasize the evolutionary emergence of an novel type of cell as a result of endosymbiotic events. According to the latter, the key transitional event leading to eukaryosis was the evolutionary acquisition of heritable intracellular symbionts, and the eucaryal branch does not represent eukaryotic cells as a whole, any more than fungal hyphae or phycobionts like the *Trebouxia* algal cells exhibit, by themselves, all the phenotypic and genetic characteristics of a lichen thallus.

Of course, antagonistic taxonomies have coexisted more or less peacefully along the history of biology. However, the urgent need to critically revise current classificatory systems cannot be underscored. Modern taxonomic schemes need to acknowledge not only the existence of three major cell lineages, but also the eukaryotic divergence patterns, which appear to be the result of rapid bursts of speciation (Sogin, 1994). Any such modifications in biological classification require the recognition of the functional and anatomical continuity between the eukaryotic cytoplasm and the intranuclear environment, as well as the likelihood that the evolution of membrane-bounded nuclei is indeed a byproduct of permanent intracellular associations. In fact, extant amitochondrial eukaryotes such as *Giardia* and *Trichomonas* appear to have had mitochondria in the past (Germont et al., 1997), and still harbor permanent intracellular bacterial endosymbionts (Margulis, 1993). These amitochondrial cells, which may include the microaerophilic, amitotic, multinucleated giant amoeba *Pelomyxa palustris*, are all located in the lowest branches of the eucarya, and contain several types of intracellular prokaryotes which may be the functional equivalents of mitochondria. The ubiquity of endosymbionts suggests that they may have played a critical role in the evolutionary development of nucleated cells. This hypothesis is amenable to observational and experimental designs, and may be supported by studying the possible bacterial affinities of membrane-bounded hydrogenosomes that are known to multiply by binary division in the *Trychomonas* cytoplasm (Müller, 1988), as well as by searching for prokaryotic endosymbionts in species of Parabasalia, Retortomonads, Diplomonads, Calonymphids, and other protist taxa, some of which may have evolved prior to mitochondrial acquisition.

3. The root of the tree or the tip of the trunk?

The construction of the unrooted rRNA tree showed that no single major branch predates the other two, and all three derive from a common ancestor. It was thus concluded that the latter was a progenote, which was defined as a hypothetical entity in which phenotype and genotype still had an imprecise, rudimentary linkage relationship (Woese and Fox, 1977). According to this view, the differences found among the transcriptional and translational machineries of eubacteria, archaebacteria, and eukaryotes, were the result of evolutionary refinements that took place separately in each of these primary banches of descent after they have diverged from their universal ancestor (Woese, 1987).

From an evolutionary point of view it is reasonable to assume that at some point in time the ancestors of all forms of life must have been less complex than even the simpler extant cells, but our current knowledge of the characteristics shared between the three lines has shown that the conclusion that the last common ancestor was a progenote was premature. This interpretation, based on rRNA-based trees for which no outgroups have been discovered, has been definitively superseded (Woese, 1993). A partial description of the last common ancestor of eubacteria, archaebacteria, and eukaryotes may be inferred from the distribution of homologous traits among its descendants. The set of such genes that have been sequenced and compared is still small, but the sketchy picture that has already emerged suggests that the most recent common ancestor of all extant organisms, or *cenancestor*, as defined by Fitch and Upper (1987), was a rather sophisticated cell with at least (a) DNA polymerases endowed with proof-reading activity; (b) ribosome-mediated translation apparatus with an oligomeric RNA polymerase; (c) membrane-associated ATP production; (d) signalling molecules such as cAMP and insulin-like peptides; (e) RNA processing enzymes; and (f) biosynthetic pathways leading to amino acids, purines, pyrimidines, coenzymes, and other key molecules in metabolism (cf. Lazcano, 1995).

Although the possibility of horizontal transfer should always be kept in mind, the traits listed above are far to numerous and complex to assume that they evolved independently or that they are the result of massive multidirectional horizontal transfer events which took place before the earliest speciation events recorded in each of the three lineages. Their presence suggests that the cenancestor was not a direct, immediate descendant of the RNA world, a protocell or any other pre-life progenitor system. Very likely, it was already a complex organism, much akin to extant bacteria, and must be considered the last of a long line of simpler earlier cells for which no modern equivalent is known.

Unfortunately, the characteristics of evolutionary predecessors of the cenancestor cannot be inferred from the plesiomorphic traits found in the space defined by rRNA sequences. Although trees constructed from such universally shared characters appear to be free of internal inconsistencies, the lack of outgroups leads to topologies that specify branching relationships but not the position of the ancestral phenotype. Thus, such trees cannot be rooted. This phylogenetic *cul-de-sac* may be overcomed by using paralogous genes, which are sequences that diverge not through speciation but after a duplication event. As noted over twenty years ago by Schwartz and Dayhoff (1978), rooted trees can be constructed by using one set of paralogous genes as an outgroup for the other set, a rate-independent cladistic methodology that expands the monophyletic grouping of the sequences under comparison.

This approach was used independently a few years ago by Iwabe et al (1989) and Gogarten et al (1989), who analyzed paralogous genes encoding (a) the two elongation factors (EF-G and EF-Tu) that assist in protein biosynthesis; and (b) the alpha and beta hydrophilic subunits of F-type ATP synthetases. Using different tree-constructing algorithms, both teams independently placed the root of the universal trees between the eubacteria, on the one side, and archaebacteria and eukaryotes on the other. Their results imply that eubacteria are the oldest recognizable cellular phenotype, and imply that specific phylogenetic affinities exist between the archaea and the eucarya.

This branching order, which was promptly adopted by Woese et al (1990), appears to be consistent with structural and functional similarities which are known to exist in the translation and replication machineries of both archaebacteria and eukaryotes (Ouzonis and Sander, 1992; Kaine et al., 1994). However, the issue is far from solved, and has in fact been further complicated by the availability of completely sequenced genomes. The situation is further aggravated by the fact that the phylogenetic analysis of sets of ancestral paralogous genes other than the elongation factors and the ATPase hydrophilic subunits has challenged the conclusion that universal trees are rooted in the eubacterial branch (cf. Forterre et al., 1993). While the sequences of the products of genes involved in the transcription/transcriptional molecular machinery of eukaryotes appear to be closer to those of the archaea than to the eubacteria, other sequences such as those encoding heat-shock proteins and several enzymes suggest the existence of phylogenetic affinities between archaebacteria and Gram positive bacteria. No support for a particular topology was detected when mean interdomain distance analysis was used to analize a set of approximately forty genes common to the three lineages (Doolittle and Brown, 1994).

The lack of congruency between different universal phylogenies may be the result not only of the statistical problems involved in the aligment and comparison of a large number of sequences that may have diverged more than 3.5×10^9 years ago, but also of even older additional paralogous duplications (Forterre et al., 1993), and of horizontal gene tranfer events (Doolittle, 1999), both of which may be obscuring the natural relationships between the lineages. Given the likelihood that microbial phylogenetic analysis will increase its reliance on paralogous duplicates to define outgroups and character polarities (Sidow and Bowman, 1991), detailed studies should be devoted to assess the validity and limits of this cladistic methodology.

Minor differences in the basic molecular processes of the three main cell lines can be distinguished, but all known organisms, including the oldest ones, share the same essential features of genome replication, gene expression, basic anabolic reactions, and membrane-associated ATPase mediated energy production. The molecular details of these universal processes not only provide direct evidence of the monophyletic origin of all extant forms of life, but also imply that the sets of genes encoding the components of these complex traits were frozen a long time ago, i. e., major changes in them are very strongly selected against and are lethal. Biological evolution prior to the divergence of the three domains was not a continuous, unbroken chain of progressive transformation steadily proceeding towards the cenancestor. However, no evolutionary intermediate stages or ancient simplified version of the basic biological processes have been discovered in extant organisms.

Nevertheless, clues to the genetic organization and biochemical complexity of the earlier entities from which the cenancestor evolved may be derived from the analysis of paralogous sequences. Their presence in the three cell lineages implies not only that their last common ancestor was a complex cell already endowed, among others, with pairs of homologous genes encoding two elongation factors, two ATPase hydrophilic subunits, two sets of glutamate dehydrogenases, and the A and B DNA polymerases, but also that the cenancestor itself must have been preceded by simpler cells in which only one copy of each of these genes existed. In other words, Archean paralogous genes provide evidence of the existence of ancient organisms in which ATPases lacked the regulatory properties of its alpha subunit, protein synthesis took place with only one elongation factor, and the enzymatic machinery involved in the replication and repair of DNA genomes had only one polymerase ancestral to the *E. coli* DNA polymerase I and II.

By definition, the node located at the bottom of the cladogram is the root of a phylogenetic tree, and corresponds to the common ancestor of the group under study. But names may be misleading. The recognition that basic biological processes like DNA replication, protein biosynthesis, and ATP production require today the products of pairs of genes which arose by paralogous duplications during the early Archean, implies that what we have been calling the root of universal trees is in fact the tip of a trunk of unknown length in which the history of a long (but not necessarily slow) series of archaic evolutionary events may still be recorded. The inventory of paralogous genes that duplicated during this previously unchacterized stage of biological evolution appears to include, in addition to elongation factors, ATPase subunits, and DNA polymerases, the sequences encoding heat shock proteins, ferredoxins, dehydrogenases, DNA topoisomerases, several pairs of aminoacyl-tRNA synthetases, and enzymes involved in nitrogen metabolism and amino acid biosynthesis. It is noteworthy that this list includes also aspartate transcarbamoyl transferase, an enzyme which together with carbamyl phosphate synthetase (whose large subunit is itself the product of an internal, i.e., partial, paralogous duplication) catalyzes the initial steps of pyrimidine biosynthesis (García-Meza et al, 1995).

Thus, prior to the early duplication events that led to what may be a rather large number of cenancestral paralogous sequences, simpler living systems existed which lacked the large sets of enzymes and the sophisticated regulatory abilites of contemporary cells. Although lateral transfer of coding sequences may be almost as old as life itself, gene duplication followed by divergence probably played a dominant tole in the accretion of complex genomes, and may have led to a rapid rate of microbial evolution. If its is assumed that the rate of gene duplicative expansion of ancient cells was comparable to today's present values, which are of 10^{-5} to 10^{-3} gene duplications per gene per cell generation (Stark and Wahl, 1984), the maximum time required to go from an hypothetical 100-gene organism to one endowed with a filamentous cyanobacterial-like genome of approximately 7000 genes would be less than ten million years (Lazcano and Miller, 1994).

Although there are no published data on the rate of formation of new enzymatic activities resulting from gene duplication events under either neutral or positive selection conditions, the role of duplicates in the generation of evolutionary novelties is well stablished. Once a gene duplicates, one of the copies may be free to accumulate non-lethal mutations and acquire new additional properties, which could lead into its specialization or recruitment into new role. Data summarized here supports the idea that primitive biosynthetic pathways were mediated by small, inefficient enzymes of broad substrate specificity (Jensen, 1976). Larger substrate ranges may had not been a disadvantage, since relatively unspecific enzymes may have helped ancestral cells with reduced genomes overcome their limited coding abilities (Ycas, 1974).

The discovery that homologous enzymes catalyzing similar biochemical reactions are part of different anabolic pathways supports the idea that enzyme recruitment took place during the early development of several basic anabolic pathways. Evolutionary tinkering of the products of duplication events apparently had a major role in metabolic evolution. This is supported by the analysis of complete genome sequences, that has shown the large proportion of gene content that is the outcome of duplication events (Tekaia and Dujon, 1999). Such high levels of redundancy represent an illuminating possibility and suggest that the wealth of phylogenetic information older than the cenancestor may be larger than realized, and its analysis may provide fresh insights into a crucial but largely undefined stage of early biological evolution during which major biosynthetic pathways emerged and became fixed.

There is a major exception to the above conclusion. True fungi, euglenids, and chrytridiomycetes synthesize lysine via an eight-step pathway in which α-aminoadipate (AAA) is an intermediate. This route is different from the seven-step diaminopimelate pathway used by bacteria, plants, and most protist (Bhattacharjee, 1985). The phylogenetic distribution of these two pathways suggest that the AAA route is the most recent one. Accordingly, if the patchwork assembly of metabolic pathways (Jensen, 1976) is valid, then it can be predicted that the enzymes catalizing the AAA-route should be homologous to those participating in other major biosynthetic routes.

The recognition that enzyme recruitment may have played a major role in metabolic evolution leads, however, to assume some caution in phylogenetic inferences. Although in some cases metabolic pathways may be sucessfully used to assess the phylogenetic relationship of prokaryotes (DeLey, 1968; Margulis, 1993), the possibility that some of the enzymes of archaic pathways may have survived in unusual organisms (Keefe et al., 1994), or that important portions of extant metabolic routes may have been assembled by a patchwork process (Jensen, 1976), suggest that considerable prudence should be exerted when attempting to describe the physiology of truly primordial organisms by simple direct back extrapolation of extant metabolism.

4. Molecular cladistics and the origin of life: is there any connection?

"All the organic beings which have ever lived on this Earth", wrote Charles Darwin in the *Origin of Species*, "may be descended from some primordial form". Although the placement of the root of universal trees is a matter of debate, the development of molecular cladistics has shown that despite their overwhelming diversity and tremendous differences, all organisms are ultimately related and descend from Darwin's primordial ancestor. But what was the nature of this progenitor?

The heterotrophic hypothesis suggested by Oparin (1938) not only gave birth to a whole new field devoted to the study of the origin of life, but played a central role in shaping several influential taxonomic schemes and different bacterial phylogenies (Margulis 1993). Although the central role of glycolysis and the wide phylogenetic distribution of at least some of its molecular components are strong indications of its antiquity (Fothergill-Gilmore and Michels, 1993), it is no longer possible to support the *ad hoc* identification of putative primordial traits to assume that the first living system was a *Clostridium*-like anaerobic fermenter or a *Mycoplasma* type of cell (cf. Lazcano et al., 1992). Like vegetation in a mangrove, the roots of universal phylogenetic trees are sumerged in the muddy waters of the prebiotic broth, but how the transition from the non-living to the living took place is still unknown.

Indeed, we are still very far from understanding the origin and attributes of the first living beings, which may have lacked even the most familiar features in extant cells. For instance, protein synthesis is such an essential characteristic of cells, that it is frequently argued that its origin should be considered synonymous with the emergence of life itself.

However, the discovery of the catalytic activities of RNA molecules has led considerable support to the possibility that during early stages of biological evolution living systems were endowed with a primitive replicating and catalytic apparatus devoid of both DNA and proteins The scheme may be even more complex, since RNA itself may have been preceded by simpler genetic macromolecules lacking not only the familiar 3',5' phosphodiester backbones of nucleic acids, but perhaps even today's bases (Lazcano and Miller, 1996).

Although molecular cladistics may provide clues to some late steps in the development of the genetic code, it is difficult to see how the applicability of this approach can be extended beyond a threshold that corresponds to a period of cellular evolution in which protein biosynthesis was already in operation. Older stages are not yet amenable to molecular phylogenetic analysis. Although there have been considerable advances in the understanding of chemical processes that may have taken place before the emergence of the first living systems, life's beginnings are still shrouded in mystery. A cladistic approach to this problem is not feasible, since all possible intermediates that may have once existed have long since vanished. The temptation to do otherwise is best resisted. Given the huge gap existing in current descriptions of the evolutionary transition between the prebiotic synthesis of biochemical compounds and the cenancestor (Lazcano, 1994), it is naive to attempt to describe the origin of life and the nature of the first living systems from the available rooted phylogenetic trees.

Nevertheless, there have been several recent attempts to use macromolecular data to support claims on the hyperthermophily of the first living organisms and the idea of a hot origin of life. The examination of the prokaryotic branches of unrooted rRNA trees had already suggested that the ancestors of both eubacteria and archaebacteria were extreme thermophiles, i.e., organisms that grow optimally at temperatures in the range 90° C and above (Achenbach-Richter et al., 1987). Rooted universal phylogenies appear to confirm this possibility, since heat-loving bacteria occupy short branches in the basal portion of molecular cladograms (Stetter, 1994).

Such correlation between hyperthermophily and primitiveness has led support to the idea that heat-loving lifestyles are relics from early Archean high-temperature regimes that may have resulted from a severe impact regime (Sleep et al., 1989). It has also been interpreted as evidence of a high temperature origin of life, which according to these hypotheses took place in extreme environments such as those found today in deep-sea vents (Holm, 1992) or in other sites in which mineral surfaces may have fueled the appearance of primordial chemoautolithotrophic biological systems (Wächtershäuser, 1990).

Such ideas are not totally without precedent. The possibility that the first heterotrophs may have evolved in a sizzling-hot environment is in fact an old suggestion (Harvey, 1924). Despite their long genealogy, these hypotheses have not been able to bypass the problem of the chemical decomposition faced by amino acids, RNA, and other thermolabile molecules which have very short lifetimes under such extreme conditions (Miller and Bada, 1988). Although no mesophilic organisms older than heat-loving bacteria have been discovered, it is possible that hyperthermophily is a secondary adaptation that evolved in early geological times (Sleep et al., 1989; Confalonieri et al., 1993; Lazcano, 1993). Such possibility is in fact strongly supported by the recent phylogenetic analysis of the G+C content of rRNA genes, which suggest that the last common ancestor was not a hyperthermophilic organism (Galtier et al., 1999).

In fact, hyperthermophiles not only share the same basic features of the molecular machinery of all other forms of life; they also require a number of specific biochemical adaptations. Any theory on the hot origin of life must address the question of how such traits, or their evolutionary precedessors, arose spontaneously in the prebiotic environment. Such adaptations may include histone-like proteins, RNA modificating enzymes, and reverse gyrase, a peculiar ATP-dependent enzyme that twists DNA into a positive supercoiled conformation (Confalonieri et al., 1993). Clues to the origin of hyperthermophily may be hidden in this list, and its evolutionary analysis may contribute to the understanding of the rather surprising phylogenetic distribution of the immediate mesophilic descendants of heat-loving prokaryotes, which shows that at least five independent abandonments events of hyperthermophilic traits took place in widely separated branches of universal trees, one of which corresponds to the eukaryotic nucleocytoplasm (García-Meza et al., 1995).

The antiquity of hyperthermophiles appears to be well established, but there is no evidence that they have a primitive molecular genetic apparatus. Thus, the most basic questions pertaining to the origin of life relate to much simpler replicating entities predating by a long series of evolutionary events the oldest recognizable heat-loving bacteria. Why hyperthermophiles are located at the base of universal trees is still an open question, but the possibility that adaptation to extreme environments is part of the evolutionary innovations that appeared in trunk of the tree cannot be entirely dismissed. The phylogenetic distribution of heat-loving bacteria is no evidence by itself of a hot origin of life, any more than the presence in the hyperthermophile archaeon *Sulfolobus solfataricus* of a gene encoding a thermostable B-type DNA polymerase endowed with 3'-5' exonuclease activity (Pisani et al., 1992) can be interpreted to imply that the first living organism had a DNA genome.

5. Final remarks

Although in the past few years the relationship between molecular biology and microbial phylogenetics has been embittered by frequent clashes and antagonism, the development of rapidly growing sequence databanks has provided a unique view of the evolution of bacterial and eukaryotic microorganisms, and has opened new perspectives in several major fields of life sciences. Molecular evolution was originally the outcome of the wedding of molecular biology with neodarwinian theory, but it has been rapidly transformed into a field of scientific enquiry in its own right. However, its full development requires not only the development of less-expensive, more rapid macromolecular sequencing techniques and more powerful computer algorithms for constructing phylogenetic trees, but also the awareness of its non-stated assumptions and more precise definitions of its conceptual framework.

As summarized by Patterson (1988), the theoretical foundations of molecular cladistics have been based on a number of central concepts, most of which were inherited from older disciplines, such as physiology, anatomy, and neodarwinism. Homology, which is one of the key concepts in evolutionary theory, was originally used by Wolfgang Goethe, Ettiene Geoffroy Saint-Hilaire, Richard Owen, and others, to describe structural resemblance to an archetype (Donoghue, 1992). In recent years it has not only been repeatedly confused with sequence similarity (Reeck et al., 1988), but is also used to describe a wider range of possible evolutionary relationships that include species- or gene-phylogeny. In fact, some classes of homology that describe phenomena at the molecular genetic level may have no exact equivalent in orthodox evolutionary analysis of morphological traits. One such case is paralogy, a term coined by Fitch (1970) to describe the diversification of genes following duplication events.

Since paralogy provides evidence of gene duplication but not of speciation events, it is the basis for infering evolutionary relationships among genes, not among species. Recognition og this distinction has led to repeated recommendations on the avoidance of paralogous sequences in phylogenetic analysis. However, the use of paralogous duplicates in outgroup analyses for determining the evolutionary polarity of character states in universal phylogenies (Gogarten et al., 1989; Iwabe et al., 1989) has rekindled keen theoretical interest in their advantageous properties. Their use, however, does pose some risks. The naive assumption that only one paralogous duplication has taken place in the set of sequences under consideratiuon may lead to incorrect topologies (Forterre et al., 1993). Indeed, the incorporation of genes that are the result of unrecognized multiple paralogous events in a tree may be even more insidious than the problem derived by convergent evolution and lateral gene transfer. The latter phenomena are much more easily identified at the molecular level.

The recognition that paralogous duplicates expand a monophyletic group of sequences raises a number of issues not encountered in classical evolutionary analysis. From a (classical) cladistic point of view, a character that is found only in outgroups is primitive. Nonetheless, in molecular phylogenetic analysis this may not be always the case. Such rule would hold if multiple paralogous duplications have taken place, and if one (or several) of the older sequences is used as an outgroup for an unrooted tree of younger sequences. This would be the case, for instance, if a myoglobin sequence is used to root alpha (or beta) haemoglobin trees. However, this rule would not hold if an alpha haemoglobin sequence (or a set of them) is used as an outgroup for the beta haemoglobin tree, or viceversa.

The same is true, of course, with universal phylogenetic trees derived from elongation factors (Iwabe et al, 1989). In this case neither set is older than its homologue. In this case, the reconstruction of ancestral character states from dichotomously varying paralogous genes does not comes from the analysis of the outgroup, but may be inferred from the realization that the root of the tree must have been preceded by an even older, more primitive condition in which only one copy of the gene existed, prior to the paralogous duplication. Recognition of this fact is likely to play a central role in future understanding of enzyme evolution during the early Archean. Although it is true that the raw material for molecular cladistic analysis is restricted to sequences derived from living organisms (or from fossil samples from which ancient preserved DNA can be retrieved) and cannot be applied to extinct groups of organisms, the construction of trees derived from archaic paralogous sequences may allow us to infer evolution prior to the ealiest detectable nodes.

The flourishing of molecular techniques has led into a proliferation not only of sequences of isolated molecular constituents of living organisms, but also of completely sequenced genomes. This is a storehouse of data that has already provided considerable insights into the phylogeny and the diversity of microbes. But because of its very nature, molecular cladistics separates clusters of adaptative characters into a nested hierarchical set which is expected to reflect the temporal sequence of their evolutionary acquisition. However fruitful, such approach has all the demerits of a reductionist one-trait approach to biological evolution chastised in early literature as "partial phylogeny", and since the birth of molecular phylogeny has rarely been used to attempt a truly integrative analysis of complete character complexes.

Such limitation may be overcomed in several ways, some of which are part of intellectual traditions deeply rooted in comparative biology. As Georges Cuvier contended in his 1805 *Lectures in Comparative Anatomy*, the appearance of the whole skeleton can be deduced up to a certain point by examination of a single bone. The success that Cuvier had in such anatomical reconstructions is legendary, and was based not only in his unsurpassed knowledge and intuition, but also on what he termed the "correlation of parts", i. e., the full recognition of a functional coordination of the parts of the body of a given animal (Young, 1992). Such correlation of parts is not restricted to bones and muscles; at subcellular levels, it underlies the functional coordination among the molecular components of multigenic traits such as metabolic pathways and protein biosynthesis. As shown by the intimate relationship between the biosyntheses of valine and isoleucine, their triplet assignments, and the phylogenetic proximity of their aminoacyl-tRNA synthetases, inquiries on the early evolution of the genetic code and other basic features of living systems should be understood not only by determining the molecular phylogenies of some of their isolated components or by mathetical discussions spiced with a distinct Pythagorean flavor, but with the integrative analysis of character complexes.

But for all its foibles, the relationship between molecular biology and evolutionary theory has opened new, unsuspected avenues of intellectual exploration. Never before has such a wealth of methodological approaches and empirical data been available to the students of life's phenomena. In part because of this prosperity, systematics and evolutionary biology, two of the most broadly oriented fields of life sciences, are now in a state of intellectual agitation. The symptoms are manifold; it is possible that the traditional species concept may not apply to prokaryotes, time-cherished concepts like that of the existence of kingdoms are under fire, the origin and taxonomic position of genetic mobile elements is unknown. There is an increased awareness that the understanding of the processes underlying the generation of evolutionary novelties and the origin of ontogenic patterns cannot be restricted by classical neodarwinian explanations. We are living in the midst of hectic times in which epoch-making debates are reshaping the future of the life sciences, and the development of a more integrated molecular biology may be a never-ending story. It is said that to wish someone to live in an interesting time is one of the most terrible of all Chinese curses. Whatever the outcome of current discussions and debates, for biology the putative Oriental curse may turn out to be nothing less than an intellectual blessing.

Acknowledgments

We are indebted to Dr. Lynn Margulis for her critical reading of the manuscript and many suggestions. Support from the UNAM-DGAPA Project PAPIIT-IN213598 is gratefully acknowledged.

6. References

Achenbach-Richter, L., Gupta, R., Stetter., K. O., and Woese, C. R. (1987) Were the original eubacteria thermophiles? *System. Appl. Microbiol.* **9**, 34-39

Bhattacharjee, J. K. (1985) α-aminoadipate pathway for the biosynthesis of lysine in lower eukaryotes, *CRC Crit. Rev. Microbiol.* **12**, 131-151

Bowler, P. J. (1990) *Charles Darwin, The man and his influence,* Basil Blackwell, Oxford

Confalonieri, F., Elie, C., Nadal, M., Bouthier de la Tour, C., Forterre, P., and Duguet, M. (1993) Reverse gyrase, a helicase-like domain and a type I topoisomerase in the same polypeptide, *Proc. Natl. Acad. Sci. USA* **90**, 4753-4758

DeLey, J. (1968) Molecular biology and bacterial phylogeny, in T. Dobzhansky,. K. Hecht, and W. C. Steere (eds), *Evolutionary Biology,* Appleton-Century-Crofts, New York, pp. 104-156

Doolittle, W. F. (1999) Phylogenetic classification and the universal tree, *Science* **284**, 2124-2128

Doolittle, W. F. and Brown, J. R. (1994) Tempo, mode, the progenote and the universal root, *Proc. Natl. Acad. Sci. USA* **91**, 6721-6728

Donoghue, M. J. (1992) Homology, in E. Fox Keller and E. A. Lloyd (eds), *Keywords in Evolutionary Biology,* Harvard University Press, Cambridge, pp. 170-179

Fitch, W. M. (1970) Distinguishing homologous from analogous proteins, *Syst. Zool.* **19**, 99-113

Fitch, W. M. and Upper, K. (1987) The phylogeny of tRNA sequences provides evidence of ambiguity reduction in the origin of the genetic code, *Cold Spring Harbor Symp. Quant. Biol.* **52**, 759-767

Forterre, P., Benachenhou-Lahfa, N., Confalonieri, F., Duguet, M., Elie, Ch., Labedan, B. (1993) The nature of the last universal ancestor and the root of the tree of life, still open questions, *BioSystems* **28**, 15-32

Galtier,, N., Tourasse, N., and Gouy, M. (1999) A nonhyperthermophilic common ancestor to extant life forms, *Science* **283**, 220-221

García-Meza, V., González-Rodríguez, A., and Lazcano, A. (1995) Ancient paralogous duplications and the search for Archean cells, in G. R. Fleischaker, S. Colonna, and P. L. Luisi (eds), *Self-Reproduction of Supramolecular Structures, from synthetic structures to models of minimalliving systems,* Klüwer, Amsterdam, pp. 231-246

Germont, A., Phillipe, H., and Le Guyader, H. (1997) Evidence for the loss of mitochondria in Microsporidia from a mitochondrial-type HSP70 in *Nosema locustae, Mol. Biochem. Parasitol.* **8**, 159-168

Gogarten-Boekels, M. and Gogarten, J. P. (1994) The effects of heavy meteorite bombardment on the early evolution of life --a new look at the molecular record, *Origins of Life and Evol. Biosph.* **25**, 78-83

Gogarten, J. P., Kibak, H., Dittrich, P., Taiz, L., Bowman, E. J., Bowman, B. J., Manolson, M. L., Poole, J., Date, T., Oshima, Konishi, L., Denda, K., and Yoshida, M. (1989) Evolution of the vacuolar H^+-ATPase, implications for the origin of eukayotes, *Proc. Natl. Acad. Sci. USA* **86**, 6661-6665

Gupta, R. S. and Golding, G. B. (1993) Evolution of HSP70 gene and its implications regarding relationships between archaebacteria, eubacteria, and eukaryotes, *J. Mol. Evol.* **37**, 573-582

Harvey, R. B. (1924) Enzymes of thermal algae, *Science* **60**, 481-482

Holm, N. G., ed., (1992) *Marine Hydrothermal Systems and the Origin of Life*, Klüwer Academic Publ., Dordrecht

Iwabe, N., Kuma, K., Hasegawa, M., Osawa, S., and Miyata, T. (1989) Evolutionary relationship of archaebacteria, eubacteria, and eukaryotes inferred from phylogenetic trees of duplicated genes, *Proc. Natl. Acad. Sci. USA* **86**, 9355-9359

Jensen, R. A. (1976) Enzyme recruitment in the evolution of new function, *Ann. Rev. Microbiol.* **30**, 409-425

Kaine, B. P., Mehr, I. J., and Woese, C. R. (1994) The sequence, and its evolutionary implications, of a *Thermococcus celer* protein associated with transcription, *Proc. Natl. Acad. Sci. USA* **91**, 3854-3856

Kandler, O. (1994) The early diversification of life, in S. Bengtson (ed), *Early Life on Earth, Nobel Symposium No. 84*, Columbia University Press, New York, pp. 124-131

Keefe, A. D., Lazcano, A. and Miller, S. L. (1994) Evolution of the biosynthesis of the branched-chain amino acids, *Origins of Life and Evol. Biosph.* **25**, 99-110

Lazcano, A. (1993) Biogenesis, some like it very hot, *Science* **260**, 1154-1155

Lazcano, A. (1994) The transition from non-living to living, in S. Bengtson (ed), *Early Life on Earth, Nobel Symposium No. 84*, Columbia University Press, New York, pp. 60-69

Lazcano, A. (1995) Cellular evolution during the early Archaean: what happended between the progenote and the cenanestor? *Microbiologia SEM* **11**, 1-13

Lazcano, A., Fox, G. E., and Oró, J. (1992) Life before DNA, the origin and evolution of early Archean cells, in R. P. Mortlock (ed), *The Evolution of Metabolic Function*, CRC Press, Boca Raton, pp. 237-295

Lazcano, A. and Miller, S. L. (1994) How long did it take for life to begin and evolve to cyanobacteria? *Jour. Mol. Evol.* **39**, 546-554

Lazcano, A. and Miller, S. L. (1996) The origin and early evolution of life: prebiotic chemistry, the pre-RNA world, and time, *Cell* **85**, 793-798

Margulis, L. (1993) *Symbiosis in Cell Evolution*, W. H. Freeman, New York

Margulis, L. and Guerrero, R. (1991) Kingdoms in turmoil, *New Scientist* **132**, 46-50

Mayr, E. (1990) A natural system of organisms, *Nature* **348**, 491

Miller, S. L. and Bada, J. L. (1988) Submarine hot springs and the origin of life, *Nature* **334**, 609-611

Müller, M. (1988) Energy metabolism of protozoa without mitochondria, *Ann. Rev. Microbiol.* **42**, 465-488

Nuttall, G. H. F. (1904) *Blood Immunity and Blood Relationship: a demonstration of certain blood-relationships amongst animals by means of the precipitation test for blood*, Cambridge University Press, Cambridge

Oparin, A. I. (1938) *The Origin of Life*, MacMillan, New York

Ouzonis, C. and Sander, C. (1992) TFIIB, an evolutionary link between the transcription machineries of archaebacteria and eukaryotes.,*Cell* **71**, 189-190

Patterson, C. (1988) Homology in classical and molecular biology, *Mol. Biol. Evol.* **5**, 603-625

Pisani, F.M., De Martino, C., and Rossi, M. (1992) A DNA polymerase from the archaeon *Sulfolobus solfataricus* shows sequence similarity to family B DNA polymerases. *Nucleic Acid Res.* **20**, 2711-2716

Reeck, G. R., de Häen, C., Teller, D. C., Doolittle, R. F., Fitch, W., Dickerson, R. E., Chambon, P., McLachlan, A. D., Margoliash, E., Jukes, T. H., and Zuckerkandl, E. (1987) "Homology" in proteins and nucleic acids, a terminology muddle and a way out of it, *Cell* **50**, 667

Rivera, M. C. and Lake, J. A. (1992) Evidence that eukaryotes and eocyte prokaryotes are inmediate relatives, *Science* **257**, 74-76

Schwartz, M. and Dayhoff, M. O. (1978) Origins of prokaryotes, eukaryotes, mitochondria, and chloroplasts, *Science* **199**, 395-403

Sidow, A. and Bowman, B. H. (1991) Molecular phylogeny, *Current Opinion Genet. Develop.* **1**, 451-456

Sleep, N. H., Zahnle, K. J., Kastings, J. F., and Morowitz, H. J. (1989) Annihilation of ecosystems by large asteroid impacts on the early Earth, *Nature* **342**, 139-142

Snel, B., Bork, P., and Huynen, M. A. (1999) Genome phylogeny based on gene content, *Nature Genetics* **21**, 108-110

Sogin, M. L. (1994) The origin of eukaryotes and evolution into major kingdoms, in S. Bengtson (ed), *Early Life on Earth, Nobel Symposium No. 84*, Columbia University Press, New York, pp.181-192

Stark, G. R,. and Wahl, G. M. (1984) Gene amplification, *Ann. Rev. Biochem.* **53**, 447-491

Stetter, K. O. (1994) The lesson of archaebacteria, in S. Bengtson (ed), *Early Life on Earth, Nobel Symposium No. 84*, Columbia University Press, New York, pp. 114-122

Tekaia, F. and Dujon, B. (1999) Pervasiveness of gene conservation and persistence of duplicates in cellular genomes, *J. Mol. Evol.* **49**, 591-600

Tekaia, F., Lazcano, A., and Dujon, B. (1999) The genomic tree as revealed from whole proteome comparisons, *Genome Research* **9**, 550-557

Wächtershäuser, G. (1990) The case for the chemoautotrophic origins of life in an iron-sulfur world, *Origins of Life Evol. Biosph.* **20**, 173-182

Wallace, D. C. and Morowitz, N. H. (1973) Genome size and evolution, *Chromosoma* **40**, 121-126

Wheelis, M. L., Kandler, O., and Woese, C. R. (1992) On the nature of global classification, *Proc. Natl. Acad. Sci. USA* **89**, 2930-2934

Woese, C. R. (1987) Bacterial evolution, *Microbiol. Reviews* **51**, 221-271

Woese, C. R. (1993) The archaea, their history and significance, in M. Kates, D. J. Kushner, and A. T. Matheson (eds), *The Biochemistry of the Archaea (Archaebacteria)*, Elsevier Science Publishers, Amsterdam, pp. vii-xxix

Woese, C. R. and Fox, G. E. (1977) The concept of cellular evolution, *Jour. Mol. Evol.* **10**, 1-6

Woese, C. R., Kandler, O., and Wheelis, M. L. (1990) Towards a natural system of organisms, proposal for the domains Archaea, Bacteria, and Eucarya, *Proc. Natl. Acad. Sci. USA* **87**, 4576-4579

Ycas, M. (1974) On the earlier states of the biochemical system, *J. Theor. Biol.* **44**, 145-160

Young, D. (1992) *The Discovery of Evolution*, Natural History Museum Publications, Cambridge

Zuckerkandl, E. and Pauling, L. (1965) Molecules as documents of evolutionary history, *J.Theoret. Biol.* **8**, 357-366

Section 4.
Study of Life in the Solar System

ASTROBIOLOGY AND THE ESA SCIENCE PROGRAMME

WILLEM WAMSTEKER,
ESA, VILSPA, P.O. Box 50727,
28080 Madrid, Spain (wwamstek@notes.vilspa.esa.es), and

AUGUSTIN CHICARRO,
ESA, ESTEC, Keplerlaan 1
NL-2201 AZ Noordwijk ZH, the Netherlands (achicarr@estec.esa.nl)

Abstract: We introduce some fundamental concepts relating to the nature of "Astrobiology" to evaluate the relevance of Space Science in these activities. It is clear that the strength of the current space missions is predominantly in establishing possible pre-biotic conditions. The overall importance of the missions of the ESA Science Programme Horizons 2000 to the general studies globally collected in the "Search for life" are highlighted through a description of the relevance of the missions *IUE, Hipparcos, HST, ISO, SOHO, Giotto, MARS EXPRESS, ROSETTA, CASSINI-HUYGENS.*

1. Introduction.

At the Ibero-American School of Astrobiology, it may be appropriate to spend a little time on the evaluation of the nature of Astrobiology (others appear to prefer the use of the term Exobiology *for the study of life beyond the earth's atmosphere, as on other planets*) and its goals. The make-up of the name "Astrobiology" suggests that in this case we are dealing with a scientific discipline in the current understanding of the word science itself. The implications of the word science in the modern world require that one deals with a body of facts and the application of generally accepted laws of tested validity. In this context one would expect Astrobiology to address one of the most important aspects of science through the generation of *verifiable predictability for natural phenomena.*

An alternative aspect of science is associated with systematic knowledge gained through observation and experimentation, which has not yet reached the maturity to produce generally valid laws. This aspect of science is the, mainly curiosity driven, human activity where wondering about our surroundings drives us to ask questions. It is however not the asking of questions and evaluation of possible answers which makes modern science. One of the first issues which has to be settled, before the question and answer activity becomes science, is the establishment of a common understanding of concepts. In science, this is generally accomplished by limiting the field of query and confinement through definitions. These definitions are not generated to limit the scope of human curiosity, but they are created to make sure that acquired knowledge can be commonly shared to the benefit of mankind. This is done through a process in which established rules, words, and symbols are related in such a way that independent individuals can fully share their common knowledge. Only then can one exchange and evaluate knowledge which is not yet part of the common heritage.

J. Chela-Flores et al. (eds.), Astrobiology, 153–162.

The activity ongoing until knowledge has reached this common heritage stage, is normally referred to as science. After that phase, the transfer and sharing of the knowledge is considered *education*.

As the concept of life is based on the single observation of Life on the planet Earth, neither of the two aspects of modern science –predictability and common heritage symbolics- is actually addressed by Astrobiology, nor Exobiology. As this could generate problems for the development of a mature scientific debate, I will try here to clarify the matter somewhat in the hope that through this, the excitement of the multi-disciplinary subject of this school can be maintained and expanded, while some of the possible confusion between philosophy, myth and rigid science disciplines is avoided.

The term Exobiology has been coined as a consequence of the fact that mankind, through its earlier scientific studies in many fields has generated a common perception that life is not unique to Earth as appears to be suggested by religious references, but could be a rather more common occurrence in the Universe. As such one can consider the activity commonly addressed as Exobiology to be rather more accurately described as the "extra-terrestrial search for familiar life forms". This avoids the rather erroneous perception that there is a factual base for the existence of such life forms beyond the Earth.

On the other hand the term Astrobiology, has been coined in similar ways as Astrophysics, which has been added to the science vocabulary to supply a more adequate description of the combination of two well established sciences: Astronomy and Physics. Each of these had reached a sufficient level of valid predictability to assure that at times predictions of one science could only (or preferably) be validated through experiments related to the other science. This has brought Astrophysics firmly into the realm of sciences and education, as outlined above. But, even in this case, a rather heated professional debate extending over years was needed to have this cross fertilization accepted in the mainstream of both sciences (DeVorkin, 1999). The current question is: Does Astrobiology fulfils the same criteria? And as is quite clear, it does not. The simple reason being that there is no predictability in biology which can be confirmed in astronomical experiments nor vice-versa. The closest we come to this is in the experiments related to the spatial orientation memory of animals, which have shown that this memory is associated with Astronomy (e.g. Emlen, 1969). It is however our perception that these experiments are not close to the essence of what is meant in the current use of Astrobiology.

Does this imply that one should discard Astrobiology, or restrict it to such phenomena, only? I dare to say that this would be very unfortunate. For the simple reason that at this school the organizers have been able to bring together a wonderful group of well defined scientists with a real desire to explore new multi-disciplinary frontiers. As is quite often the case, it is not possible to explore new frontiers if only the rigid discipline, with which we were all educated, is adhered to. Consequently, it is essential that one has to find for such extremely exciting activities new definitions which will allow us to communicate. Or a common understanding has to be developed, which will allow to bring this multidisciplinary activity to a stage where we will indeed be able to bring together Astrophysics, Biophysics and Biochemistry into a new science. This stage would represent a major breath-through in our scientific thinking: It would be the first time that ***three*** established -already multi-disciplinary- sciences would really combine into a new science and a new field of experimental science will be opened. Before we can expect to reach this stage, it is of primary importance that the first assumptions and definitions are made such that some form of common thinking can be created among the participants. As all our concepts of life are clearly based on a single observation in the sense of physical sciences, it is of importance to generate a more generic starting point with a definition of life. Of course such a definition must be able to lead to the single life form that we know: terrestrial life. But this definition must be sufficiently simple that we do not, on first approach already, impose such a large amount of restrictive information that only life -as we know it- will fit the bill. It should on the other hand, not take into consideration any requirement (source of energy) which is not currently known in the Universe, i.e. not require "deus ex machina".

Although this might not the right place here to try to do that, we think it is important for all our future space endeavors to have a concise definition of life available. The importance of this is in the fact that experiments can than be designed to accommodate well defined questions. At the same time new approaches can be evaluated to clarify more practical ways to establish the scope and field of applicability of "Astrobiology".

We would therefore like to suggest here a definition of life which I hope will be helpful for the future evolution of what might one day become the science of Astro-Chemo-Biophysics.

A definition which could serve this purpose has to be concise, complete and can not exclude any known life forms.

Generic Life:
Self-Multiplication without replication

Two Modes of Self-Multiplication have to be allowed

1. *Auto Self-Multiplication* leads to larger and complex structures (allows existence of a single complex entity)
2. *Multi Self-Multiplication* generates mobility and subsequently populations (allows a complex population of entities)

This on first sight rather skimpy definition of life is actually quite complete at the fundamental level, since it will unavoidably lead to evolutionary effects and it includes all currently known forms of terrestrial life. We would like to stress that the term *replication* is here identified as: *any non-spontaneous multiplication process in which no genetic changes will occur.* With respect to the use of the word **Evolution**, it is important to be aware that it has, in a scientific context, no single meaning (e.g. in evolution Life Sciences and evolution in Astrophysics are completely different concepts).

The above definition of Generic Life will also permit many other, not yet imagined, life forms to be included. An interesting aspect of this is that it brings the definition of life at a slightly higher level than the classical primordial soup. It does allow experimental concepts which are not related to the normal problems of the descriptions of the terrestrial life. It also does firmly de-couple life from intelligence, and can have an existence in parallel to searches for intelligence such as the, very well named, SETI project (Drake, 1993) which looks for technologically advanced intelligence, which, by itself, does *not have to* relate to life.

Attempts have been made to find, from the decomposition of genetic contents of different forms of life with a well established long presence history on Earth, the common genetic elements. For these the genetic make-up of a yeast, five bacteria (both gram-positve and gram-negative), and three fungi where gene-spliced to determine the common ancestor nature in terms of joint genetic properties. The remarkable find of this study (Tekaia, Dujon, Lazcano, 1999) is that this common ancestral environment could be best described as the "**RNA-World**". In this respect it is quite surprising that the minimalist definition of life presented above, represents an quite complete conceptual description of this RNA-World. We will not elaborate further on this quite interesting combination of the evaluation of basic concepts of life from two totally unrelated directions: A non-experimental definition of life has reached a very strong coincidence with the starting point defined for life from a genetic experiment following the strict scientific methodology of genetics.

After this introduction, we will now describe the projects of the ESA Science programme and indicate their relevance to life-related sciences, keeping in mind the issues raised in this introduction. In general the issues addressed are pre-biotic.

2. The ESA Science Programme and its relevance to pre-biotic environments. (http://sci.esa.int/index.cfm)

2.1. IUE (1978 – 1996)

Most of the relevance of the International Ultraviolet Explorer spacecraft, in this context, has been in the observations made with its UV spectrographs (120 - 320 nm) of the study of the atmospheric conditions in the major planets and especially in the ionosphere interactions between the higher layers and the Solar Wind (Ballester, 1997 and Prange and Livengood, 1997). This was also extended to the surface effects of particulate bombardment of satellite surfaces in an environment of highly energetic particles (Domingue and Lane, 1997). Also important results have been obtained on Cometary materials and composite outgassing of volatiles (Festou, 1997).

2.2. HIPPARCOS (1989 – 1993)

The Astrometry Mission Hipparcos has addressed the problems associated with the proper motions of the stars and the first step on the cosmological distance ladder beyond the Solar System (Perryman, 1997). The results of the project have importance for pre-biotic conditions in that they will allow highly improved mass estimates of stars, and allow a great improvement in the abundance and temperatures (Gomez et al., 1997). These combine to create a much improved understanding of the stellar evolution field. As a solar system must be understood as part of the life-cycle of a star this is an important contribution.

2.3 HST (1990 – present)

The Hubble Space Telescope is a joint NASA/ESA Project. It consists of an orbiting general purpose telescope with capabilities of observations at high resolution from the ultraviolet to the near infrared. As its capabilities extend to all realms of Astrophysics and the instrumentation is regularly changed through an in-orbit replacement, a general summary of all result would clearly extend beyond the purpose of this paper. I will here therefore, only touch upon one of the more direct results associated with the formation of planetary systems, which is a typical pre-biotic condition. Important new results have been obtained by a detailed study of the dusty disk around the star Beta Pictoris. Kalas and his collaborators found that the nature of the disk is not really a stable configuration, on astronomical time scales, but is actually a resonance phenomenon associated with the passage of another star some 100,000 years ago (Kalas et al., 2000). This process is quite reminiscent of the phenomena needed to explain the orbital distributions of the comets in our solar system.

2.4. ISO (1995 – 1998)

The Infrared Space Observatory was a short duration mission for observations in the infrared domain (Kessler, 1999). The duration of the mission was determined by the evaporation of the liquid helium used to cool the telescope and detector systems.
The relevance to pre-biotic conditions was mostly related to the reconfirmation of the prevalence of water (H2O) in many shapes and forms and the complex organic molecules collected under the common denominator PAH's (Polycyclic Aromatic Hydrocarbons containing about 50 carbon atoms). Other long chains and non-aromatic groups were found to be much less abundant (Tielens et al., 1999). Within our Solar System new lines of H2O lines were observed on all four giant planets and Titan. These detections have allowed a much better clarification of the physical chemistry of water in these environments since its structure could be determined. Both water vapour as well as ices and frosts were identified. On another front many results were obtained on circumstellar dust and disks (e.g. Tsiji, et al., 1999 on H2O; Robberto et al. , 1999 on disks; and Rosenthal et al.,1999 on molecular hydrogen flows).

2.5. GIOTTO (1985 – 1992)

The extremely successful Giotto mission – ESA's first Deep-Space mission – was developed to have a close encounter with a Comet. This was made possible using the unique opportunity (once every 76 years) supplied by the close approach of the periodic Comet Halley in 1986 to study a periodic Comet, its physics, astrophysics and pre-biotic physics (Grewing, Praderie, and Reinhard, 1987). Its relation to pre-biotic activities are clearly associated with the fact that Comets were considered

1. To be the left over materials from the formation of the solar system
2. To contain significant quantities of water.
3. To show evidence of simple carbon chemistry.

It is therefore clear that this mission had a high pre-biotic relevance. Also the in-situ studies of the environment would give insights in the energetic conditions and would allow a first evaluation if life bearing conditions were fulfilled.

Many previously unsuspected plasma phenomena were found in the cometary environment, especially through the plasma interaction with the Solar Wind. The major success of the mission was of course the capability to make actual close up images of a Comet for the first time in history. The images confirmed the model of Comets as '*dirty snowballs*' with a quite irregular shape. The precise determination of the size of the nucleus (14.9x8.2 km) of the Comet, together with the exceptional resolution on the cometary surface of as small as 100 m has allowed to determine that the surface of Comet Halley is one of the darkest surfaces in the Solar System. It was found that the ejection of matter from the Comet is very irregularly distributed over the surface of the Comet with localised evaporation sites from where high velocity jets are driven by the increase in temperature as a consequence of the close approach to the Sun. Strong variations in the evaporation process, following the rotation period of 7.35 days (or its double 14.7 days) of the nucleus, were found. These are caused by the strong temporal changes in temperature over the surface of the Comet. Combination of these results with those in the ultraviolet (IUE) allowed to determine that Comet Halley lost, at it closest approach to the Sun, in the order of 10 tons of water per second. The GIOTTO mission also generated co-ordination of the scientific community in a world-wide effort without precedent, allowing even more important advances in our understanding of the physical nature of these Celestial messengers from the early formation days of the Sun and Planets. It gave very important new insights in pre-biotic conditions which are considered to be closely related to the formation of the Earth.

2.6. SOHO (1995 – present)

The ESA/NASA Solar and Heliospheric Observatory plays an important role since it has given for the first time an extensive and long duration exposure of the Sun in the Ultraviolet and X-ray domain allowing a detailed evaluation of the interaction between the thermal phenomena of the Sun, more associated with the Chromoshpere of the Sun and the very hot tenuous outer Corona. Paricle acceleration and ejection mechanisms can now be studied directly and in detail.The long duration of the mission extending over the current Solar Maximum has given completely new insights in the variations and mechanisms driving the deposition of highly energetic particles in the Earth Magnetospere and can be expected to strongly chance our views of the Solar influence on the development and possibly even the evolution of terrestrial life. Together with the mission CLUSTER II (launch in June and July 2000) which will directly observe the interactions between the Solar Wind and the the outer interfac between the terrestrial Magnetospere and the general Heliosphere.

A important contribution of the SOHO mission in this context is the discovery of more than 90 Sungrazing Comets, many of which do not survive their close encounters with the Sun. Most of these are thought to be the result of a breakup of a major Comet described 2000 years ago.

2.7. HUYGENS/CASSINI (1997 – present)

Although this mission has already been launched it will not enter science operations phase until 2004 when the spacecraft reaches the planet Saturn. ESA's contribution to Huygens consists of a lander to go to the satellite Titan.

(*http://sci.esa.int/huygens/*). This mission present one of the most tantalising expectations of those related to the pre-biotic conditions anywhere in our Solar System at this moment. The known facts about Titan, the only large satellite of the planet Saturn, suggest that it has an atmosphere rich in simple hydrocarbons and has a strong thermal inversion in its atmospheric temperature profile. Also the discovery of Water vapour on Titan. All these combine to suggest the closest conditions similar to those on the primitive Earth anywhere in the Solar system.

2.8. ROSETTA

The International Rosetta Mission was selected as the first Planetary Cornerstone Mission in ESA's long-term Science Programme. The mission goal is a rendezvous with comet 46 P/Wirtanen. On its eight-year journey to the comet, the spacecraft will pass close to two asteroids (Otawara and Siwa). Rosetta will study the nucleus of comet Wirtanen and its environment in great detail for a period of nearly 2 years, the near-nucleus phase starting at a heliocentric distance of about 3.25 AU, with continued observation activities leading ultimately to close observation from about 1 km distance. Rosetta will be launched in January 2003 by an Ariane-5 Launcher.

The prime scientific objective of the Rosetta mission is to study the origin of comets, the relationship between cometary and interstellar material and its implications with regard to the origin of the Solar System. Cometary nuclei, the prime target bodies of the Rosetta mission, and - to a lesser extent - asteroids, represent the most primitive solar-system bodies. They are assumed to have kept a record of the physical and chemical processes that prevailed during the early stages of the evolution of the solar system. It is the abundance of volatile material in comets that makes them particularly important and extraordinary objects: they demonstrate that comets were formed at large heliocentric distances and have been preserved at low temperatures since their formation. The only alteration processes since that time were induced by irradiation close to the surface and, later, by the thermal wave experienced by the comet when it was transferred from the Oort cloud to a Jupiter family trajectory, a few thousand years ago. Cometary material therefore represents the closest we can get to early condensates in the solar nebula. This may also be true to a lesser extent of primative (carbonaceous) asteroids.

Cometary material has been submitted to the lowest level of processing since its condensation from the protosolar nebula. It is therefore, considered likely that presolar grains may have been preserved in comets. As such, cometary material should constitute a unique repository of information on the sources which contributed to the protosolar nebula, as well as on the large condensation processes that resulted in the formation first of planetismals, then of larger planetary bodies. While tantalising results were obtained in situ from cometary grains, and from interplanetary dust particles collected on Earth, these cannot be considered as fully representative, in particular in terms of the organic and volatile complement (i.e. pre-biotic conditions). Direct evidence on cometary volatiles is particularly difficult to obtain, as species observable from Earth, and even during the Halley flybys, result from physico-chemical processes such as sublimation, interaction with solar radiation and the solar wind. Information currently available on cometary material gained from in situ studies and ground-based observations demonstrates the low level of modification of cometary material. The tremendous potential of cometary material for providing on the constituents and early evolution of the solar nebula has yet to be exploited.

The Rosetta mission has specifically been designed to achieve this important scientific goal by focusing on in situ investigations of cometary material. The surface-science package will provide information on the chemical and physical properties of a selected area. It should also be possible to determine the isotopic composition. This information will be used as ground-truth for high-resolution coverage of the nucleus by remote-sensing investigations. In situ analyses will be performed on the dust grains and the gas flowing out from the nucleus. The physical and chemical processes that link material in the coma to volatile and refractory species in the nucleus will also be investigated. To achieve these goals, the spacecraft will remain for most of the mission within a few tens of kilometres of the nucleus, where the analysed dust and gas is likely to present minimal alterations relative to surface material, and where it can be traced back to specific active regions on the cometary nucleus. The physics of the outer coma and the interaction with the solar wind will also be studied. Rosetta will constitute a major step towards a better understanding of the formation and composition of planetesimals in the outer solar system, as well as their evolution over the last 4.57 X 109years. The global characterisation of a cometary nucleus and one or two asteroids will provide essential information about the provenance of meteorites and interplanetary dust from which we have obtained most of our present knowledge about the formation of the solar system. Last, but not least, the Rosetta mission has important and exciting astrophysical implications, as it will provide a link between the solar system and nucleosynthetic processes, the formation and evolution of molecular clouds, and their further collaps into protostars and planets.

2.9. MARS EXPRESS

The European Space Agency and the scientific community have performed concept and feasibility studies for more than ten years on potential future European missions to the red planet (*Marsnet, Intermarsnet*), focusing on a network of surface stations complemented by an orbiter, a concept which is being implemented by the CNES-led *Netlander* mission to be launched in 2005. Before that, however, the ESA *Mars Express* mission includes an orbiter spacecraft and a small lander module named Beagle-2 in remembrance of Darwin's ship Beagle. The mission, to be launched in 2003 by a Russian Soyuz rocket, will recover some of the lost scientific objectives of both the Russian *Mars-96* mission and the ESA *Intermarsnet* study, following the recommendations of the International Mars Exploration Working Group (IMEWG) after the failure of *Mars-96*, and also the endorsement of ESA's Advisory Bodies that *Mars Express* be included in the Science Programme of the Agency.
The scientific objectives for the orbiter spacecraft include: global high-resolution remote sensing at 10 m resolution and super-resolution imaging at 2 m/pixel of selected areas, global mineralogical mapping at 100 m resolution, global atmospheric circulation and mapping of the atmospheric composition, subsurface structure at km scale down to the permafrost, surface-atmosphere interactions and interaction of the upper atmosphere with the interplanetary medium. Current design estimates allow for an orbiter scientific payload of about 106 kg compatible with the approved mission scenario. The orbiter instruments include a super/high resolution stereo color imager (HRSC), an IR mineralogical mapping spectrometer (OMEGA), an atmospheric Fourier spectrometer (PFS), a subsurface-sounding radar/altimeter (MARSIS), an energetic neutral atoms

analyzer (ASPERA), an UV and IR atmospheric spectrometer (SPICAM), a radio science experiment (MaRS) and a lander communications relay (MARESS).

The Beagle-2 small lander concept of 60 kg total lander mass (at launch) was selected for its innovative scientific goals, which include geology, geochemistry, meteorology and exobiology (see definition in Introduction) of the landing site. Beagle-2 includes a suite of imaging instruments, organic and inorganic chemical analysis instruments, robotic sampling devices and meteorological sensors. It will deploy a sophisticated robotic-sampling arm, which could manipulate different types of tools and retrieve samples to be analyzed by the geochemical instruments mounted on the lander platform. One of the tools to be deployed by the arm is a 'mole' capable of subsurface sampling to reach soil unaffected by solar-UV radiation.

A Soyuz-Fregat launcher will inject a total of about 1100 kg into Mars transfer orbit in early June 2003, which is the most favorable launch opportunity to Mars in terms of mass in the foreseeable future. The orbiter will be 3-axis stabilized and will be placed in an elliptical martian orbit (250 10142 km) of 86.35 degrees inclination and 6.75 hours period, which has been optimized for communications with Beagle-2, the *Netlanders*, as well as NASA landers to be launched both in 2003 and 2005. The Beagle-2 lander module, will be independently targeted from separate arriving hyperbolic trajectory, enter and descend through the martian atmosphere in about 5 min, and land with an impact velocity <40 m/sec and an error landing ellipse of 100 20 km. A preliminary Beagle-2 landing site has been proposed in the Maja Valles area (19° N, 51.5° W). The nominal mission lifetime of one martian year (687 days) for the orbiter investigations will be extended by another martian year for lander relay communications and to complete global coverage. The Beagle-2 lifetime will be of a few months.

ESA will provide the launcher, the orbiter and the operations, while the Beagle-2 lander will be delivered by an UK-led consortium of space organizations. ASI will play a major role in the operations of the mission with the provision of a lander communication package and the use of a 64 m antenna in Sardinia later in the mission. The orbiter instruments, selected in June 1998 from proposals in response to an international announcement of opportunity, are being provided by scientific institutions through their own funding. The *Mars Express* mission is now in Phase-C/D, with Matra Marconi Space in Toulouse, France, as its Prime Contractor and involving a large number of European companies.

International collaboration, either through the participation in instrument hardware or through scientific data analysis is very much valued to diversify the scope and enhance the scientific return of the mission, such as NASA's major contribution to the subsurface-sounding radar. Also, arriving at Mars at the very end of 2003, *Mars Express* will be followed by the Japanese *Nozomi* spacecraft a few days later. Both missions are highly complementary in terms of orbits and scientific investigations; *Nozomi* focusing on the study of the upper atmosphere of Mars as well as the interaction of the solar wind with the ionosphere from a highly elliptic equatorial orbit. Close cooperation, including scientific data exchange and analysis, is foreseen by the *Nozomi* and *Mars Express* teams within a joint ESA-ISAS programme of Mars exploration. (Web sites: http://sci.esa.int/marsexpress/ and http://www.beagle2.com/)

3. References

Ballester, G.E., 1997, in Ultraviolet Astrophysics beyond the IUE Final Archive, Eds. W.Wamsteker and R. Gonzalez-Riestra, ESA SP-413, pg 21.

DeVorkin, D.H., 1999, in "*The American Astronomical Societey's first Century*", Editor David H.DeVorkin, AAS Special Publication, pg 20.

Domingue,D.L., Lane, A.L., 1997, in Ultraviolet Astrophysics beyond the IUE Final Archive, Eds. W.Wamsteker and R. Gonzalez-Riestra, ESA SP-413, pg 13.

Drake,F., 1993, 3rd Decennial US-USSR Conf. On SETI, Ed. G.Shostak, A.S.P. Conf. Ser.Vol. 47, pg. 11.

Emlen, S.T., 1969, Sky and Telescope, 38, 4

Festou, M.C.,1997, in Ultraviolet Astrophysics beyond the IUE Final Archive, Eds. W.Wamsteker and R. Gonzalez-Riestra, ESA SP-413, pg 45.

Gomez, A.E., Luri,X.,Mennessier, M.O., Torra,J., Figueras,F., in HIPPARCOS Venice '97, Eds. M.A.C. Perryman, P.L.Bernacca, ESA SP-402, pg 207.

Grewing, M.,Praderie,F., Reinhard,R., 1987, Proceedings of Exploration of Halleys's Comet, Springer Verlag, Berlin, also Astron. Astrophys, 187, pg 1- 936

Kessler, M.F., 1999, in the Universe as seen by ISO, Eds. P.Cox and M.F.Kessler, ESA SP-427, pg 23

Kalas, P., Larwood,J., Smith,B., Schultz, A., 2000, Ap.J.Lett., *in press*

Prange, R., and Livengood, T.,1997, in Ultraviolet Astrophysics beyond the IUE Final Archive, Eds. W.Wamsteker and R. Gonzalez-Riestra, ESA SP-413, pg 29.

Robberto, M., et al., 1999, in the Universe as seen by ISO, Eds. P.Cox and M.F.Kessler, ESA SP-427, pg 195

Rosenthal, D., et al., 1999, in the Universe as seen by ISO, Eds. P.Cox and M.F.Kessler, ESA SP-427, pg 561

Tekaia, F., Dujon, B., and Lazcano, A., 1999, 12th International Conference on the Origin of Life & 9th ISSOL Meeting (San Diego, California, USA, July 11-16, 1999), Abstract c4.6, p. 53

Tsuji, T., et al., 1999, in the Universe as seen by ISO, Eds. P.Cox and M.F.Kessler, ESA SP-427, pg 219

Tielens,A.G.G.M., Hony,S., van Kerckhoven, C., Peeters, E., in the Universe as seen by ISO, Eds. P.Cox and M.F.Kessler, ESA SP-427, pg 579

THE CHEMICAL COMPOSITION OF COMETS

HUMBERTO CAMPINS
University of Arizona, Lunar and Planetary Laboratory, Tucson, Arizona 85721, USA and
Research Corporation 101 N. Wilmot Rd Suite 250, Tucson Arizona 85711, USA

Abstract: Comets are believed to have played an important role in the origin and evolution of life on Earth, the chemical composition of comets can provide significant insights into this role. Our understanding of the composition of comets has improved significantly since the recent apparitions of comets Hale-Bopp (1995 O1) and Hyakutake (1996 B2), the two brightest comets in the past 20 years. New information on the composition of cometary molecules and on the deuterium to hydrogen ratio (D/H) in cometary water provides evidence for presolar chemical signatures in cometary ices. D/H ratios and noble gas abundances also suggest that some, and possibly most, of Earth's water has a cometary origin. Comets also contain organic molecules that may have contributed to the inventory of prebiotic organic matter on Earth.

1. Introduction

The extent to which comets influenced the origin and evolution of life on Earth has been a topic of great interest in recent years (e.g., Oró and Lazcano 1997; Delsemme 1997, 1999). Comets are the most primitive members of our solar system. Because of their orbits and small size, comets have undergone relatively little processing, unlike larger bodies such as the Moon and the Earth, which have been modified considerably since they formed. This pristine nature of comets is evidenced in their high abundance of volatile compounds. The composition of comets contains a wealth of information on their origin and evolution, as well as about the origin and evolution of the rest of our solar system; hence, comets are often called cosmic fossils. Two sources of comets have been identified. The Oort cloud is a spherical shell of comets located approximately 50,000-100,000 astronomical units (AU) from the Sun. The Edgeworth-Kuiper belt, or Kuiper belt, is a disk-like distribution of icy objects beyond the orbit of Neptune (30 AU).

The study of comets benefited significantly from the recent apparitions of comets Hale-Bopp (1995 O1) and Hyakutake (1996 B2), which were the two brightest comets in the past 20 years. Some of the observations of these two comets are still under analysis, and Comet Hale-Bopp is still being observed in the southern sky. However, many

J. Chela-Flores et al. (eds.), Astrobiology, 163–176.

exciting results with profound implications have emerged from studies of these two comets and of Comet Halley.

Comet Hale-Bopp was discovered at 7.15 Astronomical Units (AU) from the Sun on July 23, 1995, 21 months pre-perihelion. This advanced notice allowed for considerable preparation and resulted in extensive observations, some of which are still being carried out. Closest approach to Earth (1.3 AU) occurred on March 21 and perihelion (0.91 AU) on April 1, 1997. This is the most productive and dustiest comet ever observed; its gas and dust production were, respectively, 20 and 100 times greater than those for Comet Halley (Schleicher *et al.* 1997).

Comet Hyakutake was discovered at 2 AU from the Sun, on January 30, 1996, less than two months before closest approach to Earth (0.1 AU) on March 25. This is the closest approach by a comet since 1983, when comets IRAS-Araki-Alcock and Sugano-Saigusa-Fujikawa came to 0.03 and 0.07 AU of Earth, respectively. Comet Hyakutake reached its perihelion (0.23 AU) on May 3, 1996. This comet had a high gas-to-dust ratio, i.e., it is considered a "gassy" comet (Schleicher *et al.* 1996).

In this paper, recent results on the chemical and isotopic composition of comets are reviewed. The implications of these results on the origin and evolution of cometary and terrestrial volatile material are discussed, with an emphasis on the origin of cometary ices as well as terrestrial water. In addition to water, comets are also rich in complex organic molecules. The possible delivery of this organic material to the early, prebiotic Earth is also discussed.

2. Composition of Neutral Gas

Parent molecules sublimate directly from ices in the comet's nucleus. Daughter molecules are photodissociation fragments of parent molecules. The most prominent daughter fragments in the visible spectra of comets are OH, CN, C_2, C_3, and NH. H_2O is the most abundant volatile species in comets; however, because of the H_2O in Earth's atmosphere it is difficult to detect cometary H_2O. The first direct detection of H_2O vapor in a comet was achieved in 1985, on Comet Halley. In the solar radiation field H_2O will photodissociate mainly into OH plus H. OH is relatively easy to detect from Earth's surface, and is used as a tracer of the H_2O in comets where H_2O can not be observed directly.

2.1. CHEMICAL CLASSIFICATION

Emission bands from OH, CN, C_2, C_3, and NH dominate the visible and near ultraviolet spectrum of most comets. Filter photometry has been used to study the absolute and relative abundances of these species in approximately 100 comets (A'Hearn *et al.* 1995). This has resulted in the first chemical classification of comets into two main groups, those with "typical" abundances and those depleted in C_2 and C_3 with respect to OH and CN. All the comets likely to have originated in the Oort cloud show typical abundances; Hyakutake and Hale-Bopp are among these (Schleicher *et al.* 1996, and

1997). Comets likely to have originated in the Kuiper belt can show either typical abundances or C_2 and C_3 depletions (A'Hearn *et al.* 1995). The origin and significance of these two compositional groups are difficult to establish at this time because we do not yet know the parent molecules of these fragments.

2.2. NEW PARENT MOLECULES

All the chemical species discussed above, except for H_2O, are photodissociation fragments of parent molecules. It is preferable to study parent molecules whenever possible because they are indicative of the composition of the ices. The detection of these parent molecules has been difficult, partly because most of them do not have transitions observable at visible wavelengths. The development of new instrumentation for spectroscopy at infrared and microwave wavelengths, coupled with the apparition of these two bright comets, has resulted in an explosion in the identification and study of new parent molecules. Recent reviews of this subject are given by Bokelée-Morvan (1997), Bokelée-Morvan and Crovisier (2000), Despois *et al.* (1999), and Irvine *et al.* (2000). Table 1 is based on Bokelée-Morvan and Crovisier (2000) and presents the chronology of the identification of parent molecules in comets; a question mark indicates a tentative detection. Note that each molecule is listed only once, under the comet where it was discovered. Table 2 is also based on Bokelée-Morvan and Crovisier (2000), and it gives the abundances of the main parent molecules in Comet Hale-Bopp.

2.3. ORTHO-PARA H_2O RATIO

H_2O is divided into ortho and para species, according to the hydrogen nuclear spins being parallel or antiparallel, respectively. The ortho-para ratio yields a spin temperature, which is used to infer the physical temperature at which the nuclear spins were last equilibrated. This equilibrium requires a sufficiently long time to be reestablished (greater than three years at atmospheric pressure) that observations of the H_2O gas reflect the ortho-para ratio of the ices in the comet nucleus. Four comets have been observed in sufficient detail to detect emission lines from both ortho and para H_2O, and in three of these cases, estimates of the ice temperature have been possible. These temperature estimates are: 29 K in comet Halley (Mumma *et al.* 1993), 35 K in comet Hartley 2 (Crovisier *et al.* 1999), and 25 K for Comet Hale-Bopp (Crovisier *et al.* 1997). The implications of these temperatures will be addressed in the next section.

2.4. ISOTOPIC RATIOS

Isotopic abundances can provide information on the formation conditions and subsequent processing of cometary material. The measurement of isotopic ratios in comets using remote sensing is difficult because it requires high resolution spectroscopy. Consequently, isotopic ratios have been measured for few elements and in few comets (Vanysek 1991). Observations of comets Hyakutake and Hale-Bopp have contributed greatly to this area.

Table 1: Chronology of the direct detection of cometary parent molecules

Year	Comet	First detection
1973	C/Kohoutek 1973XII	HCN (?)
1976	C/West 1976VI	CO
1983	C/I-A-A 1983VII	NH_3(?), S_2
1985	P/Halley	H_2, HCN, H_2CO, CO_2
1990	C/Austin 1990V C/Levy 1990XX	CH_3OH, H_2S
1996	C/1996 B2 Hyakutake	CH_4, C_2H_2, C_2H_6 HNC, CH_3CN, NH_3 OCS, HDO, HNCO (?) $H^{13}CN$
1997	C/1995 O1 Hale-Bopp	SO, SO_2, H_2CS HC_3N, HNCO, NH_2CHO HCOOH, CH_3OCHO DCN, $HC^{15}N$

Table 2: Molecular abundances in Comet Hale-Bopp near perihelion[1]

Molecule	Abundance[2]	Methods	Comments	References
H_2O	100	IR		[Crovisier *et al.* 1997]
CO	20	radio, IR, UV	extended	[Biver *et al.* 1997]
CO_2	6[3]	IR		[Crovisier *et al.* 1996, 1997]
CH_4	~0.6	IR		[Mumma *et al.* 1996]
C_2H_2	~0.1	IR		[Brooke *et al* 1996; Weaver *et al.* 1999]
C_2H_6	~0.5	IR		[Mumma *et al.* 1996]
CH_3OH	2	radio, IR		[Biver *et al.* 1997]
H_2CO	~1[4]	radio	extended	[Biver *et al.* 1997]
HCOOH	0.06	radio		IAUC 6599
$HCOOCH_3$	0.06	radio		IAUC 6645
NH_3	0.6	radio		[Bird *et al.* 1997]
HCN	0.2	radio, IR		[Biver *et al.* 1997]
HNCO	0.06	radio		IAUC 6566
HNC	0.04	radio	extended?	[Biver *et al.* 1997]
CH_3CN	0.02	radio		[Biver *et al.* 1997]
HC_3N	0.02	radio		IAUC 6566
NH_2CHO	0.01	radio		IAUC 6614
H_2S	1.6	radio		[Biver *et al.* 1997]
SO	0.6	radio	from SO_2 ?	IAUC 6573
CS	0.2	radio, UV	from CS_2 ?	[Biver *et al.* 1997]
SO_2	0.15	radio		IAUC 6591
OCS	0.5	radio, IR	extended?	IAUC 6573, 6683
H_2CS	0.02	radio		IAUC 6607
S_2	0.005[1]	UV		[Weaver *et al.* 1996]

1. The abundance given for S_2 is that measured in Hyakutake (Weaver *et al.* 1996). S_2 has not been detected in Comet Hale-Bopp
2. Abundances relative to water. Except for CO_2,H_2O, S_2 and hydrocarbons, they are derived from radio measurements
3. Assuming the CO_2/CO abundance ratio measured at 3 AU from the Sun (Crovisier *et al.* 1997)
4. Abundance in the coma. Abundance in the nucleus may be smaller.

2.4.1. *Deuterium-to-Hydrogen Ratio*

So far, the deuterium-to-hydrogen, or D/H, ratio is the most diagnostic isotopic ratio in comets. Almost all of the deuterium is believed to have formed in the early universe. In our solar system, there is evidence for two distinct reservoirs of deuterium (Owen *et al.* 1986). Jupiter and Saturn, which likely obtained most of their hydrogen directly from the solar nebula gas, have a D/H ratio of about $2.6 \pm 1.0 \times 10^{-5}$ (Irvine *et al.* 2000, and references therein). The second reservoir is enriched in deuterium compared with the solar nebula, and contributed to bodies that accreted from solid grains (i.e., comets and asteroids), and to those bodies that may have acquired a significant fraction of their atmospheres from asteroid and comet bombardment (i.e., Earth, Mars, Titan, and Venus).

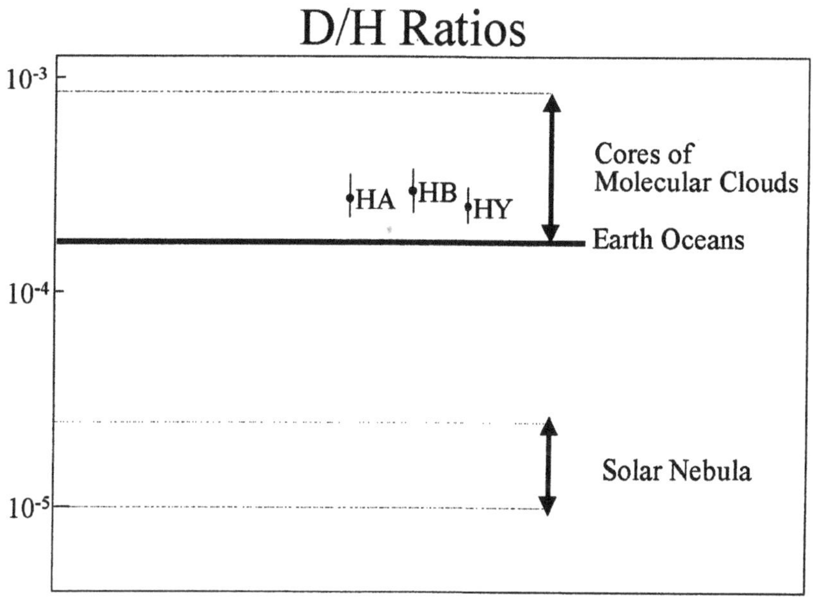

Figure 1. The D/H ratios are plotted for Earth's oceans, the three comets observed so far, the cores of molecular clouds, and the solar nebula. These values suggest that cometary ices formed in dense molecular clouds, and that cometary impacts have contributed significantly to Earth's water.

We now have D/H ratios from H_2O in three comets, Halley ($3.16 \pm 0.34 \times 10^{-4}$; Eberhardt *et al.* 1995), Hyakutake ($2.9 \pm 1.0 \times 10^{-4}$; Bokelée-Morvan *et al.* 1998), and Hale-Bopp ($3.3 \pm 0.8 \times 10^{-4}$; Meier *et al.* 1998). These are all about twice the value for terrestrial water (1.5×10^{-4}; Hageman *et al.* 1970), about ten times the solar nebula value (Jupiter and Saturn), and consistent with the values for "hot cores" of dense molecular clouds ($2\text{-}6 \times 10^{-4}$; Gensheimer *et al.* 1996). These results are illustrated in Figure 1, they suggest that cometary ices formed in dense molecular clouds, and are consistent with the idea that cometary impacts have contributed significantly to Earth's ocean water. Both of these implications will be discussed in more detail in the next two sections.

2.4.2. *Heavier Elements*

Values for $^{12}C/^{13}C$ (Lis *et al.* 1998, Matthews *et al.* 1997), $^{14}N/^{15}N$ and $^{32}S/^{34}S$ (Matthews *et al.* 1997) measured in Comet Hale-Bopp were all consistent with solar system values. Isotopic fractionation in the heavier elements is more difficult to achieve due to the smaller difference in mass between the isotopes. Hence, this agreement in the isotopic composition of the heavier elements does not contradict the view that ices in comets may have originated in dense molecular clouds (see below).

3. Interstellar Signatures in Cometary Ices

The connection between comets and the presolar interstellar environment out of which the solar system formed has been discussed for many years (e.g., Greenberg 1982). Recent advances in our understanding of the composition of comets have reinforced the view that comets are the most primitive bodies in the solar system (e.g., Mumma *et al.* 1993). However, clear evidence that comets retain a record of the presolar environment has only emerged in the past four years. This section summarizes the main results linking comets with the molecular cloud environment.

3.1. D/H RATIOS

As discussed above, the refined value for the D/H ratio in H_2O for Comet Halley and those measured for comets Hyakutake and Hale-Bopp, are virtually identical to each other and a clear presolar signature. If the H_2O in cometary ices had equilibrated with the gaseous hydrogen in the solar nebula, then the D/H ratio in cometary H_2O should be like that in Jupiter and Saturn, and not an order of magnitude greater. So where did the higher D/H in cometary ices come from? It is believed that ion-molecule reactions in dense molecular clouds at temperatures near 35 K can produce these D/H enhancements in icy grains mantles (Millar *et al.* 1989). These icy mantles evaporate in the "hot cores" (temperatures near 200 K) of these clouds and are thus observed with spectroscopic techniques in the gas phase. Hence, the D/H abundance in hot cores allows a direct comparison between molecular cloud and cometary ices. The preservation of this enhanced D/H ratio in comets through the formation of the solar system is evidence for a gentle formation process and the primitive nature of comets.

Just how unprocessed do ices have to be in order to retain this D/H signature? Lunine *et al.* (1991) argue that water ice can sublimate during the free fall of grains into the solar nebula from a surrounding molecular cloud. However, practically all the H_2O would recondense at heliocentric distances beyond 10 AU, and the D/H ratio would not be significantly altered. On the other hand, the presolar ices may never sublimate in the process of solar system and cometary formation; this second scenario is favored by some of the evidence discussed later in this section.

3.2. COMPOSITIONAL ANALOGIES

The similarity between the list of cometary parent molecules and those molecules found in molecular clouds is remarkable (e.g., Shalabiea and Greenberg 1994). Even complex molecules first detected in interstellar clouds, such as NH_2CHO, HCOOH, and CH_3OCHO have now been detected in comets Hyakutake and Hale-Bopp with comparable abundances (Bocklée-Morvan 1997); this strengthens the case for an interstellar origin of these parent molecules in comets. The only molecular species found in comets and not yet in interstellar clouds are S_2 and C_2H_6 (ethane). Both of these cases are discussed below.

3.3. DETECTION AND ABUNDANCE OF CERTAIN SPECIES

3.3.1. *The Ethane (C_2H_6) to Methane (CH_4) Ratio*

In Comet Hyakutake, Mumma *et al.* (1996) detected both ethane and methane, with a surprisingly high abundance of ethane (C_2H_6/CH_4 of 0.53). This implies that the ices in this comet did not originate in a thermochemically equilibrated region of the solar nebula. Instead, Mumma *et al.* argue that the C_2H_6/CH_4 ratio is consistent with production of C_2H_6 in icy grain mantles in a molecular cloud, either by photolysis of CH_4-rich ice, or by hydrogen addition reactions to C_2H_2 (acetylene). This proposed origin for the abundant C_2H_6 appears to constrain the formation temperature of the ices in Comet Hyakutake to 65 K (Notesco *et al.* 1997).

3.3.2. *The HNC/HCN Ratio*

The relatively large abundance of interstellar hydrogen isocyanide (HNC) has been attributed to gas phase ion-molecule chemistry in molecular clouds. HNC was detected in Comet Hyakutake, with a ratio to hydrogen cyanide (HCN) of 0.06-0.15, similar to that characterizing the non-equilibrium conditions in dense interstellar clouds (Irvine *et al.* 1996). If the HNC has an interstellar origin, its preservation through the formation of the solar system and the comet would provide important contraints on this process. However, observations of the HNC/HCN ratio in Comet Hale-Bopp at a range of heliocentric distances is consistent with production of HNC by chemical reaction in the coma (Irvine *et al.* 2000). Nevertheless, the application of the same coma chemistry model to Comet Hyakutake cannot produce all the observed HNC. It may be that in an extremely active comet such as Hale-Bopp most or all of the observed HNC is produced by coma chemistry, but in less active comets such as Hyakutake, there is at least some interstellar HNC. Further study of this process is underway.

3.3.3. *S_2 in Comets Hyakutake and IRAS-Araki-Alcock*

S_2 is a parent molecule, and was observed for the first time in any astrophysical object in Comet IRAS-Araki-Alcock (A'Hearn *et al.* 1983). On May 11, 1983, this comet approached Earth at a distance of 0.03 AU, the closest in about 200 years. A'Hearn and Feldman (1985) argued that S_2 can be produced by irradiation of ices on interstellar grains. If so, this would require that cometary ices have accreted from interstellar grains with icy mantles that were never vaporized, and maybe never reached more than a few tens of degrees K. Objections to this mechanism for production of S_2 in Comet IRAS-Araki-Alcock were raised by Russel *et al.* (1987), who suggested that the detection of S_2 may have been related to an unusual alignment of the interplanetary magnetic field at the time of the observations. The detection of S_2 in Hubble Space Telescope spectra of Comet Hyakutake (Weaver *et al.* 1996) rules out the objection of Russel *et al.* (1987). S_2 has now been observed in the last two extreme near-Earth comets. The lifetime of the S_2 molecule is so short that only spectra having spatial resolution of order a few hundred kilometers can succeed in detecting S_2 in the presence of a strong continuum signal. Given that the S_2 abundance in Comet Hyakutake is similar to that observed in Comet IRAS-Araki-Alcock, it seems likely that S_2 is present in all cometary nuclei, and its lack of detection in most cases is due to an observational selection effect (Weaver *et al.* 1996).

The production of S_2 in the interstellar medium has been questioned because laboratory work (Grim and Greenberg, 1987; Moore *et al.* 1988) indicated it would also produce more SO and SO_2 than had been observed (Kim and A'Hearn 1991); however, this discrepancy has essentially been solved. SO and SO_2 were both observed in Comet Hale-Bopp. A reanalysis of the SO using new oscillator strengths is consistent with the interstellar origin, and reanalysis of the SO_2 is underway (Kim *et al.* 1999). Hence, S_2 has returned to the status of a cosmogonically interesting species worthy of detailed study and potentially quite diagnostic of the formation conditions for comets.

3.3.4. *Jupiter's Atmosphere*

Recent observations of the noble gas content in Jupiter's atmosphere by the Galileo probe indicate that the icy planetesimals (comets) that contributed to Jupiter's mass were formed at temperatures lower than those predicted by present models of giant-planet formation (Owen *et al.* 1999). This conclusion is based on the observed abundance of argon, krypton, and xenon. The noble gas abundances in Jupiter are significantly different from those in the Sun (and the solar nebula). The enrichment in argon, krypton, and xenon is the result of comets contributing a significant fraction to Jupiter's inventory of elements heavier than hydrogen and helium. In order for comets to be able to trap sufficient argon, krypton, and xenon to produce the observed abundances in Jupiter, Owen *et al.* argue that they had to have formed at temperatures below 30 K.

3.4. EVIDENCE FOR AMORPHOUS ICE

Evidence for amorphous ice in Comet Hale-Bopp has been found in spectroscopic observations in the 1 to 2 micron region (Davies *et al.* 1997) and from models of the CO and H_2O activity versus heliocentric distance (Enzian *et al.* 1997). The presence of amorphous ice in comets has been suspected for many years (e.g., Smoluchowski 1985). The formation of amorphous ice is favored when condensation occurs at temperatures below about 150 K. At temperatures above 150 K amorphous ice crystallizes. The detection of amorphous ice in Comet Hale-Bopp constrains the thermal history of its ices. Although the presence of amorphous ice does not constrain the ice temperature to levels as low as some of the other observations discussed above, it contributes to the view that cometary ices have not been altered significantly since their formation.

3.5. ORTHO-PARA H_2O TEMPERATURES

As discussed in section 2, the temperature of the nuclear H_2O ice has been estimated for comets Halley, Hale-Bopp, and Hartley 2 to be approximately 29 K, 25 K, and 35 K, respectively. These temperatures are certainly consistent with the low formation and processing temperatures suggested by the other observations discussed in this section. All these results are in agreement with models that form cometary ices in cold molecular clouds (e.g., Greenberg and Hague 1990) and keep them cold since their formation. In other words, comets preserve records of presolar chemistry.

4. Origin of Earth's Water

The observed D/H ratio in the three comets discussed above, and the abundance of noble gases on Earth, suggest that comets contributed some or most of Earth's water. As discussed in section 2, the D/H ratio in sea water is about six times greater than the value for the solar nebula (as indicated by the values in the atmospheres of Jupiter and Saturn). This is clear evidence for at least two distinct reservoirs of deuterium in our solar system. The three comets are enriched in deuterium by a factor of about 12 compared with the solar nebula, and a factor of two compared with Earth's oceans. Hence, it has been argued that comets are the "deuterium rich" reservoir, and that they may have contributed much or most of the water to the primitive Earth.

Because the D/H ratio observed so far in comets is twice the value for Earth's oceans, it has been argued that comets cannot be the only source of ocean water (e.g., Despois 1999). While this argument may be true, Delsemme (1999) favors comets being the main source of Earth' atmosphere, sea water, and volatile carbon compounds. It is generally accepted that an intense comet bombardment of the primitive Earth occurred, and that these comets were brought into the inner solar system by orbital scattering due to the growth of the giant planets. Delsemme believes most of Earth's impactors were comets that accreted in Jupiter's zone. In this zone, water vapor convecting from the hot, inner solar nebula, condensed onto icy material from the outer regions. This process, argues Delsemme, diminished the deuterium enrichment of comets coming from Jupiter's zone. Hence, these comets with moderate deuterium enhancement (similar to that currently observed in sea water), may have been the main source of Earth's water. Delsemme's model implies that the D/H ratio in some, yet to be observed, comets should be at least a factor of two lower than that observed so far in the three comets. He also argues that because the three comets with measured D/H originated in the Oort cloud, they probably formed beyond Jupiter's zone, therefore, retaining the higher D/H signature.

The observed deuterium enrichment in the cores of molecular clouds is attributed to ion-molecule reactions in the icy mantles of interstellar grains. This process is favored over the exchange of neutral atoms or molecules because of long time scales for neutral exchanges at low temperatures. Interstellar ices are the only known source for the deuterium enrichment in cometary water. Some meteorites are also enriched in deuterium, and the source of this enrichment is believed to be the same as that in comets. Could meteorites be the source of Earth's water? The main argument against meteorites being a major contributor to Earth's water is the noble gas abundance in the atmospheres of Earth, Mars, and Venus, which we discuss below.

The D/H evidence in favor of a cometary origin of Earth's water is in agreement with an argument made by Owen and Bar-Nun (1995) based on the abundance of noble gases. The noble gas abundances in the atmospheres of Earth, Mars, and Venus are significantly different from those observed in the Sun (Figure 2). This difference indicates that these three planetary atmospheres were not formed directly from solar nebula gases. The noble gas patterns in the atmospheres of Venus, Earth, and Mars are also significantly different from those in volatile rich meteorites (H and C chondrites, Figure 2). The noble gas evidence does not rule out a modest meteoritic contribution to Earth's atmosphere, but it does argue strongly against meteorites being the main source. Although noble gas abundances in comets have not been measured, laboratory studies of how noble gases are trapped in ices forming at low temperatures predict abundances, including that of Xenon, similar to those observed on Earth (Owen and Bar-Nun 1995). In fact, Owen and Bar-Nun use a mixture of three types of comets, formed at different temperatures, to account for the observed volatile inventories of Venus, Earth, and Mars; with the caveat that impact erosion is necessary to explain the current condition of the Martian atmosphere.

The number of comet impacts necessary to provide all the water in Earth's oceans (10^{21}kg) is about 10^6, assuming a typical comet contains 10^{15}kg of water ice. This number is consistent with an independent estimate of the cometary impacts that must have occurred early in the history of the solar system. The scattering of comets in all directions, including that of Earth, is a natural byproduct of the formation of the giant planets. This independent estimate of cometary impacts is based on the current number of comets, the cratering record on the Moon, and models of giant planet formation (e.g., Delsemme 1999, Lunine 1999).

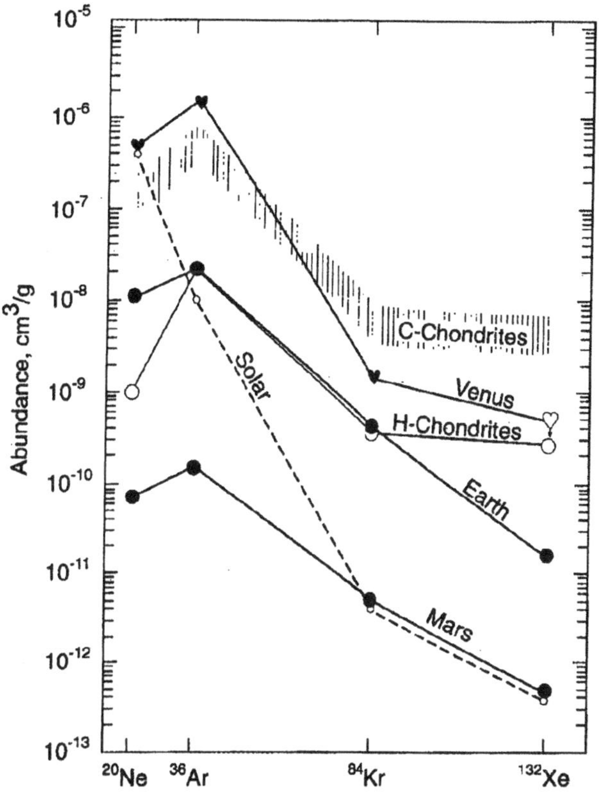

Figure 2. Abundances of noble gases in the atmospheres of Venus, Earth, and Mars compared with solar abundances and those in two kinds of meteorites (from Owen and Bar-Nun 1995). Note that the noble gas abundances in the three planets differ significantly from solar abundances, indicating that these atmospheres were not derived directly from the solar nebula. The large argon-to-xenon and krypton-to-xenon ratios on Earth relative to those in meteorites suggest that comets were an important source of Earth's atmosphere and Oceans.

5. Dust Composition

The dust component of cometary material also contains important information about the origin and evolution of comets. In this section the composition of the silicate and organic components of cometary dust is discussed.

5.1. SILICATES

5.1.1. *Ground Based 8-13 μm Spectra*

This spectral window includes the best studied emission from silicate grains in astrophysical sources. In circumstellar and interstellar sources, a broad emission or absorption feature near 9.7 μm was attributed to the stretching of the Si-O bound in silicate grains; other features due to the O-Si-O bending mode at longer wavelengths were also identified. This silicate emission has also been observed in several comets. The first identification of a specific silicate was made in Comet Halley. Observations of Halley at two heliocentric distances (Bregman *et al.* 1987, Campins and Ryan 1989) showed another peak near 11.3 microns which was identified with crystalline olivine, $[Mg, Fe]_2SiO_4$ (Campins and Ryan 1989). This identification was somewhat of a surprise because at that time no structure in the spectra of circumstellar or interstellar silicates had been observed, and it was believed that astrophysical silicates were amorphous. Comets Hyakutake and Hale-Bopp both showed a prominent silicate emission, with a well defined olivine signature (Hayward and Hanner 1997). Furthermore, a detailed study of the silicate emission by Wooden *et al.* (1998) has now identified pyroxene from features near 9.3 and 10.5 microns, specifically enstatite, which is magnesium-rich pyroxene ($MgSiO_3$).

5.1.2. *Infrared Space Observatory Spectra*

The Infrared Space Observatory (ISO) made extensive infrared observations of Comet Hale-Bopp (Crovisier *et al.* 1997). Additional emission features near 19.5, 23.5, 27.5, and 33.5 microns allowed the identification of forsterite, which is magnesium-rich olivine (Mg_2SiO_4). Crystalline silicates were also observed by ISO in shells around oxygen-rich evolved stars, which are the formation sites of interstellar silicates (Waters *et al.* 1996). This result is very significant and solves an important issue involving cometary silicates. If all circumstellar and interstellar silicates had been amorphous, how did the silicates in comets crystallize, and how did this happen without vaporizing the ice mantles? Now that crystalline silicates have been identified in their formation site, there is no need to heat amorphous olivine or pyroxene to get crystalline silicates after formation of ice mantles. If so, then the silicates grains are also likely to contain presolar signatures. Such presolar isotopic signatures in cometary silicates should be detected by spacecraft such as NASA's Stardust mission.

5.2. ORGANICS IN COMETARY DUST

Delsemme (1982) had predicted that a significant fraction of the carbon in comets was probably tied up in solids. The organic component in cometary solids was first detected by spacecraft observations of the dust in comet Halley. Some of the mass spectra obtained by the Giotto and VEGA spacecraft were dominated by the elements H, C, N, and O. In the other mass spectra, these elements were essentially absent, instead the rock forming elements Mg, Si, and Fe were most abundant. This led to the designation of "CHON" particles to those grains that contained mainly these four light elements (e.g., Gruen and Jessberger 1990). The molecular composition of organic solids in comets is not well defined, mainly because the mass spectra yielded only atomic compositions. The likely molecular composition of cometary organic solids has been discussed recently by Huebner and Boice (1997) and by Kissel, Krueger, and Roessler (1997).

The survivability of cometary organics during impacts with Earth has been questioned by some and defended by others (e.g., Chyba and Sagan 1997). Cometary organics are not expected to survive the impacts of comets greater than a few hundred meters in diameter. However, smaller objects typically explode in the atmosphere prior to reaching the surface and are capable of delivering most of the organic molecules to the surface. In addition, cometary solids in the form of interplanetary dust particles (IDPs) constantly reach Earth's surface.

Acknowledgments
This paper benefited from discussions with M. A'Hearn, D. Bokelée-Morvan, R. Navarro, J. Oró, D. Osip, S. Lederer, and H. Weaver. This work was supported by NASA.

References

A'Hearn, M.F., Feldman, P.D., and Schleicher, D.G. 1983. *Astrophys. J.* 274, L99.
A'Hearn, M.F. and Feldman, P.D. 1985. In *Ices in the Solar System* (J. Klinger *et al.*, eds.), Dordrecht: Reidel, 463.
A'Hearn, M.F., Millis, R.L., Schleicher, D.G., Osip, D.J., and Birch, P.V. 1995. *Icarus* **118**, 223.
Bird, M.K., *et al.* 1997. *Astron. & Astrophys.* **325**, L5.
Biver, N., *et al.* 1997. *Science* **275**, 1915.
Bockelée-Morvan, D. 1997. In *Molecules in Astrophysics: Probes and Processes* (E.F. Van Dishoeck ed.), Dordrecht: Reidel, 219.
Bockelée-Morvan, D. and Crisp, D. 1996. *Nature* **383**, 606.
Bockelée-Morvan, D. and Crovisier, J. 2000. In *Planetary Systems: The Long View*, Blois, France, in press.
Bockelée-Morvan, D., *et al.* 1998. *Icarus* **133**, pp. 147-152.
Bregman, J.D., *et al.* 1987. *Astron. & Astrophys.* **187**, 616.
Brooke, T.Y., Tokunaga, A.T., Weaver, H.A., and Crovisier, J. 1996. *Nature* **383**, 606-608.
Campins, H. and Ryan, E.V. 1989. *Astrophys. J.* **341**, 1059.
Chyba, C.F. and Sagan, C. (1997). In *Comets and the Origin and Evolution of Life*. P. J. Thomas *et al.* (eds.), Springer-Verlag, New York, pp. 147-173.
Crovisier, J., *et al.* 1996. *Astron. & Astrophys.* **315**, L385.
Crovisier, J., *et al.* 1997. *Science* **275**, 1904.
Crovisier, J., *et al.* 1999. In *The Universe As Seen by ISO*. European Space Agency SP **427**, 161.
Davies, J., *et al.* 1997. *Icarus* **127**, 238.
Delsemme, A.H. (1982). In *Comets*. L.L. Wilkening (ed.), Univ. of Arizona Press, Tucson, pp. 85-163.
Delsemme, A.H. (1997). In *Comets and the Origin and Evolution of Life*. P. J. Thomas, *et al.* (eds.), Springer-Verlag, New York, pp. 29-67.
Delsemme, A.H. 1999. *Planet. & Space Sci.* **47**, 125-131.
Despois, D. 1999. *Earth, Moon, and Planets* **79**, 103-124.
Eberhardt, P., Reber, M., Krankowsky, D., and Hodges, R.R. 1995. *Astron. & Astrophys.* **302**, 301.
Enzian, A., Cabot, H., and Klinger, J. 1997. *Astron. & Astrophys.* **319**, 995.
Gensheimer, P.D., Mauersberger, R., and Wilson, T.L. 1996. *Astron. & Astrophys.* **314**, 281.
Greenberg, J.M. 1982. In *Comets*. L.L. Wilkening (ed.), Univ. of Arizona Press, Tucson, 131.
Greenberg, J.M. and Hague, J.I. 1990. *Astrophys. J.* **361**, 260.
Grim, R.J.A. and Greenberg, J.M. 1987. *Astrophys. J.* **181**, 155.
Guen, E. and Jessberger, E. K. (1990). In *Physics and Chemistry of Comets*. W. F. Huebner (ed.), Springer-Verlag, New York, pp. 113-176.
Hagemann, R., Nief, G., and Roth, E. 1970. *Tellus* **20**, 712.
Hayward, T.L. and Hanner, M.S. 1998. *Science* **275**, 1907.
Huebner, W.F. and Boice, D.C. (1997). In *Comets and the Origin and Evolution of Life*. P. J. Thomas *et al.* (eds.), Springer Verlag, New York, pp. 111-129.
Irvine, W.M. 1996. *Nature* **383**, 418.
Irvine, W. M., Schloerb, F.P., Crovisier, J., Fegley, B., and Mumma, M. (2000). In *Protostars and Planets IV*, V. Mannings, *et al.* (eds), Univ. Arizona Press, Tucson, in press.
Kim, S.J. and A'Hearn, M.F. 1991. *Icarus* **90**, 79.
Kim, S.J., Bockelee-Morvan, D., Crovisier, J., and Biver, N. 1999. *Earth, Moon, and Planets* **78**, 65.

Kissel, J., Krueger, F.R., and Roessler, K. (1997). In *Comets and the Origin and Evolution of Life*. P.J. Thomas *et al.* (eds.), Springer Verlag, New York, pp. 70-109.
Lis, D.C., *et al.* 1999. *Earth, Moon, and Planets* **78,** 13.
Lunine, J.I. (1999). *Earth: Evolution of a Habitable World*, Cambridge University Press, Cambridge, UK.
Lunine, J.I., Engel, S., Rizk, B., and Horanyi, M. 1991. *Icarus* **94**, 333.
Matthews, H.E., *et al.* 1997. IAU Circular 6353.
Meier, R., *et al.* 1998. *Science* **279**, 842.
Millar, T.J., Bennett, A., and Herbst, E. 1989. *Astrophys. J.* **340**, 906.
Moore, M.H., Donn, B., and Hudson, R.L. 1988. *Icarus* **74**, 399.
Mumma, M.J., Weissman, P.R., and Stern, S.A. 1993. In *Protostars and Planets III*, 1177.
Mumma, M.J., *et al.* 1996. *Science* **272**, 1310.
Notesco, G., Laufer, D., and Bar-Nun, A. 1997. *Icarus* **125**, 41.
Oró, J. and Lazcano, A. (1997). In *Comets and the Origin and Evolution of Life*. P. J. Thomas *et al.* (eds.), Springer-Verlag, New York, pp. 3-27.
Owen, T., Lutz, B.L., and De Bergh, C. 1986. *Nature* **320**, 224.
Owen, T. and Bar-Nun, A. 1995. *Icarus* **116,** 215.
Owen, T., *et al.* 1999. *Nature*, **402**, pp. 269-270.
Russel, C.T., Luhmann, J.G., and Baker, D.N. 1987. *Geophys. Res. Lett.* **14,** 991.
Schleicher, D.G., Millis, R.L., Osip, D.J., and Lederer, S.M. 1996. *Bull. Amer.Astron. Soc.* 28, No.3, 1089.
Schleicher, D.G., Lederer, S.M., Millis, R.L., and Farnham, T.L.1997. *Science* **275**, 1913.
Shalabiea, O.M., Greenberg, J.M. 1994. *Astron. & Astrophys.* **290**, 266.
Smoluchowski, R. (1985). In *Ices in the Solar System* J. Klinger *et al.* (eds.), Dordrecht:Reidel, 397.
Vanysek, V. 1991. In *Comets in the Pos-Halley Era*, vol. 2. Newburn *et al.* (eds) Dordrecht:Kluwer, 879.
Waters, L.B.F.M., *et al.* 1996. *Astron. & Astrophys.* **315**, L361.
Weaver, H.A., *et al.* 1996. *Bull. Amer. Astron. Soc.* **188** #62.01.
Weaver, H.A., *et al.* 1999. *Earth, Moon, and Planets* **78**, 71-80.
Wooden, D.H., Harker, D.E., Woodward, C.E., Kioke, C., and Butner, H.M. 1999. *Earth, Moon, and Planets* **78**, 285.

Section 5.
Origins of cognitive systems

INFORMATION, LIFE AND BRAINS

JUAN G. ROEDERER
Geophysical Institute
University of Alaska-Fairbanks
Fairbanks, AK 99775, USA

1. Life and Information

As we look around us we realize that much of the human environment appears to be made up of discrete clumps of matter, objects with clearly defined, sharp boundaries. It is our experience from daily life (and from precise observations in the laboratory) that the presence of one object may alter the state of other objects in some well-defined ways. We call this process an "*interaction*", and consider it one of the fundamental, primary concepts, like space, time and measurement, with which science, and physics in particular, works.

Consider the two examples of interactions sketched in *Fig. 1a*. On the left we show several satellites orbiting the Earth, on the right we show insects "in orbit" around a light bulb. Both have in common some well-defined patterns that are followed at a regular, well-defined pace. The motion of a satellite is governed by the force of gravitational interaction *f*, which is a function of position and masses—and nothing else. The motion of an insect is governed by the force of propulsion imparted by the wings, which is controlled by a sensory system with a complicated mechanism of light detection and pattern recognition—a process that involves *information acquisition and processing*. In other words, we have the chain: light emission → detection → pattern analysis → muscle activation. No such set of algorithms appears in the case depicted on the left side: the concept of information is totally alien to gravitational interactions—and to *all* purely physical interactions between inanimate objects. The latter just "happen": they don't require intermediate operations of information detection and processing. A second pair of examples is shown in

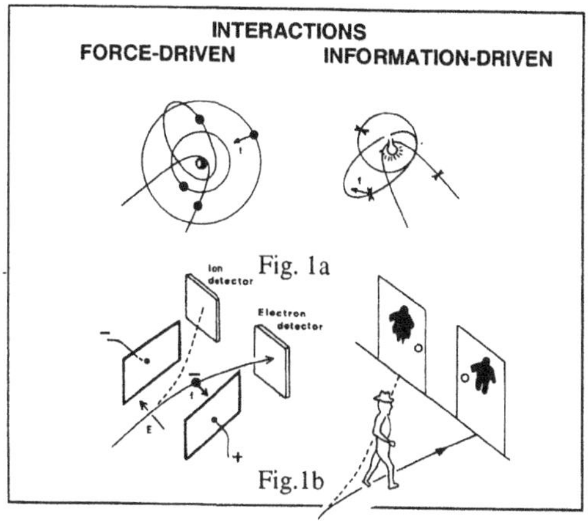

Fig. 1a

Fig.1b

J. Chela-Flores et al. (eds.), Astrobiology, 179–194.

Fig. 1b. At left we have a schematic view of an "electric charge separator" in a particle detector, on the right, a "sex separator".

We call the interactions between inanimate objects "*force-field driven interactions*", and those between a living object and other objects (living or not) "*information-driven interactions*". The latter, of course, ultimately are coupled to physical interactions; the key aspect, however, is their *control* by information processing operations. While in the former the energy involved is tapped from the potential energy reservoir of the interactive force field, in the latter the energy is provided by a reservoir *external* to the interaction mechanism. Let us summarize: inanimate interactions respond to force fields, interactions of living things are controlled by information fields. We offer the following definition: *a life system is a natural (not-man-made) system exhibiting interactions that are controlled by information processing operations* (Roederer, 1978) (the proviso in parentheses is there to emphasize the exclusion of computers and robots!). Information is thus the defining concept that separates the living world from the inanimate world[1]. Note that in the two examples of *Fig. 1* involving living organisms, the electromagnetic waves (light) themselves do not drive the interactions—it is the *information content* in the patterns of the wave trains, not their energy that plays the controlling role. Quite generally, an information-driven interaction behaves like a *trigger* of physical/chemical processes, but does not participate in them.

Let us turn to the scene sketched in *Fig. 2*, and group the depicted objects into different classes, according to their origin and, where applicable, purpose. We begin with the snow-capped mountain, a truly inanimate object unaffected by life (if we disregard biologically generated minerals). It appeared and grew through the action of tectonic forces and was further shaped by weather erosion. No information processing intervened in its emergence; as time goes on, the mountain will gradually disintegrate in an irreversible way—its organization decays and its entropy increases.

Fig. 2

[1] Science in general defines information as "a statement that describes the outcome of expected alternatives". This is the definition of what is usually called "semantic information". The statement itself that represents the information in question may be given in numerical form, digital when the number of alternatives is finite (e.g., the faces of a die) or real numbers when there is an infinite number of alternatives (e.g., the coordinates of a mass point). The numerical form is used in communication theory, which introduces a mathematical expression for the *amount* of information: for n equally probable alternatives, knowledge of the outcome represents $\log_2 n$ "bits" (binary digits) of information (the outcome of a coin toss thus represents 1 bit of information). Note that for information to have a concrete scientific meaning, it *must* refer to a given, identifiable set of alternatives or choices (even if infinite in number).

Next we consider the tree, the bird, the dog and the person, all living organisms. As long as they are alive, they do not disintegrate; moreover, in general they grow following lines of *increasing* organization. They maintain themselves based on interactions involving information processing in its multiple forms (a growing leaf that follows blueprints in the genes, an ameba moving in response to changes in saline concentration, an animal recognizing food or avoiding danger, etc.). All living organisms are the product of *evolution*, a long-term process of mutation, multiplication, natural selection and diversification. In this process, information about the environment is gradually incorporated in the genetic memory structures of the organism: the genetic blueprints in the DNA change, enhancing the species' performance and survival probability in a gradually changing environment.

Finally, in *Fig. 2* we have the nest, the house, the newspaper and the cruise missile. They are all inanimate *products* of life systems: they are the result of some quite specific information processing, they all carry information, and they all serve a specific purpose [2]. But there are fundamental differences among them. The nest is an instinctive, "automatic" result of a blueprint imprinted during the long process of evolution in the genes of the bird and "hardwired" into its brain; the house is the result of a blueprint conceived "on the spur of the moment" by a human brain that had a vision of the future—a *plan* (what is imprinted in the human genes is the *capability* of making blueprints) [3]. The newspaper represents a *repository* of information deliberately planned for later use by other humans [4]; and the cruise missile is a device deliberately constructed by a human with information processing capabilities (recognition of terrain and target) for its ultimate mission.

To understand the unique role of information in biological systems we must clarify its role in physics. Furthermore, to understand the unique role of "future time" in human brain function we also must clarify in general terms the role of time in physics, which appears intertwined with that of information. This is the purpose of the next Section.

---

[2] Note that nothing in the non-living world "serves a specific purpose" (if something does—say, a cave in the mountain—it is because a living being has found it to be useful for its own survival, or because we, humans, interpret it this way!).

[3] Note that a bird from a given species will always build the *same* nest (only slightly modified to adapt it to the immediate environment) whereas a human can develop an unlimited variety of blueprints for a house and change them at will before it is actually built.

[4] Animals, of course, also leave "imprints " of information in the environment for some future use (see the dog in Fig. 2!). Yet there is a crucial difference. If we were to suddenly destroy all animal imprints (tracks, nests, food reserves, etc.), animal life would still continue essentially as it was today: the key information for survival and preservation is genetically stored in the organism, and relics of animal "cultures" are not used by later generations as a source of information. But if we were to destroy at once all human imprints (our "externalized" memory storage), civilization as we know it would disappear and future generations would be left in a state similar to where humanity was tens of thousands of years ago.

2. Physics, Information and Time

Classical physics starts with the formulation of basic laws for bodies idealized as "*mass points*" (i.e., it works with *models* of real bodies whose dimensions are negligible compared to the distances between them). One particular feature of classical dynamics of interacting mass points is that knowledge of the positions and velocities at a given time t_0 (the "initial conditions") *completely determines* the position and velocities of the mass points for *all* times $t > t_0$ and $t < t_0$. All fundamental equations remain unchanged if the time t is run backwards: a movie of interacting mass points run backwards in time represents the correct dynamic picture that would ensue if all velocities of the mass points were to be reversed at a given instant of time. In other words, the system is fully deterministic and "invariant with respect to time reversals". This means that there is no qualitative distinction between past and future. The concept of "cause and effect relationship" does not enter into the picture of basic physics in that we can think of cause and effect going either way: we can deduce what *will* happen at $t > 0$ from knowledge of what happens at $t = 0$, but we can also deduce what *did* happen at $t < 0$ from that same knowledge. This state of affairs is even true for the atomic domain (with the exception of nuclear decay processes): quantum mechanics is both deterministic (in the probabilistic sense, though) and invariant with respect to time reversals.

The bodies with which we deal in daily life are not individual mass points, but assemblages of 10^{23}-plus molecules. Our senses respond to macroscopic features, and in order to describe the dynamics of macroscopic bodies (solids or fluids), we must work with macroscopic variables such as temperature, density, pressure, internal energy and entropy, which are statistical averages over huge ensembles of particles. This is the realm of thermodynamics and, when explicitly expressed in terms of the behavior of the constituent particles, statistical mechanics. In these disciplines, like in real life, the indistinguishability of future and past time disappears: heat can only flow in *one* direction from a hot to a cold body, but not in reverse; a gas can only diffuse toward regions of lower concentration; the entropy of a closed system can never decrease, etc. Films of macroscopic happenings run in reverse look "funny": the pieces of a broken cup assembling together on the floor and jumping up into your hand, a cloud of water vapor being sucked back into the kettle, etc.! The familiar *arrow of time* appears, time-ordered cause and effect relationships can now be clearly identified, and the future is fundamentally different from the past.

Why is this so, despite the fact that macroscopic bodies are made up of mass points, which, if contemplated individually, behave in the "time-reversible" way of basic physics? For instance, let us consider a perfect gas enclosed in a volume V_1 and let it expand by opening a valve into an empty vessel of volume V_2. Once equilibrium has been reached, the gas occupies a volume V_1+V_2 and the change of entropy ΔS of the system (assumed thermally isolated) is $\Delta S = n\,\mathrm{k}\ln[(V_1+V_2)/V_1] > 0$ [5]; this increase tells us that the process is irreversible, that it cannot proceed by itself in the opposite direction, and that therefore it defines a unique direction of time. By opening the valve,

[5] n: total number of molecules; k: Boltzmann constant (= 1.38×10^{-23} Joule °K^{-1}).

we have suddenly created an unstable initial condition with *reduced* uncertainty about the microscopic state (we know that all molecules are in V_1 and *not* in V_2), but which is highly *unlikely* to occur all by itself (when the valve is open). The final state is one of increased uncertainty (a given molecule could be *anywhere* in V_1+V_2) but highest likelihood (equilibrium). The direction of time is thus given by the natural evolution of a thermodynamic system toward a microscopically more uncertain but macroscopically more probable state.

But now consider this: if after the expansion we were to reverse the velocities of *all* molecules at some instant *t*, they would retrace *exactly* their paths and assemble again in the original, smaller, volume V_1! Leaving aside the technical impossibility of reversing the velocities with present experimental means, what's wrong here? Nothing—provided we clearly understand what it means "to reverse the velocities of all molecules"! First of all, we note that we would obtain a very exceptional initial condition for all 10^{23}-plus molecules. There is an infinite number of possible positions and velocities of the molecules (i. e., microscopic states) compatible with one and the same macroscopic state of a gas (described by its pressure, volume and temperature). Among all these possible microscopic states there is, indeed, a tiny subset of "bad" substates (still infinite in number!) that do lead to a "funny" macroscopic behavior (like the gas getting "sucked" into the smaller vessel of volume V_1). But the probability of occurrence of such "bad" states is so small that we would have to wait very long (Poincaré's "recurrence time"—longer than the age of the Universe!) to see them arise naturally. Still, this shows us that, in a sense, the direction of time as determined by the course of irreversible processes does have a certain "statistical flavor"—fluctuations of macroscopic behavior are not prohibited, they are just awesomely rare!

There is an "ultimate reason" why most macroscopic processes are irreversible, thus introducing a fundamental physical difference between future time and past time. Let us take the hypothetical case of a system of molecules whose initial conditions (the initial microscopic state) are *completely known* (zero entropy!). We said that since molecular interactions obey basic physical laws, we should be able to predict all future positions and velocities exactly—the system (it's microscopic state) should remain "completely known" at all times. Leaving aside all practical difficulties involved, the main problem with this is that all particles are subjected to *unavoidable* interactions with "what is outside". For a gas in a vessel, it will be collisions with molecules of the walls; for a gas cloud in outer space there are no walls but there still remain the gravitational interactions—however weak—with nearby planets and stars (Layzer, 1975). All these *random perturbations* "from the Cosmos" will gradually destroy any information we initially had on the system—the uncertainty about the system will increase, and so will its entropy until it reaches the traditional equilibrium value.

Now we are in a better position to understand why all cause-and-effect relationships in the macroscopic world define a common direction of time. Consider two events *A* and *B*, of which *A* is said to be the cause of *B*, e.g., a golfer striking a ball on the putting green, and the ball rolling into the hole. Golf player, club, ball and the green are all made of molecules, each one of which follows basic physical laws that are strictly time-reversible. Why, then, is the macroscopic cause-and-effect sequence not reversible, too? In the macroscopic world, any cause A will have a *multiplicity* of other effects C, D, E, ... besides B. In the case of the golfer, the ball, while rolling on the grass, will bend the blades of the grass, thereby doing mechanical work and slowing down. To conform with basic physical laws in the microscopic, molecular picture, an exact time reversal of a macroscopic process would require that not only B but *all* the concomitant events C, D, E, ... *also* occur in time-reversed order. In our example, not only would the ball have to jump out of the hole, but the molecules of each grass blade would have to move cooperatively in such a way as to make it "unbend" at the appropriate time to impart the appropriate kick to the ball and accelerate it toward the golfer. The occurrence of all these conditions by chance (fluctuation), while not prohibited, is very, very unlikely to the point it has never been observed—this is why a film shown in reverse looks "funny" and why, again, in the macroscopic world there is a "preferential" direction of time!

The so-called Gibbs' paradox offers a good view of the association between entropy and information. Suppose that we have two vessels, each of volume V filled with n molecules of the same gas at the same pressure p. Both vessels are connected by a tube with a valve, initially closed. The total entropy of the system will be (see above) $S = S_1 + S_2 = 2\, n\, \mathrm{k} \ln V$. When we open the valve, nothing "macroscopic" is expected to happen and the entropy of the system of course will remain constant. Yet we do know that, from the microscopic point of view, the molecules in vessel 1 will expand into vessel 2, and vice versa. Each process, contemplated in isolation, would imply a change of entropy from $n\, \mathrm{k} \ln V$ to $n\, \mathrm{k} \ln 2V$; the total entropy would thus change by $\Delta S = 2\, n\, \mathrm{k} \ln 2$—which thermodynamically is indeed nonsense, because "nothing has happened" from the macroscopic point of view by opening the valve! This is, in essence, Gibbs' paradox. It is resolved by realizing that in the microscopic picture (which led to the apparently nonsensical entropy change), we have, albeit only mentally, *tagged* all molecules (as "initially in vessel 1" or "not in vessel 1"), but have *lost* these tags once the molecules from both vessels start getting mixed (because we cannot follow their individual fate). Erasing the tag represents a one-bit informational process for each molecule; such a procedure *increases* the uncertainty about the initial system, hence increases its entropy! We can reconcile the microscopic view with macroscopic thermodynamics if we assume that the erasing operation increases the entropy of the initial system by $2\, n\, \mathrm{k} \ln 2$, or $\delta S = \mathrm{k} \ln 2$ *per bit of lost information* [6] (there are $2n$ molecules). The total entropy change stands now at zero, as it should [7]!

In the transition from "just a few" interacting mass points to "very many", somehow the concept of information (or loss thereof) has crept in! In the examples with gas

---

[6] The expression $\delta S = -\,\mathrm{k} \ln 2$ is well-known in communication theory as the *entropy decrease per bit of gained information*.

[7] Gibbs' paradox illustrates the fickle character of the entropy concept: in reality, we must distinguish at least two classes of entropy: one is defined as a function of the macroscopic variables exclusively (Clausius' approach), the other one is defined on the basis of the microstates (Boltzmann's approach). Both are equal only for an *equilibrium* case. The second approach requires explicit accounting for any information gained or lost, a fact that lies at the center of the difficulty of defining the entropy of a living, information-driven organism!

molecules discussed above, it seems as if information is playing a controlling role in how a thermodynamic system behaves. A loss of information is always accompanied by a gain in entropy and vice versa. Does this mean that our definition of "life" in Section 1 is no longer valid because the role of information has lost its exclusiveness? This is not so, because in our definition of life, information controls the *interactions per se*. In a thermodynamic system, "information", whenever it appears, relates to the observer, experimenter or thinker, *not* to the system per se. It does not influence the system, whose molecules do what they are supposed to do according to basic physical laws in which information plays no role. What does influence the behavior of a system is *what we humans do to it*, such as deliberately changing the initial conditions of its microscopic state without altering the macroscopic one (e.g., tagging molecules, or reversing molecular velocities). It is *us humans* who prepare a system, set up its boundaries and initial conditions—even if only in thought experiments. And it is those human-induced actions that condition the response of a physical, inanimate thermodynamic system, whether in reality or only in thought: in Maxwell's demon paradox, the demon (or its robot surrogate (Leff and Rex, 1990)) is placed there *by a human*, it has no chance of appearing naturally! In thermodynamics, the concept of information appears because we humans imagine or fabricate *unnatural* situations. In life systems, instead, the concept of information is an active participant in their *natural* evolution and behavior.

3. Evolution of the Animal Brain, Instincts and Neural Codes

The formation of large, complex organic molecules was the primary condition for the emergence of life on Earth; the early ocean-atmosphere system provided the appropriate medium. Chemical reactions are governed by physical interactions in the quantum domain, in which, according to our previous discussion, "information as such" should play no role—these interactions just "happen". In the process of molecular synthesis, however, large and complex polymer-like macromolecules emerged as *code-carrying templates* whose effect on other molecules in their environment is to bind them through a catalysis-like process into conglomerates according to patterns represented in the code. Some of these polymers might also have served as templates for the production of others like themselves—those more efficient in this process would multiply. Replication and natural selection may indeed have begun already in a pre-biotic chemical environment. Concerning these macromolecules, we can say that beyond a certain degree of complexity, *information as such* (on the existence of certain patterns, like the code in the sequence of nucleotides in RNA) begins to play the decisive role in organizing the chemical environment. It is because of the supremacy of information as the controlling agent that these molecules can catalyze chemical reactions that would be highly improbable to occur naturally under the same environmental conditions and energy sources. In summary, the perhaps *first "grand moment"* in the evolution of life occurred when molecules of sufficiently high complexity appeared, for which the information expressed in their structural patterns began to take a highly selective control over their interaction with the surrounding medium [8].

Another "*grand moment*" in the chemical evolution of life is the appearance of topologically closed two-dimensional molecular structures or *membranes* separating distinct regions of space from the rest to keep groups of interacting entities (molecules, multi-molecular bodies) together through time. Information-driven interactions and transmission of information from one place to another (communication) inside the shielded spaces make it possible to move and expend energy to create order and to prevent an increase in entropy. With the emergence of closed membranes in conjunction with other fundamental processes such as *reproduction*, the first living organisms appeared in the form of *prokaryotic cells* over 3 billion years ago, followed by the appearance of oxygen-producing photosynthesis in blue-green algae, and the emergence of the *eukaryotic cell* between 2 and 1.4 b.y.a.

The first multicellular organisms appeared about 670 m.y.a.—another of the "*grand moments*" in the evolution of life. Vascular plants appeared about 440 and the first land animals about 400 m.y.a. With the appearance of *locomotion*, especially in land animals, the number of relevant environmental variables affecting the organism increased drastically, with time-scales of change down to a fraction of a second. Sophisticated sensory organs developed, and with them, nervous systems. What started out as a simple environmental signal conversion and transmission apparatus evolved into the central nervous system of the higher vertebrates, with sophisticated input information analysis and behavioral response-setting capabilities. The *brain* emerged as the "central processor" to carry out the fundamental operations of sensory information processing, cognitive operations of object recognition and environmental representation, and the planning of motor response based on the momentary state of the environment, on the state of the organism, on innate instructions (instincts) and on learned information (another *"grand moment"* in evolution!).

In its evolution, the animal brain developed in a way quite different from the development of other organs of the vertebrate body. Separate layers appeared, with distinct functions, *overgrowing* the older structures but not replacing them, thus preserving "older functions"—the "hard-wired" memories or *instincts* that a species has acquired during evolution. The outermost layer is the cerebral *cortex*, which executes all higher order cognitive operations. These functions are, listed here in oversimplified form: to (i) analyze the information received from the senses and the detectors that monitor the posture of the body in the environment; (ii) sort out that which is of relevance for the organism's well-being and intentions; (iii) construct new or improved "mental maps" or representations of the surrounding space; (iv) determine and store in memory relevant cause-and-effect relationships; and (v) activate the musculature to elicit the most appropriate behavioral response. The human brain, of course, has some

[8] No doubt that we are dealing here with a "threshold" period in the evolutionary schedule in which a transition takes place to information-driven interactions. It is interesting to note that in the molecular domain this transition is conceptually similar to the boundary that separates the domains of quantum and classical behavior of matter, the so-called region of wave function de-coherence.

very distinct additional capabilities. In this Section we will discuss some aspects common to animal and human brain function (e.g., Roederer, 1995, and references therein).

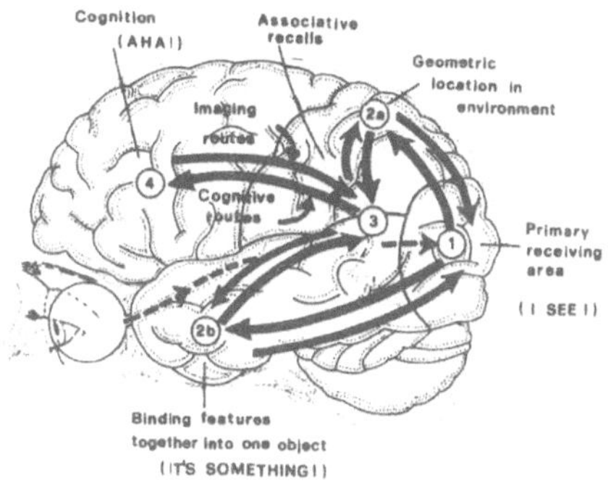

Fig. 3

We first turn briefly to perception. The most refined senses are vision and audition and we will refer mostly to these in what follows. The neural circuitry in the periphery and afferent pathways up to and including the so-called primary sensory receiving area of the cortex (*Fig. 3*) carries out some basic preprocessing operations mostly related to *feature detection* (e.g., detection of edges and motion in vision, spectral pitch and transients in hearing). Recordings of individual neuron activity with microelectrodes have shown that in the primary areas one indeed finds "feature detectors", cells that respond only to well-defined patterns in the original sensorial stimulus (e.g., dark or light bars inclined at a certain angle, a blue dot moving in a certain direction; the "meow detectors" in a cat's auditory cortex). The next stage is *feature integration* or *binding*, needed to sort out from an incredibly complex input those features that belong to one and the same spatial or temporal object (i.e., binding those edges together that define the boundary of the object; sorting out those resonance regions on the basilar membrane that belong to one musical tone). At this stage the brain "knows" that it is dealing with an object, but it does not yet know *what* the object is. This requires a complex process of comparison with existing, previously acquired information. The recognition process can be "automatic" (associative recall) or require a further analysis of the full sensory input in the frontal lobes. As one moves up the stages of *Fig. 3*, the information processing becomes less automatic and more centrally controlled; more *motivation-controlled* actions and decisions are necessary, and increasingly the previously stored (learned) information will influence the outcome.

The motivational control deserves special attention. One of the lower, phylogenetically much older parts of the vertebrate brain is the so-called *limbic system*. It comprises several subcortical structures including the amygdala, the ventral tegumental area or VTA, and the hippocampus. In conjunction with the hypothalamus (the region of the brain that receives and integrates signals from the autonomic nervous system and regulates the neuro-chemical hormonal information system), the limbic system "polices" sensory input, selectively directs memory storage according to the relevance of the information and mobilizes motor output. In these tasks, it communicates directly with the prefrontal regions of the cortex. The aim of this system is to ensure a behavioral response that is most beneficial to the organism according to genetically acquired information—the so called *instincts* and *drives*. Motivation and emotion are integral manifestations of the limbic system's guiding principle to assure that all cortical processes are carried out to maximum benefit of the organism and the propagation of the species [9]. The limbic system constantly challenges the brain to find solutions to alternatives, to probe the environment, and to perform certain actions even if not needed at that moment (e.g., animal play for the purpose of training in skilled movement). It also motivates the brain to avoid danger.

To carry out its functions the limbic system works in a curious way by dispensing sensations of reward or punishment; pleasure or pain; love or anger; happiness or sadness or fear [10]. Of course, only we humans can report to each other on these feelings, but on the basis of behavioral and neurophysiological studies we have every reason to believe that higher vertebrates experience similar feelings. What kind of evolutionary advantage was there to this mode of operation? Why does pain hurt? Why do we feel pleasure scratching a mosquito bite or eating chocolate? How would we program similar reactions into a robot? [11] Obviously this has to do with evoking an *anticipation* of pain or pleasure whenever certain constellations of environmental events are expected to lead to something detrimental or favorable to the body, respectively. Since such anticipation comes *before* any actual harm or benefit could arise, it helps guide the organism's behavior in the direction of maximum chance of survival. In short, the limbic system directs a brain to *want* to survive and to find out the best way of doing so given current, genetically unprogrammable circumstances. Plants cannot respond quickly and plants do not exhibit emotions; their defenses (spines, poisons) or insect-attracting charms (colors, scents) develop only through the long process of evolution.

How does the brain work? The new non-invasive techniques of neural activity imaging (functional nuclear magnetic resonance or fNMR, positron emission tomography or PET) are providing a wealth of information. For instance, observations clearly confirm the processing stages sketched in *Fig. 3*, and, moreover, demonstrate that they do unfold exactly in *reverse* order during the process of mental imaging (imagining things—more on this later). Tomographic techniques detect neural activity only indirectly by mapping brain regions that show enhanced metabolic interchange. Therefore, they cannot provide a detailed picture of what goes on at the neural network level; in this quest one must resort to invasive techniques of recording the electrical activity of individual neurons or groups of neurons in laboratory animals or during neurosurgery in humans. Such measurements are yielding data on how the neural

[9] The functions of the limbic system are sometimes referred to as "the four F's": Feeding, Fighting, Fleeing and F... reproducing!

[10] Experiments with mice have shown that electrical stimulation of the VTA causes such a sensation of pleasure that animals taught to self-stimulate will do so continuously until they die of exhaustion, even if given the alternative of obtaining food. Conversely, electrical stimulation of the amygdaloid body can instantly induce fear and convert a gentle monkey into a raging beast. The intimate neural integration with the hippocampus, which is the part of the brain involved in long-term memory storage, reaffirms the role of the limbic system as a controller of the flow and storage of somatically relevant information.

[11] We could program a robot to emit a crying sound whenever it loses a part, to tighten loose screws, or to seek an electrical outlet whenever its batteries are running low, but how do we make it feel "pain" or "pleasure"?

circuitry works at the "microscopic", neuron-to-neuron level. Still, many difficulties remain to be resolved, especially in the attempts to match results obtained with single cell recordings (neurophysiology) with those obtained on global brain function (neuropsychology).

The picture that is emerging is summarized in the following paragraphs (e.g., Roederer, 1995, and references therein). Sensory information, or *any* other kind of information, is encoded in the brain in the form of a specific spatio-temporal distribution of electrical impulses [12]. The mental representation or image of an object (visual, acoustic, olfactory, tactile) appears in the form of a specific distribution of electrical signals in the neural network of the cerebral cortex that is in *one-to-one correspondence* with the specific features sensed during the perception of this object. By "one-to-one" we do not mean a "geometric" or isomorphic correspondence but, rather, a distribution which, however complex, is always the *same* (within limits) whenever information on that particular object is involved in brain processing. According to this result, "cognition" is nothing else but the occurrence of a specific neural activity in certain areas of the cortex that is in one-to-one correspondence with the object that is being recognized, remembered or imagined. For instance, the distributions of neural activity displayed on the intervening cortical areas by the perception of the following things—a big red apple, an apple tree, a piece of an apple pie—though widely different, would all bear in common some subset of neural activity, namely the one that appears in correspondence with, and defines the cognition of "apple". Every time a dog hears or sees the face of its master, no matter under what circumstances, some unique distribution of neural activity will occur in its brain that is specific to the dog's recognition of its master.

It takes some time to get used to the meaning and relevance of this "specific spatio-temporal activity distribution" and the information it represents. We could describe such a distribution mathematically with a function $\psi(\boldsymbol{r}, t)$, representing the neural activity (e.g., neural firing rate) at point $\boldsymbol{r}$ in the brain tissue at the time t. Unfortunately, this function would be completely *discontinuous*: two neighboring neurons may participate in totally different tasks, and thus fire completely unrelated electrical impulses. We could use a vector function $\psi = [f_1(t), f_2(t), \ldots\ldots f_m(t), \ldots\ldots f_n(t)]$, where $f_m(t)$ is the firing rate of the m-th neuron in an ensemble of n cells. For a given distribution, $f_m(t)$ would be zero if the corresponding neuron does not participate in the particular cognitive task represented by ψ (even if it is firing); on the other hand, there are many neurons which participate in several processing tasks at the same time. Another, even more frustrating problem is that in each cognitive task there may be hundreds of millions if not billions of neurons involved! In any case, if we *could* determine ψ, there would be *one specific* vector function for each mental image (and *that* number is, of course, beyond comprehension!) and a typical duration of a triggered imaging event would be only 50-200 ms. Although it seems hopeless to use this description in a

---

[12] It is important to point out that there is also a *neurochemical* information system (brain peptides) which regulates the general state of the brain and organism (a sort of "volume control" of general brain activity). This system is "slow" and only *transmits*, but does not process, information.

mathematical sense, it does help organize one's ideas about the way information is encoded in the brain. In particular, it helps understand the *categorical* and *hierarchical* way in which information is treated in brain processing: the representation (ψ) of more complex concepts can be thought of as the sum of the representations of its simpler parts (e.g., apple orchard $\Rightarrow$ apple tree $\Rightarrow$ apple all having ψ_{apple} in common). This is precisely how memory storage and recall works.

The act of remembering, or memory recall, consists of the re-elicitation or "replay" of that particular distribution of neural activity which is specific to the object or concept that is being remembered. This pattern can be triggered by external sensory information: when a dog looking at a group of people suddenly recognizes his master, the corresponding "ψ_{master}" was triggered. A memory recall can also be triggered internally by the perception (or, for humans, imagination) of *correlated* events: a hungry feeling may trigger ψ_{master} in that dog's brain and he may run to his master to beg for food. On the other hand, during a given learning phase the association areas of the cortex and the frontal lobes display patterns of neural activity triggered by a series of specific stimulus constellations arriving simultaneously through different channels (for instance vision and hearing, when a dog is taught to use its doghouse). As this happens repeatedly, changes occur in the connections between neurons, the *synapses* [13]. In particular, new synapses are established between neurons that fire simultaneously during the learning process in such a way that *after* the learning procedure, the stimulus constellation in only one input channel (the conditioned stimulus or "key word") will suffice to trigger the full pattern of neural activity specific to the entire original event ("associative recall"). These examples show that one must recognize the neural activity distribution as the fundamental "physical quantity" that represents "information" in the working brain.

In other words, learning is not represented by the storage of information or images per se, but by appropriate modifications of the information-processing network, and memory emerges as the "storage" of information-processing *instructions*. This is why the brain is called a "self-organizing" system. There is no software: memory, instructions, and operations are all based on appropriate changes in the hardware (the architecture and efficiency of synaptic connections between neurons). This kind of self-organization is also the principle on which today's "neural computers" operate . The study of computer-simulated neural networks is leading to improved understanding of neural networks in the brain. For instance, one finds that a pattern recognition network is capable of correctly recognizing *incomplete* patterns that have not been fed in at all during the training process. This ability of interpreting novel (unlearned) situations is a fundamental property of brain function, and plays a crucial role in the human brain (see next Section).

[13] Recently, synaptic growth has been observed with optical microscopy *while it was happening*!

4. Human Brain, the Concept of Future Time, and Intelligence

Another "*grand moment*"—the last and grandest of all on Earth—came over two million years ago when the evolution of the *human* brain began. Interestingly, from the neurophysiological and neuroarchitectonic points of view the human brain today is not particularly different from the brain of a primate like the chimpanzee. It just has a cortex with larger prefrontal lobes, more neurons and some of the intercortical fasciculae have more fibers. But the difference in numbers is barely one order of magnitude—except for the number of synapses in the adult brain, which in humans is several orders of magnitude higher. Is the difference in information handling capabilities only one of quantity but not one of substance? Many scientists (but not all!) believe that there are some very *basic* differences between primate brain function and human brain function.

Aristotle already recognized that "animals have memory and are able of instruction, but no other animal except man can recall the past at will". More recently, J. Z. Young (1987) stated this in the following terms: "Humans have capacity to rearrange the "facts" that have been learned so as to show their relations and relevance to many aspects of events in the world with which they seem at first to have no connection". More specifically, the most fundamentally distinct operation that the human, and only the human, brain can perform is to recall stored information, images or representations (i.e. ψ's), manipulate them, and re-store modified or amended versions thereof *without any concurrent external sensory input*. The acts of information recall, alteration and re-storage without any external input represent the *human thinking process* or *reasoning*. Intimately related to this is the emergence of *self-consciousness*, i.e., feeling totally in control of one's own brain function and free-will decisions [14].

All the aforesaid had vast consequences. The capability of re-examining, rearranging and altering images led to the discovery of previously overlooked cause-and-effect relationships (creation of *new* information!), to a quantitative concept of elapsed time, and to the awareness of future time. Along came the possibility of *long-term prediction* and *planning* [15], the postponement of behavioral goals and the

---

[14] The reader may notice that nowhere do we mention the famous "brain-mind problem". Most neuroscientists today consider this "problem" as dead: there are no scientific grounds for having to assume the existence of *two* separate entities!

[15] Experiments with conditioned reflexes to carefully timed sequences of cause-and-effect relationships show that animals have a very short "window of time" for prediction-making (a few tens of seconds). This seems to contradict many observations of animal behavior: one may say that "animal hunting strategies involve sophisticated long term planning", or that "my dog knows very well when I am expected to come home tonight". However, hunting follows instincts and memory of past experience and the decisions and actions proceed as time goes on—there is no evidence that a lion decides *today* to follow a different strategy *tomorrow*. And the dog "knows" the time of his master's return using momentaneous cues such as its internal clock, darkness and hunger. It may well be that higher primates have "bursts" of self-consciousness during which internally recalled images are manipulated and a longer-term future is briefly "illuminated". But there is no clear and convincing evidence that any outcome is stored in memory for later use. In other words, it is conceivable that some higher mammals

capability to overrule the dictates of the limbic system (the instincts): the body started serving the brain instead of the other way around. Mental images could thus be created that had no relationship with sensory input; abstract thinking and artistic creativity began; this also brought the development of beliefs and values. Concurrently came the ability to encode complex mental images into specific auditory utterances (concurrent with the development of a more sophisticated muscular control of the larynx) and the emergence of *human language*. This was of such decisive importance to the development of human intelligence that certain parts of the auditory and motor cortices began to specialize in verbal image coding (in the left cerebral hemisphere in 97% of all persons), and the human thinking process began to be controlled by the language networks [16]. Finally, though much later in human evolution, came the deliberate storage of information in the environment; this externalization of memory led to the documentation of feelings and events through written language, visual artistic expression, and science—to human culture as such.

In Section 1 we briefly alluded to genetic blueprints and the blueprints of a house. Genetic blueprints are the result of a long process of evolution during which key information, such as how to build a nest, was incorporated into the genetic code. While building a nest, the bird follows the blueprints of this code, adapting its actions to fit the local terrain. The code was an "automatic" result of evolutionary circumstances; it is not the result of a "designer" (as it would be the case if we were to construct a nest-building robot), but, instead, the experimental outcome of past processes of mutation and survival of the fittest; no thinking, no decisions, no "vision", no planning were involved [17]. The future was not considered, because nobody having a concept of future time was involved. In the case of the house blueprint, however, something drastic happened: a *time reversal* in the cause-and-effect relationship! The blueprint is the *result* at time t_0 of something which only will occur at a later time $t > t_0$: the house is the cause, the blueprint is the effect [18]. We quote E. Squires (1990): " ... *the idea of purpose might be regarded as inverting the usual order of cause and effect. I want a particular outcome and it is my concept of the "final state" that makes me adjust things so that it is achieved. The future has thus determined the present"*. According to this view, in human planning the preferential direction of time and cause-effect relationships run backwards: blueprints and plans are "pictures of the future". But isn't this true for the nest-building bird, too? No, because the genetic blueprints used by the bird evolved

---

may exhibit bursts of human-like thinking, but they seem not to be able to do anything with the results. There is a contentious debate on this issue between animal psychologists and brain scientists.

[16] It does not mean that one always thinks in words; however, the language centers' participation in thinking is essential: injecting an anesthetic into the carotid artery that feeds the speech hemisphere leaves the patient incapable of thinking—or outright unconscious.

[17] "Anticipation" (which animals can experience) is not the same as "prediction".

[18] Rubbish! some will say, this is just a dialectic twist of words! The "cause" of a blueprint is my hand picking up a pencil and committing to paper some ideas that have come to my mind just a few minutes *before*, not in the future. However, what counts is that these ideas (brain information processing) *are* about something in the future—the cause for these ideas lies in the future!

gradually and automatically through trial and error in the past—there were no predictions and decisions about the future involved at any stage [19].

The question of "information traveling backwards in time" is not such an unusual thought as it may sound: just remember that positrons can be described as (negative) electrons traveling backwards in time (which does not violate any laws of basic physics) [20]! It would have many consequences for the interpretation of time—in biology and even in physics. Behavior normally consists of a temporal chain of elementary steps, each one involving specific action triggered by decision-making to reach some intermediate goal. At the end of the chain lies the "ultimate" goal. In animals, at each step, information from the past (both, genetic and learned) and the present (sensory perception and information on the state of the organism) is processed in the brain (associative recall plays a fundamental role here) and triggers motor output leading to action. The chain is genetically preprogrammed and modified according to current circumstances (remember the nest building in Section 1). The course of long-term purpose-oriented *human behavior* involving a plan and a schedule looks very different: at each step the decision is based on the convergence of information from the past, the present *and the future*. The information from the future (the immediate and distant goals) can be amended as a function of the information from the present.

It is important to remark that the information from the past and present is always "filtered" by the limbic system. For animals, this is a most decisive factor in decision-making. In humans, as stated before, limbic dictates may be overruled (think of a diet), but they also may be stimulated by human thought about the future (think of wanting to listen to music, or wanting to communicate with extraterrestrial intelligence). As a matter of fact, it is unlikely that there could be art, culture, and even science in a society of intelligent organisms with brains that have no limbic system to control motivation and dispense sensations of reward and punishment. Human concern for the long-term future, specifically about "what will happen after death", contributed to the development of religious beliefs.

To make a final point, let us return to *Fig. 2*. A lifeless world of mountains, land, water, air will gradually decay—order will decrease and entropy increase. A world with only sub-human life will have some "pockets" exhibiting increased order and decreased entropy compared with the rest (e.g., nests, beaver dams, calcium deposits, etc.) In a world with human-like intelligence, the *rate* of changes corresponding to goal-directed activity is vastly increased (think of urbanization, agriculture, technology, use of energy sources that have lain dormant in unstable equilibrium for eons, etc.). To satisfy the second principle, a decrease of entropy *must* be compensated by an even bigger increase elsewhere (think of waste and pollution). In other words, by building regions of "unnaturally" high order and consistently low entropy, human activity is accelerating

[19] It is *us*, humans, who in our anthropomorphic approach tend to interpret animal behavior as being "future-oriented". This bias has plagued animal behavior research from the very beginning.
[20] Anyone who abhors the idea of information going backwards in time should consider this description as nothing else but a paradigm to help organize thoughts!

the increase of the total entropy of the Universe. This is not a curiosity, but likely the most general cosmological consequence of *any* human-like intelligence!

References

Layzer, D. (1975). The arrow of time, *Sci. Am.* December 1975, 56-69.

Leff, H. S. and Rex, A. F., eds. (1990). *Maxwell's Demon, Entropy, Information, Computing*, Adam Hilger, Bristol, New York.

Roederer, J. G. (1978). On the relationship between human brain functions and the foundations of physics, science and technology, *Found. of Phys. 8*, 423-438.

Roederer, J. G. (1995). *Physics and Psychophysics of Music*, Springer-Verlag, New York (Spanish translation: *Acústica y Psicoacústica de la Música*, Ricordi Americana, Buenos Aires, 1997).

Squires, E. (1990). *Conscious Mind in the Physical World*, Adam Hilger, Bristol, New York.

Young, J., Z. (1987). *Philosophy and the Brain*, Oxford Univ. Press, Oxford, New York.

THE ORIGIN OF THE NEURON: THE FIRST NEURON IN THE PHYLOGENETIC TREE OF LIFE

RAIMUNDO VILLEGAS, CECILIA CASTILLO and GLORIA M. VILLEGAS
Instituto de Estudios Avanzados • IDEA
Apartado 17606, Caracas 1015-A, Venezuela

Abstract: This contribution can be considered as an attempt to provide some kind of basic anchorage to look for neurons and nervous systems in possible new forms of life in the Universe. The peculiarities exhibited by the late protozoans and early groups of metazoans, led us to conclude taking into consideration the knowledge available at present, that the first neuron appeared in the Coelenterates, including Cnidarias and Ctenophores, and the first cerebral ganglia, as central component of a nervous system, appeared in the Platyhelminthes and Nematodes.

Introduction

Two important subjects of interest to both Neuroscience and Astrobiology, such as the neuron and how it works, and the search for the first neuron in the phylogenetic tree of life, are considered here.

1.- The neuron and how it works

1.1. NEURON

The nerve tissue is constituted by individual units, the nerve cells or neurons and the neuroglia or glial cells. This organization was known only upon the advent of the compound microscope and the studies of Ramón y Cajal towards the end of the XIX century. Ramón y Cajal (54), using the silver impregnation method or "reazione nera", developed by Golgi (18) was able to observe the microscopical anatomy of the whole neuron: cell body and processes, as well as the arrangement of the neurons in a sort of intrincate network of individual cells connected one to another by specialized contacts, and not forming a syncytium. Sherrington (60) coined the word synapse for the functional yuxtaposition of the neurons and later the term was extended to design a structural and functional entity.

The microscopical observations allowed Ramón y Cajal to propose the classical concepts of the neuron doctrine and of the neuron dynamic polarization, which have been confirmed and enriched throughout this present century and constitute the fundamental principles of the organization and function of the nervous system.

Moreover, Ramón y Cajal remarked that the fundamental fact is that neurons are the structures generating and conducting the nerve wave (nerve impulse), and therefore, their morphology is adapted to fulfill both functions.

As shown in Fig. 1, the neuron has an asymmetric morphological feature constituted by a cell body (soma) bearing a large nucleus and by two types of processes: multiple, ramified dendrites and a single axon. At the dendrite surface, there exist receptors for different chemicals known as neurotransmitters and the information received through them is processed, integrated and stored in the dendrite. From there, information is sent to the cell body where it is transformed into a binary signal: all-or-none, which is the so-called action potential, and this potential, travelling along the axon, is then sent either to other neurons or to peripheral target cells.

J. Chela-Flores et al. (eds.), Astrobiology, 195–211.

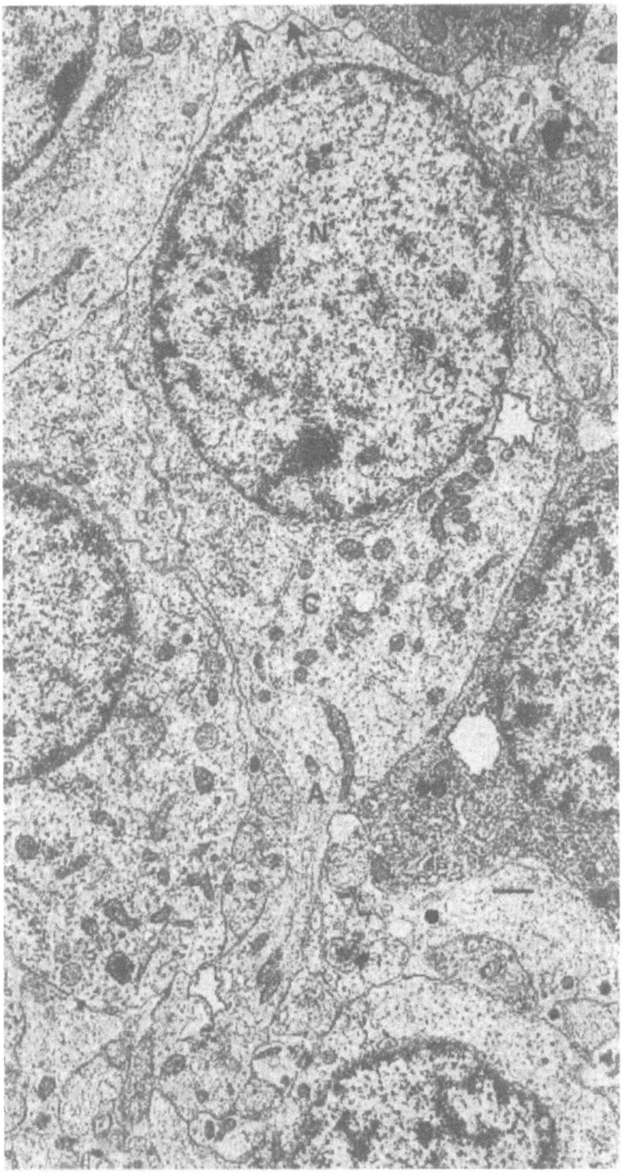

Figure 1.- Electron micrograph of a neuron of frog brain. A large, pale nucleus (N) occupies a good part of the cell body. In the cytoplasm (C), several different organelles are observed. The emergency of the axon with the characteristic bundles of microtubules is also seen (A). Arrows indicate emergency of dendrites. Bar= 1µm.

1.2. SYNAPSES AND NEUROTRANSMITTERS

Upon the arrival of the potential to the synaptic region sited at the axon terminal, neurotransmitters are released in minute amounts from the synaptic microvesicles existing in that region. Neurotransmitters then diffuse across the interspace separating the synaptic terminal from the target cell, thus creating in the membrane of this latter cell the conditions for generating and transmitting a new action potential.

Neurotransmitters are of different chemical nature and work, either as excitatory, inhibitory or regulatory of the ionic channels involved in excitability. The first attempt to demonstrate the synaptic chemical transmission corresponds to Elliot (16) who showed in 1904 that mammalian muscles innervated by sympathetic nerve responded to adrenaline. Then, in 1914 Dale (11) reported that acetylcholine (ACh) has effects on various body functions, such as lowering blood pressure, inhibiting cardiac beat and contracting intestinal muscle, and concluded that ACh works as an antagonist of adrenaline. Corresponded to Loewi (38) to conclusively demonstrate that ACh was released from vagus nerve endings and its effects were prevented with atropine. Other families of neurotransmitters were added afterwards: amino acids, such as γ-amino butyric acid (GABA) and l-glutamic and l-aspartic acids working in the central nervous system and also exerting opposed actions (inhibitory or excitatory actions) respectively; and the neuronal existing peptide hormones, named neuropeptides, which also work as signaling neurotransmitters in the nervous system (32). Among them, the opioids or endorphins, the intestinal peptides, such as substance P and vasoactive intestinal peptide (VIP), angiotensin and the pituitary peptides such as vasopressin, prolactin, corticotropin (ACTH) and others. Usually the neuropeptides coexist in the same neurons with the monoamines and are also coreleased, thus suggesting them as neuromodulators instead of neurotransmitters. (For a general review see ref. 36).

1.3. NEUROGLIA

The other cellular element of the nerve tissue, the neuroglia, was first visualized by Virchow (77) as a soft, medullary, interstitial material constituted by stellate or spindle-shaped cells. The present view on the role of the glial cells transcends the glue function originally attributed by Virchow. In fact, as revealed by recent research (33, 67, 71), they perform several important functions directed towards the maintenance of the neurons and their ambiance homeostasis, guidance of embryonary neurons and their processes, formation of the insulating myelin sheath around the axons, synthesis of neurotransmitters and the production of trophic factors.

1.4. EXCITABILITY, ELECTRICAL POTENTIALS AND ION CHANNELS

In certain species, the axon, the part of the neuron specialized in the transmission of information from one part to another of the cell, has dimensions that allow it to be seen with the naked eye. These characteristics have mainly contributed to the using of axons as the ideal preparation for the study of excitability, electrical potentials and ionic channels. The work of Young (83), Cole (10), Hodgkin , Huxley and Katz (22-25), and Keynes (29) on the squid giant axon permitted the understanding of those phenomena.

At rest, the neuron as any cell, exhibits a potential or voltage difference across its plasma membrane. This resting potential is due to the unequal distribution of charges at both sides of the cell membrane, caused by some ions moving more easily than others. When the cell is stimulated with a depolarizing stimulus that exceeds threshold, the permeability to the ions changes, thus giving origin to the action potential or nerve impulse. Both, the ionic permeability of the membrane at rest and the change during the action potential are associated to the existence of membrane proteins forming channels for each different ion, these proteins changing their ionic permeabilities during activity. The action potential shape and size, as well as the pattern of firing, can vary among different neurons and are related to differences in the membrane ion channels.

The existence of hydrophilic pores in cell membranes (61), including the nerve excitable membrane (46, 69 ,70 ,72), and the possibility that ion currents might flow through hydrophilic pores or channels was suggested in the 1950-1960s (10, 22-25, 29). The protein nature of these channels in the excitable membrane and the ion flow through them was demonstrated in the 1970s (73-75). Similar results obtained utilizing the patch clamp method in the same nerve membrane has been reported (6-8).

A breakthrough in the study of ion channels occurred in the early 1980's, when a novel method, patch clamp recording, permitted the measurement of ionic currents of single cells or single channels (19), in either the cell membrane or after the incorporation of the channels into artificial liposomes or into planar phospholipid bilayers. The single channel recording of all studied channels revealed transitions between an open and a close state, these transitions being considered as conformational changes in the structure of the channel protein. Following the first successful cloning of voltage-gated ion channels in 1984 (48), the combination of molecular biological and electrophysiological techniques has contributed to clarify the molecular basis of the electrical excitability of cells (63).

The channels are usually named according to the ions allowed to pass through the open channel and are classified according to their ion selectivity, gating properties (voltage-dependent or modified by neurotransmitters, hormones or other molecules), pharmacology (sensitivity to neurotoxins and other compounds that block the channel pore or modify other channel properties), and conductance. The ion channels are the most important membrane components related to electrical potentials. (For reviews see refs. 2, 5, 21, 36, 66, 76.)

1.4.1. *Sodium and Potassium Channels*

The action potentials in the squid axon are related to the presence of two different cationic channel populations: sodium channels and potassium channels. When the membrane is suddenly depolarized the probability of the sodium channels being open increases very rapidly and this is known as activation. Activation is followed by a progressive decrease in open sodium channels or inactivation. Potassium channels exhibit also activation at a slower rate and prolonged as long as the membrane depolarization is maintained. The potassium channel inactivation is usually also slower than that of sodium channels.

Voltage clamp studies which allow the measurements of the current flowing through the ionic channels at different fixed voltages, have provided a detailed understanding of the sequence of changes in channel activity that gives rise to action potentials. The course of action potentials varies according to the cells, the whole set of events taking in the squid axon only few milliseconds.

1.4.2. *Sodium Channel and Potassium Channel Pharmacological Probes*

Several molecules have been found to bind selectively and with high affinity to the sodium channel, thus modifying its function. Some of them have been utilized to detect the presence of sodium channels in cells, isolated membranes, and in solubilized membrane proteins during sodium channel purification. They have also been used to investigate the functioning of the channel in intact nerves and when incorporated in liposomes or in planar lipid bilayers. Among these molecules, the best studied are tetrodotoxin (TTX), saxitoxin (STX) and μ-conotoxin that block the sodium channel from outside; local anesthetics that block the channel, mostly from inside; batrachotoxin (BTX), grayanotoxin (GTX) and aconitine that activate the sodium channel; sea anemone toxins and α-scorpion toxins that slow the rate of inactivation; β-scorpion toxins that shift voltage dependence of inactivation to more negative potentials.

Among the compounds that selectively block the potassium channel are the organic cations tetraethylammonium (TEA) and 4-aminopyridine, and the inorganic cations cesium and barium.

1.4.3. *Sodium-Potassium Pump*

When membrane currents flow, ions move following their concentration gradients. The currents are small at rest, but much larger during the action potential. An active ion transport mechanism, the sodium-potassium pump consisting of a sodium-potassium ATPase, is then needed for mediating the pumping of sodium out and of potassium into the cells to maintain the ion concentration gradients.

1.4.4. *Calcium Channels*

In most neurons, a depolarizing voltage clamp step elicits an inward current with kinetics different from that of sodium in the squid axon. The current reaches its peak more slowly and inactivates partially and far more slowly. When any sodium contribution to this inward current is eliminated, a smaller inward current with the same kinetic characteristics remains. Further ion substitution experiments have revealed that most neurons exhibit a voltage-dependent calcium current, which in some cases is partially or totally responsible for the depolarization during the rising phase of the action potential.

1.4.5 *Calcium Exchange Mechanisms*

Cytoplasmic ionic calcium has a critical role in intracellular signaling. Two mechanisms are important for the maintenance of intracellular calcium level. They are the calcium ATPases and the Na/Ca exchanger, the latter being a calcium transport system that can move calcium at a high rate from the cell to the medium or viceversa This exchanger allows the rapid return of cytoplasmic calcium levels to the rest value during the post-activity period of the presynaptic neuron (13).

1.4.6. *Calcium Channel Pharmacological Probes*

Drugs and toxins have been important in discriminating among the different calcium channels. As examples, we can mention the dihydropyridines (DHP) that bind to the L-type channels that exhibit long-lasting calcium current. DHP antagonists cause the L-type channels to have shorter opening times, whereas the agonists cause longer channel openings. Other channels resistant to DHP are sensitive to peptide toxins, such as ω-conotoxin GVIA, which inhibits the N-type channels in neurons. Another peptide toxin, ω-agatoxin IVA, inhibits the P-type calcium channel responsible for most of the calcium current in cerebellar Purkinje neurons.

1.5. BIOCHEMISTRY AND MOLECULAR BIOLOGY OF IONIC CHANNELS

1.5.1. *Sodium Channels*

Two possibilities about the nature of the voltage-dependent sodium channels, though hypothetical by 1970s, were considered at that time. One, the existence of individual and permanent molecular structures in the cell membrane corresponding to the voltage-gated sodium channels and the other, the transitory occurrence of structural rearrangements in the membrane components induced by potential changes across the membrane. The answer was offered by our research group working in cooperation with Racker in 1977 (75), by incorporating lobster nerve functional sodium channels present in crude nerve membrane preparations, into phospholipid liposomes by the freeze-thaw-sonication method. Similar reconstitution experiments were then successfully carried out with partially purified sodium channel preparations of *Electrophorus electricus* (43, 56, 57).

The sodium channel protein from the electric organ of *E. electricus* was the first to be purified, using TTX as a marker, and identified with a glycoprotein of 260 kDa (1). Later, a cDNA encoding its complete sequence was obtained (48).

When the rat brain sodium channel protein was purified, in addition to a large 260 kDa component named α-subunit, two other subunits were found and named β_1 of 39 kDa and β_2 of 33 kDa (20). A different approach revealed the existence of distinct mRNAs in rat brain corresponding to the large polypeptide of three distict sodium channels named I-III. The cDNA clones derived from two out of the three mRNAs, permitted to deduce their complete amino acid sequences (49). Then, mRNAs generated by transcription of the cloned cDNAs corresponding to the large polypeptides, when injected in *Xenopus* oocytes were able to direct the formation of functional sodium channels (50). This was confirmed when purified mRNA obtained from a cDNA clone encoding only α-subunit of sodium channel II was expressed in the oocyte, though in this latter case inactivation occurred at less negative potential (63). This change in inactivation could be related to the lack of β-subunits in the purified α-subunit mRNA preparation.

The protein predicted from the cDNA sequence contains around 2000 amino acids that accounts for 208 kDa out of the 260 kDa, the remainder corresponding to carbohydrates. The hydrophobicity plot for the α-subunit amino acids reveals the presence of 24 possible transmembrane segments made of about 23 hydrophobic amino acids each. The amino acid sequence of the α-subunit reveals the existence of four internal similar domains identified as I-IV, each one of them having six possible transmembrane segments, S1 to S6 (49).

1.5.2. *Calcium Channel*

As in the case of sodium channels, a large subunit, 175-230 kDa, and named α_1, was found to be sufficient to reconstitute a functional voltage-dependent calcium channel. One or more smaller subunits, β, γ, α_2 and δ, also present, interact with α_1 and modulate the kinetic properties of the calcium channels.

Molecular cloning, based on protein sequence, has revealed the existence of at least six distinct genes for α_1-subunits and four for β-subunits, being all the α_1 and the majority of the β mRNA transcripts able to undergo alternative splicing. This makes the number of α_1-subunits expressed in the brain, as well as the number of channels product of their combinations, fairly large. As a consequence, an overwhelming diversity of calcium current kinetics and pharmacological properties can be observed.

The membrane organization of the calcium channel protein, predicted from the α_1-subunit amino acid sequence, is closely similar to that of sodium channel, i.e., 24 transmembrane segments, forming four internal similar domains of six segments each.

In addition to its contribution to the action potential, calcium ions play important intracellular roles, such as regulating some ion channels gating, interacting with calcium binding proteins for enzyme regulation, and participating in the control of the release of some chemical neurotransmitters. (For reviews see refs. 9, 21, 36, 45, 66, 68.)

1.5.3. *Potassium channel*

The diversity of voltage-gated potassium channels, as revealed by single channel and molecular biology studies, outnumber that of known calcium channels (21, 78). Among them the most common are described below.

A type existing in the cell body of many molluscan neurons has been used for voltage clamp studies of membrane ionic currents. In those neurons, the total outward current carried by potassium shows a characteristic N-shape current-voltage curve in the range of depolarized voltages. When the intraaxonal calcium is lowered or eliminated, the N-shape disappears and the curve becomes equal to the delayed rectifier potassium current observed in the squid axon (K-Ca channel). The current removed is considered a calcium-dependent potassium current (K channel). There is also a transient potassium current or A-current, that activates rapidly and then inactivates as the sodium current does (K_A channel). In some cells also exists a potassium current activated by hyperpolarization that causes a decrease in the slope of the current-voltage curve, a phenomenon known as anomalous or inward rectification (K_{IR} channel).

There are other channels gated by ligands and intracellular metabolites that may also have voltage dependence. These channels include those modulated by the neurotransmitters muscarine and serotonin, known as M-current and S-current potassium channels.

The lack of a rich source of channel protein precluded to carry out molecular cloning by sequence homology. One alternative was provided by studies of mutant *Drosophila melanogaster* fruit flies, that shake their legs, wings and abdomen when anesthetized with ether (78). Among the strains of these mutant flies, there are some with defects in the gene identified as the Shaker locus. The availability of a precise genetic map in *Drosophila* made possible to define the relative location of the Shaker mutation and to identify cloned segments of DNA, including the mutant locus. The protein encoded by the Shaker locus was found to be only one-quarter the size of sodium or calcium channel proteins, and to have only one domain with six transmembrane regions, S1 to S6. It is interesting to point out that four Shaker subunits, equal or distinct, have to come together in the membrane to form functional tetramers, i.e., potassium channels that resemble the sodium and calcium channel α-subunits. (For reviews see refs. 16, 21, 36, 66.)

2. Searching for the first neuron

2.1. THE PHYLOGENETIC TREE OF LIFE

Until the late 1970s, scientists considered that the living organisms were separated in two basic groups: Prokaryotes and Eukaryotes, the Prokaryotes comprising the eubacteria and the archaebacteria. At that time Woese et al. (17, 81, 82), proposed the existence of three domains, Archaea (archaebacteria), Bacteria (eubacteria), and Eucarya (eukaryotes). The rationale behind this new division came from evidences , in particular from rRNA phylogenies, that the archaebacteria deserves the same taxonomic status as eubacteria and eukaryotes. This new division takes into consideration the molecular structures and sequences and therefore, is considered more revealing of evolutionary relationships than the phenotypes, particularly among microorganisms.

Some of these studies done with the small subunit of rRNA, revealed special structural features at certain regions of the sequence which, in one case are common to archaebacteria and eukaryotes and different from the one found in eubacteria, and in another case are common to archaebacteria and eubacteria but not found in eukaryotes.

Taking into consideration these findings, Woese et al. (82) proposed the universal tree of life, which as modified by Doolittle (14), is depicted in Fig. 2A. The model considers that life on this planet, would be seen as comprising three domains, the Bacteria, the Archaea, and the Eucarya, each containing two or more kingdoms. The Eukarya, for example, contains Animalia, Plantae, Fungi, and a number of others yet to be defined. The common ancestor of the three domains is still unknown.

More recently, Doolittle (14) has questioned the sufficiency of these sequencing comparative analyses as the unique basis for a "natural" hierarchical classification of all living things. He remarks that most archaeal and bacterial genomes (and the inferred ancestral eukaryotic nuclear genome) contain genes from multiple sources. This lead him to conclude that if chimerism and lateral gene transfer can not be dismissed as trivial or limited to certain genes, no hierarchical classification can be taken as natural. He proposed then to adopt a reticulated tree of life or net.

2.2. ARCHAEA, BACTERIA AND EUKARYA UNICELLULAR ORGANISMS

The molecular studies mentioned above have also revealed that the three domains Archaea, Bacteria and Eukarya, are similarly ancient, being the Eukarya also more than three billion years old, much older than it was estimated in the past (82).

Bacteria and Archaea are single-celled organisms consisting of cytoplasm and genetic material bounded by a membrane. They do not have a nucleus, nor membrane-bounded internal organelles, such as mitochondria. The genetic material is arranged in a circular chromosome that lacks histones and segregates into daughter cells. Due to the lack of mitochondria or an equivalent organelle, the energy metabolism of these cells must take place in the plasma membrane, being the energy produced by the chemiosmotic proton-motive force stored in the same membrane.

In relation to the domain Eukarya, the early eukaryotes are thought to be also single-celled organisms that lived anaerobically and had the genetic material in a nucleus surrounded by a nuclear membrane. They had several chromosomes

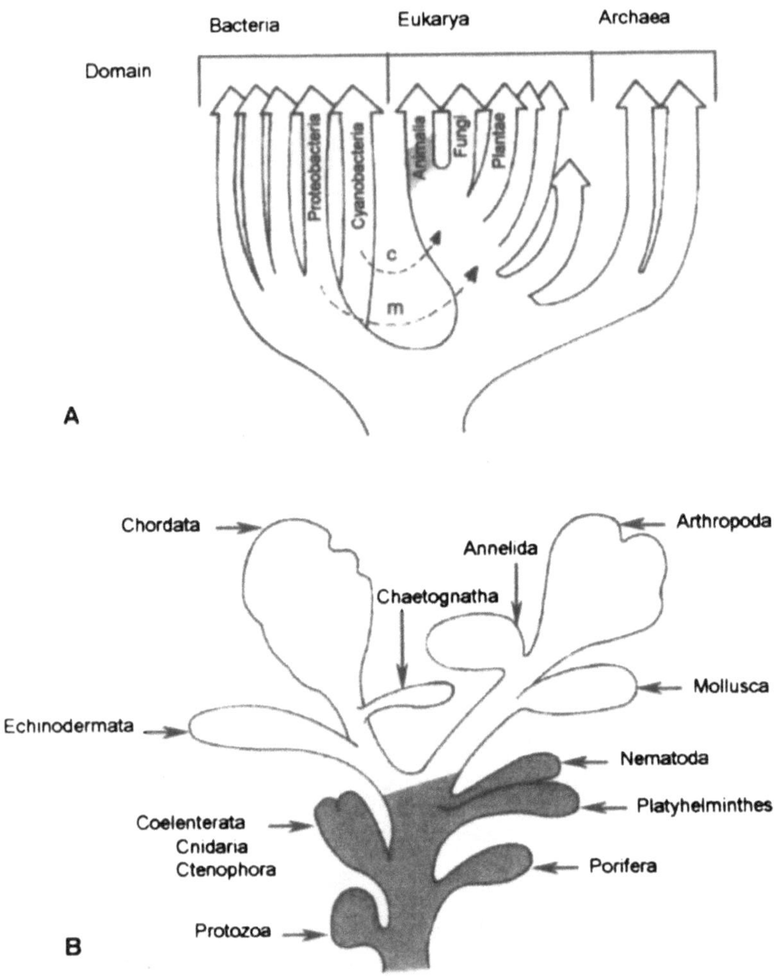

Fig. 2.- A.- Simplified version of the original tree of life standard model, supported by molecular and phenotypic characteristics and exhibiting the three domains proposed by Woese et al. (81), as interpreted by Doolittle (14). The arrows represent the kingdoms of each domain, the curved, dash-arrows indicate the chloroplast (c) and mithocondrial (m) endosymbioses. In the domain Eukarya, the kingdoms Animalia, Fungi and Plantae and in Bacteria, the kingdoms Proteobacteria and Cyanobacteria are identified. The shaded portion in the Animalia arrow corresponds to the phyla studied in the present work. B.- Version of the kingdom Animalia, modified from the original by Hille (21). The shaded portion corresponds to the phyla analyzed.

containing histones and a nucleolus, the chromosomes being segregated by mitosis or meiosis organized by spindle fibers and centrioles, at cell division. In agreement with the possibility of early eukaryotes being anaerobic, there exists nowadays the case of *Trichomonas* and *Giardias* that lack mitochondria. Once the unicellular eukaryotes obtained mitochondria (or chloroplast), the plasma membrane was set free of storing energy. The incorporation of other organelles, such as endoplasmic reticulum, Golgi apparatus and lysosomes, and also of some proteins like tubulin, actin, myosin and calmodulin, allowed these unicellular eukaryotes the sorting and packaging of proteins in vesicles and their secretion by exocytosis. They also could build an endoskeleton for changing shape and generating movements, and the use of calcium ion as a second messenger.

A question remains: Do these unicellular organisms exhibit some neuron-like characteristics? Though lacking a nervous system, these organisms must be able to solve survival and reproduction problems, same that face all living organisms. For this purpose they need to have some receptor and effector components with the capacity to receive information and respond to the sensory input.

Bacteria sense their environmental conditions (pressure, nutrients, toxins, pH, temperature and others) through receptors, usually transmembrane proteins, or additional cytoplasmic proteins. Many bacterial species swim by means of a flagellum or set of flagella that allow them to control their swimming direction. Upon sensing the environmental conditions, the bacteria move towards the most favorable environment. Sensing involves transducing membrane stress into an electrochemical response that affects the flagella. Recent studies have revealed that bacteria also integrate signals through common pathways to produce a balanced flagellar response. The mechanisms underlying such a behavior has been recently reviewed (3, 65), although how the sensory information is processed to give an integrated response remains an open question.

Pores formed by porin proteins located in the outer membrane of bacteria (47), and mechanosensitive channels involved in osmoregulation and located in the inner membrane (51, 64), are subjects of numerous investigations. Voltage-dependent porin-like ion channels (4) and mechanosensitive ion channels (35) have been described in the archaea *Haloferax volcanii*.

In Eukarya unicellular organisms and particularly in the ciliated protozoan *Paramecium*, biochemical, electrophysiological and genetical research have offered valuable information on the mechanisms of sensing and responding to a variety of stimuli. Normally, *Paramecium* swims forward by a continual beating of cilia in the posterior direction and when it strikes an object the cilia reverse their beating direction and the *Paramecium* moves backwards. After a fraction of a second, the cilia revert to the normal beating orientation, the *Paramecium* tumbles and moves forward. The frontal impact activates mechanosensitive calcium channels located in the surface membrane and the resulting depolarization activates voltage-gated calcium channels in the membrane of each cilium which elicits a regenerative calcium spike that depolarizes the whole *Paramecium*. The entry of calcium activates an organelle sited at the base of each cilium that changes its movement direction. Cell repolarization occurs when potassium channels are activated and calcium channels shut (15). Work done with mutants defficient in distinct channels has permitted to identify the channels involved in the *Paramecium* movements (28, 53).

2.3. MULTICELLULAR EUKARYA: SPONGES, COELENTERATES, PLATYHELMINTHES AND NEMATODES

Following the changes in the unicellular eukaryotes described above, a fast phase of evolution started about 700 million years ago, leading them to become multicellular organisms or metazoa, such as fungi, plants and animals. After the Protozoans, as shown in Fig. 2B, four at present living groups among the animals: Sponges, Coelenterates, Platyhelminthes and Nematodes appear to be the most interesting in relation to the search for the first neuron.

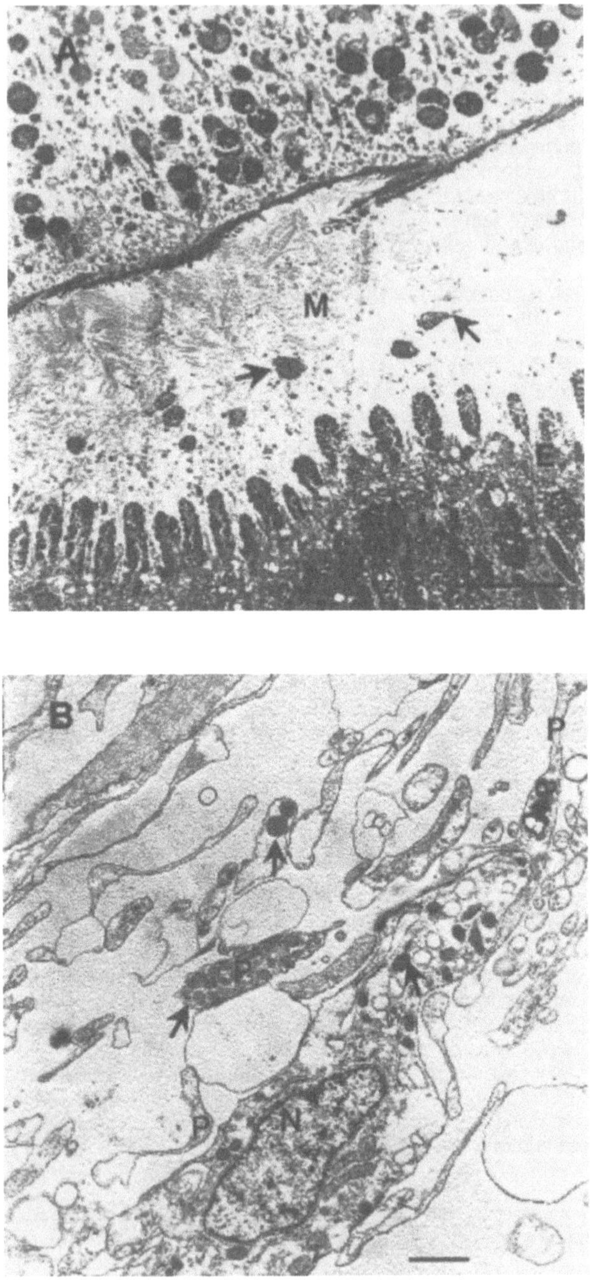

Figure 3.- Composite picture of sea anemone tubular wall. A.- Light micrograph showing the two epithelial layers, internal (I) and external (E), separated by the mesoglea (M), in which neuronal bodies are seen (arrows). Bar=10μm. B.- Electron micrograph of the mesoglea region. A neuron with a central nucleus (N) is seen, as well as several neurites sections (P). In the cytoplasm of the cell body, as well as in the processes, dark granules among other organelles are observed (arrows). Bar=1μm.

2.3.1. *Sponges / Porifera*

Sponges or Porifera are multicellular organisms with certain degree of cellular differentiation, although lacking well-defined organs. In the simplest type of sponge, the body remains as a simple tubular unit with the central cavity apically opened. Some remain solitary, while others made colonies. There seems to be no differentiation of functions among individuals within a colony. The body walls of the individual sponges appear interrupted by canals and are formed by an outer and inner cell layer named pinacoderm (formed by pinacocytes) and choanoderm (formed by choanocytes), respectively, separated by a gelatinous substance termed mesohyl, that contains disperse cells known as amoebocytes. Most sponges secrete mineralized skeletal spicules that serve as support of the body walls. Due to the resemblance of the choanocytes to the choanoflagellates, some authors consider that the sponges evolved from those protozoans. Considering their peculiarities and the possibilty that sponges have evolved independently from other animal groups, sometimes they are classified as Parazoa.

Conduction of contractile responses over long distances and the ability to regulate overall water flow have been demonstrated in certain sponges. It has been established that the hexactinellid sponge *Rhabdocalyptus dawsoni* is capable of arresting its exhalant water current in response to mechanical and electrical stimuli and also that this response is coordinated by an unpolarized conduction system (34). However, since attempts to record electrical activity were unsuccessful, it was concluded that the existence of a nervous system in sponges was doubtful.

Recently, Leys et al. (37) using the same sponge *R. dawsoni* were able to record all-or-none propagated electrical impulses from lumps of aggregated sponge tissue grafted onto the surface of pieces of the same sponge. Impulses were evoked by electrical shocks and recorded externally. They also report that the water flow through the sponge was arrested following the passage of an impulse, presumably as result of the cessation of beating of the flagella. Impulses continued to propagate during arrest, indicating that the conduction and effector systems were independent. About the mechanism of the electrical response they found that sodium deficient solutions had little effect on the action potential, whereas propagation was blocked by cobalt and manganese ions or by nimodipine, strongly suggesting that propagation depends on calcium influx. TEA ions also blocked propagation without prolonging the action potential.

Since at present it is generally agreed that sponges lack nerves (52), it is necessary to look for a tissue forming a continuos pathway through the whole sponge and linked to the flagellated epithelium (choanoderm). According to early and recent studies, the only tissue distributed in this manner is the syncytial flat layer of homogeneous cytoplasm constituting the mesohyl and that connects through trabecular portions the pinacoderm to the choanoderm. There is no evidence for the existence of specialized sensory structures associated with this trabecular reticulum. However, the system resembles the excitable epithelia of hydromedusae and amphibian tadpoles.

The electrical signaling system found in *R. dawsoni* and not yet reported in other group of sponges, shows many parallels with non-nervous pathways carrying action potentials in other metazoans. This may has also characterized the early metazoans. Recent molecular data suggest that the hexactinellids, that includes *R. dawsoni*, are the most ancient sponges and therefore, the most ancient metazoans (59).

2.3.2. *Coelenterates and the Appearance of the Neuron*

The Coelenterates are probably the simplest metazoans with neurons and nervous systems. The phylum Coelenterata embraces Cnidaria and Ctenophora (30), these latter being sometimes treated as a separate phylum (26). However, apart from how they are classified, their origin appears to be tied up with that of the Cnidaria, with which they share numerous common features. (For a review see ref. 58.)

Cnidaria are constituted by the classes Hydrozoa, Scyphozoa and Anthozoa. The body of these diploblastic animals is a sac formed by an outer and an inner columnar epithelia standing on basement membranes with mostly a cellular mesoglea in between.

The Ctenophora (or comb jellies) are biradially symmetrical coelenterates, structurally more advanced than Cnidaria, but also having epidermal and endodermal layers standing on basement membranes with mesoglea in between.

The organization of the nervous systems of Cnidaria and Ctenophora is very similar and could be described as a nerve net formed by fusiforn bipolar neurons located in the mesoglea, as shown in Fig. 3, A and B. Ultrastructural studies have revealed neurons bearing neurosecretory dense granules and with isopolar processes termed neurites connected by synaptic-like junctions. The processes of these neurons, are simple isopolar units, not differentiated into dendrites or axons. Synapses exhibit bidirectionality and occur where the dendrites make contact. A cnidarian nerve net forms a single functional unit that combines sensory, associative and motor functions with no evidence of central control. Action potentials arriving from either direction can evoke excitatory postsynaptic potentials and conventional action potentials on the other side of the synapses. The conduction velocity and the refractory periods are like those from unmyelinated neurons in other systems.

We would like to illustrate with few references, some of the advances in the characterization of the coelenterates that permit to consider this group of animals as the first to evolve recognizable neurons and nervous systems. In the giant axons of the jellyfish *Aglantha digitale,* order Hydromedusae, two sorts of action potentials were found by Mackie and Meech (39), a low-threshold, low-amplitude, calcium-dependent spike that mediates normal swimming, and a higher-threshold, overshooting sodium-dependent action potential that mediates the escape response. This was the first report of an axon capable of two kinds of impulse propagation. Later, they confirmed these findings and further characterized the ionic basis of the action potentials (40). It is interesting to note that the calcium-dependent low-threshold spikes and the sodium-dependent high-threshold action potentials, though coexisting in the same axon, may be elicited separately taking into account their particular properties. In addition, three distinct potassium channels, according to their inactivation kinetics (fast, intermediate, or slow) have been detected (40, 41) in the same axons of the jellyfish. More recently, Meech and Mackie (42) characterized the synaptic potentials and threshold currents underlying action potential production in the motor giant axons and found that the calcium and sodium spikes arise from different excitatory postsynaptic potentials which originate separately in units of the inner and outer nerve rings sited at the base of the bell.

A morphological, comparative study of the synapses of all classes of Cnidaria has permitted to define an early neuronal synapse as one with paired electron dense plasma membranes separated by a filament-containing, 13-25 nm cleft, and with vesicles at one or both sides of the cleft (79). The neuromuscular synapses resemble neuronal synapses and lack the postsynaptic specialization of higher animals. Dense vesicles at both ends of the interneuronal synapse, as well as some specific neuromuscular synapses of sea anemones, can be labeled with antisera to the neuropeptides Antho-RFamide and Antho-RWamide I and II. As noted by Moosler et al. (44), the presence of RFamide neuropeptides in all major cnidarian classes suggested that this type of substance was among the first neurotransmitters appearing in evolution.

Recent work on the voltage-gated ion channels found in the nerve net has permitted to confirm their presence and to advance in the knowledge of their pharmacological and molecular biological characterizations (27, 62, 80).

2.3.3. *Platyhelminthes, Nematodes and the Cerebral Ganglion*

Many authors consider that a group of flatworm-like animals, the Platyhelminthes, that evolved after the Coelenterates, gave origin to the remaining higher animals (21). They would be the first triploblastic animals, having a well developed mesoderm in addition to the ectoderm and endoderm. They have bilateral symmetry and the major internal organs, but no true coelom. Among the organs are those of the nervous system which resembles a hybrid between the classical nerve net of the coelenterates and the more centralized nervous system of higher animals.

In recent years the most studied flatworm has been the marine species *Notoplana acticola.* This animal already has a cerebral ganglion or primitive brain that receives inputs from sensory organs and delivers outputs to muscles via nerve cords (31). The cerebral ganglion has many types of neurons, including typical invertebrate monopolar neurons, bipolar cells and also vertebrate-like multipolar neurons. In the neurons, there exist sodium channels, both sensitive and insensitive to TTX, and the major types of potassium channels. Some of these channels have been already cloned and sequenced. Although the cerebral ganglion functions to coordinate peripherally based neuronal circuits, these animals are able to perform many fragments of normal behavior after its removal. The peripheral system consists of a series of plexuscs or nerve nets. Neuropil and cell bodies are located in both the central and peripheral systems together with variable amounts and kinds of neuroglial sheaths (21, 55).

Close to the Platyhelminthes are the Nematodes or unsegmented round-worms, represented by *Ascaris lumbricoides* and *Caenorhabditis elegans.* Their nervous system is made of neurons, with the same type of ion channels as the Platyhelminthes, and neuroglial cells. They also have a cerebral ganglion that receives signals from sensory organs and sends output signals, via nerve cords, to muscles through cholinergic synapses. Inhibitory transmission involves a gabaergic-induced increase in chloride permeability. The nervous system has also octopamine, dopamine and serotonine (21). Both groups of animals, Platyhelminthes and Nematodes, have highly developed peptidergic components in their nervous systems, though with key differences in the structure and action of the neuropeptides (12).

2.4. HIGHER ANIMALS

About 570 million years ago, two major groups of coelomate animals had diversified as Deuterostomes and Protostomes. Deuterostomes include Echinoderms and Chordates, and Protostomes include Annelids, Molluscs and Arthropods. The nervous system of Chordates, Molluscs and Arthropods are considered well developed, according to their capacity of giving support to sensorial discrimination, learning, communication and social behavior. Underlying these capacities exist structural, biochemical, molecular-biological, physiological and pharmacological characteristics present in all of them.

3. Concluding remarks

Out of the invertebrates considered in the preceding paragraphs, we want to point out five of them: protozoans, sponges, coelenterates, platyhelminthes and nematodes, because they represent the most interesting groups in relation to the search for the first neuron in the phylogenetic tree of life.

Single-celled Eukarya organisms like the ciliated protozoan *Paramecium,* have to deal with problems of survival like higher animals, which imply having receptors for sensing the environment, a signal integration mechanism and effectors for responding to the external stimuli. It is worthy to note that Bacteria and Archaea oganisms, also single-celled, exhibit similar behaviors.

In the *Paramecium,* mechanosensitive calcium channels, as well as voltage-gated calcium and potassium channels in charge of the cilia movements, constitute the essential components of the mechanism for sensing, integrating and responding to the stimuli. Above-threshold stimuli cause a depolarization that spreads in an all-or-none mode. This and the associated successive steps, share basic characteristics with the functioning of a hypothetical simple neuron.

Sponges, the most ancient metazoans present a higher degree of organization than the protozoans, being the multicellularity one of their important features. This multicellularity allows the sponges to increase in size and also to have a high degree of coordination. All-or-none propagated electrical impulses have been recorded in the sponge *R. dawsonii* in response to electrical shocks. However, the electrical signaling system of this species appears to be similar to other non-nervous pathways carrying action potentials in other organisms.

The next group corresponds to Coelenterates, comprising Cnidarias and Ctenophores, in which the first neurons and a true nervous system exist. This is constituted by individual interconnected neurons transmitting, chemically or electrically, in both directions and forming a sort of nerve net in the mesoglea. Low-threshold calcium-dependent and high-threshold sodium-dependent channels, as well as three distinct potassium channels have been found in *R. dawsonii*. Some neuromuscular synapses and interneuronal synapses, labeled with antisera to neuropeptides Antho-RFamide and Antho-Rwamide, have revealed these peptides being among the first neurotransmitters appearing in evolution.

The Platyhelminthes that evolved after the Coelenterates, together with the Nematodes posses a nervous system resembling a hybrid between the coelenterates nerve net and the centralized nervous system of higher animals. They have a cerebral ganglion that represents a central nervous system with many types of neurons, including vertebrate-like multipolar neurons, and neuroglial cells. They also have several nerve plexuses representing the peripheral nervous system. The neurons have sodium channels, both sensitive and insensitive to TTX, and the major types of potassium channels. Some of these channels have been already cloned and sequenced. In addition, an advanced degree of synaptic plasticity is suggested by selective excitation and inhibition of the neuronal pathways. Several neurotransmiters have been found in both groups of animals, including neuropeptides.

4. Acknowledgements

Authors would like to thank Profs. J. Chela, J. Villegas and G. Whittembury for critical reading of the manuscript, and to Miss Carola Ono and Mr. A. Bonelli for the preparation of the figures.

5. References

1. Agnew WS, Moore AC, Levinson SR and Raftery MA (1978) Purification of the tetrodotoxin-binding component associated with the voltage-sensitive sodium channel from *Electrophorus electricus* electroplax membranes. *Proc Nat Acad Sci U.S.A.* **75**: 2606-2610.
2. Armstrong CM and Hille B (1998) Voltage-gated ion channels and electrical excitability. *Neuron* **20**:371-80
3. Armitage JP (1999) Bacterial tactic response. *Adv Microb Physiol* **41**: 229-289.
4. Besnard M, Martinac B and Ghazi A. (1997) Voltage-dependent porin-like ion channels in the archaeon Haloferax volcanii. *J Biol Chem* **272**: 992-995.
5. Bezanilla F and Stefani E (1994) Voltage-dependent gating of ionic channels. *Ann Rev Biophys Biomol Struct.* **23**: 819-46
6. Castillo C, Piernavieja AC and Recio-Pinto E. (1996) Anemone toxin II unmasks two conductance states in neuronal sodium channels. *Brain Research* **733**: 231-242.
7. Castillo C, Piernavieja AC and Recio-Pinto E. (1996) Interactions between anemone toxin II and veratridine on single neuronal sodium channels. *Brain Research* **733**: 243-252.
8. Castillo C,Villegas R and Recio-Pinto E (1992) Alkaloid-modified sodium channels from lobster walking leg nerves in planar lipid bilayers. *J Gen Physiol* **99**: 879-930.
9. Catterall WA (1998) Structure and function of neuronal Ca^{++} channels and their role in neurotransmitter release. *Cell Calcium* **24**: 307-323.
10. Cole KC (1949) Dynamic electrical characteristics of the squid axon membrane. *Arch Sci Physiol* **3**: 253-258
11. Dale HH (1914) The action of certain esters and ethers of choline, and their relation to muscarine. *J Pharmac Exp Ther* **6**: 147-190.
12. Day TA and Maule AG (1999) Parasitic peptides! The structure and function of neuropeptides in parasitic worms. *Peptides* **20**: 999-1019.

13. DiPolo R and Beauge L (1999) Metabolic regulation of the Na/Ca exchange, the role of phosphorylation and dephosphorilation. *Biochim Biophys Acta* (review in Biomembranes) **1422**: 57-71.
14. Doolittle WF (1999) Phylogenetic classification and the universal tree. *Science* **284**: 2124-2128.
15. Eckert R and Brehm P (1979) Ionic mechanisms of excitation in *Paramecium*. *Ann Rev Biophys Bioeng* **8**: 353-383.
16. Elliot TR (1904) On the action of adrenaline. *J Physiol, (London)* **31**: 20p.
17. Fox GE, Magrum IJ, Balchm WF, Wolfe RS and Woese CR (1977) Classification of methanogenic bacteria by 16S ribosomal RNA characterization. *Proc Natl Acad Sci USA*. **74**: 4537-4541.
18. Golgi C (1873) Súlla struttura délla sostanza grigia del cervello. *Gaz Med Ital Lombardía* **6**: 244-246
19. Hamill OP, Marty A, Neher E, Sakmann B and Sigworth FJ (1981) Improved patch-clamp techniques for high resolution current recording from cells and cell-free membrane patches. *Plügers Arch* **391**: 85-100.
20. Hartshorne RP and Catterall W.A (1984) The sodium channel from rat brain: purification and subunit composition. *J Biol Chem* **259**: 1667-1675.
21. Hille B (1992) *Ionic channels of excitable membranes*. 2nd ed., Sinauer Associated Inc., MA.
22. Hodgkin AL, and Huxley AF(1952) Currents carried by sodium and potassium through the membrane of the giant axon of *Loligo*. *J Physiol (London)* **116**: 449-472.
23. Hodgkin AL, Huxley AF and Katz B (1949) Ionic currents underlying activity in the giant axon of squid. *Arch Sci Physiol*. **3**: 129-150.
24. Hodgkin AL and Katz B.(1949) The effect of sodium ions on the electrical activity of the giant axon of the squid. *J Physiol (London)* **108**: 37-77.
25. Hodgkin AL Huxley AF and Katz B (1952) Measurements of current-voltage relations in the membrane of the giant axon of *Loligo*. *J Physiol (London)* **116**: 424-448.
26. Hyman LH (1940) *The invertebrates: Protozoa through Ctenophora* Vol. 1 pp. 662-96 Mc Graw-Hill, New York.
27. Jegla T and Salkoff L (1994) Molecular evolution of K^+ channels in primitive eukaryotes. *Soc Gen Physiol Ser* **49**: 213-222.
28. Jegla T and Salkoff L (1995) A multi gene family of novel K^+ channels from *Paramecium tetraurelia. Receptor Channels* **3:** 51-60.
29. Keynes RD (1951) The ionic movements during nervous activity. *J Physiol (London)* **114**: 119-150.
30. Komai T (1963) A note on the phylogeny of the Ctenophore. In: *The lower metazoa, comparative biology and phylogeny*, EC Dougherty et al., Eds. University of California Press PP 181-188.
31. Koopowitz H (1982) Free-living platyhelminthes. In: *Electrical conduction and behaviour in simple invertebrates*. Shelton, G.A.B. (Ed.) Clarendon Press, Oxford, pp. 359-392.
32. Kuffler SW and Edwards C (1958) Mechanism of gamma-aminobutyric acid (GABA) action and its relation to synaptic inhibition. *J Neurophysiol* **21**: 589-610.
33. Kuffler SW and Nicholls JG (1966) The physiology of neuroglial cells. *Ergebn Physiol* **57**: 1-90.
34. Lawn ID, Mackie GO and Siver G (1981) Conduction system in a sponge. *Science* **211**: 1169-1171.
35. Le Dain AC., Saint N, Kloda A, Ghazi A and Martinac B. (1998) Mechanosensitive ion channels of the archaeon *Haloferax volcanii*. *J Biol Chem* **273**: 12116-12119.
36. Levitan IB and Kaczmarek LK (1997) *The neuron, cell, and molecular biology* 2nd edition. Oxford University Press Inc. New York
37. Leys SP, Mackie, GO and Meech RW (1999) Impulse conduction in a sponge. *J Exp Biol* **202**: 1139-1150.
38. Loewi O (1921) Über humorale Übertragbarkeit der Hertznervenwirkung. *Pflügers Arch Ges Physiol* **189**: 239-242.
39. Mackie GO and Meech RW (1985) Separate sodium and calcium spikes in the same axon. *Nature* **313**: 791-793.

40. Meech RW and Mackie GO (1993) Ionic currents in giant motor axons of the jellyfish Aglantha digitale. *J Neurophysiol* **69**: 884-893.
41. Meech RW and Mackie GO (1993) Potassium channel family in giant motor axons of Aglantha digitale. *J Neurophysiol* **69**: 894-901.
42. Meech RW and Mackie GO (1995) Synaptic potentials and threshold currents underlying spike production in motor giant axons of Aglantha digitale. *J Neurophysiol* **74**: 1662-1670.
43. Miguel V, Balbi D, Castillo C and Villegas R (1992). Reconstitution of sodium channels in large liposomes formed by the addition of acidic phospholipids and freeze-thaw sonication. *J Memb Biol* **129**: 37-47.
44. Moosler A, Rinehart KL and Grimmelikhuijzen CJ (1997) Isolation of three novel neuropeptides, the Cyanea-Rfamides IIII, from Scyphomedusae. *Biochem Biophys Res Commun* **236**: 743-749.
45. Moreno Davila H (1999) Molecular and functional diversity of voltage-gated calcium channels. *Ann NY Acad Sci* **30**: 102-117.
46. Mullins LJ (1960) An analysis of pore size in excitable membranes. *J Gen Physiol* **43** (5, Part 2) 105-117.
47. Nikaido H (1994) Porins and specific diffusion channels in bacterial outer membranes. *J Biol Chem* **269**: 3905-3908.
48. Noda M, Shimizu S, Tanabe T, Takai T, Kayano T, Ikeda T, Takahashi H, Nakayama H, Kanaoka Y, Minamino N, Kangawa K, Matsuo H, Raftery MA, Hirose T, Inayama S, Hayashida H, Miyata T and Numa S (1984). Primary structure of Electrophorus electricus sodium channel deduced from cDNA sequence. *Nature* **32**: 121-127.
49. Noda M, Ikeda T, Kayano T, Susuki H, Takeshima H, Kurasaki M, Takahashi H, and Numa S (1986) Existance of distinct sodium channel messenger RNAs in rat brain. *Nature* **320**: 188-192.
50. Noda M, Ikeda T, Susuki H, Takeshima H, Takahashi T Kuno M and Numa S. (1986) Expression of functional sodium channels from cloned cDNA. *Nature* **322**: 826-828.
51. Oakley AJ, Martinac B and Wilce MC (1999) Structure and function of the bacterial mechanosensitive channel of large conductance. *Protein Sci* **8**: 1915-1921.
52. Pavans de Ceccatti M (1989) Les éponges, à l'aube des communications cellulaires. *Pour la Science* **142**: 64-72.
53. Preston RR (1990) Genetic dissection of Ca^{2+} -dependent ion channel function in Paramecium *Bioessays* **12**: 273-281.
54. Ramón y Cajal S (1894) La fine structure des centres nerveux. Croonian Lecture. *Proc R Soc (London)* **55**: 444-468.
55. Reuter M and Gustafsson MK (1995). The flatworm nervous system pattern and phylogeny. *EXP* **72**: 25-59.
56. Rosenberg, RL, Tomiko S A and Agnew WA (1984 a). Reconstitution of neurotoxin-modulated ion transport by the voltage-regulated sodium channel isolated from the electroplax of Electrophorus electricus. *Proc Natl Acad Sci U.S.A.* **81**: 1239-1243.
57. Rosenberg RL, Tomiko S A and Agnew WA(1984 b). Single channel properties of the reconstituted voltage-regulated Na channels isolated from the electroplax of Electrophorus electricus. *Proc Natl Acad Sci U.S.A.* **81**: 5594-5598.
58. Shelton GAB. (Ed.) (1982) *Electrical conduction and behaviour in simple invertebrates.* Clarendon Press, Oxford, 567pp.
59. Schütze J, Krasko A, Custodio MR, Efremova SM, Müller IM and Müller WEG (1999) Evolutionary relationships of Metazoa within the eukaryotes based on molecular data from Porifera. *Proc R Soc Lond B* **266**: 63-73.
60. Sherrington CS (1897) *The Central Nervous System.* A Text Book of Physiology, Macmillan, London.
61. Solomon AK (1960) Red cell membrane structure and ion transport. *J Gen Physiol* **43** (5, Part 2) 1-15.
62. Spafford JD, Spencer AN and Gallin WJ (1998) A putative voltage-gated sodium channel a subunit (PpSCN1) from the hydrozoan jellyfish, Polyorchis penicillatus: Structural comparisons and evolutionary considerations. *Biochem Biophys Res Commun* **244**: 772-780.

63. Stühmer W, Methfessel C, Sakmann B, Noda M, and Numa S. (1987) Patch clamp characterization of sodium channels expressed from rat brain cDNA. *Eur Biophys J* **14**: 131-138.
64. Sukharev S (1999) Mechanosensitive channels in bacteria as membrane tension reporters. *FASEB J.* **13** Suppl: S55-61.
65. Taylor B.L, Zhulin IB and Johnson MS (1999) Aerotaxis and other energy-sensing behavior in bacteria. *Ann Rev Microbiol* **53**: 103-128.
66. Terlau H and Stühmer W (1998) Structure and function of voltage-gated ion channels. *Naturwissenschaften* **85**: 437-444.
67. Treherne J.E (1981) Glial-neuron interactions. *J Exp Biol* **95**: 1-20.
68. Ugarte G, Perez F and Latorre R (1998) How do calcium channels transport calcium ions? *Biol Res.* **31**: 17-32
69. Villegas GM and Villegas R (1968) Ultrastructural studies of the squid nerve fibers. *J Gen Physiol* **51**: 44s-60s
70. Villegas GM and Villegas R (1984) Squid axon ultrastructure. In: *Current Topics in Membranes and Transport.* vol **22**, The squid axon. P.F. Baker, ed. Academic Press, London. pp 3-37.
71. Villegas J (1995). Learning from the axon-Schwann cell relationships of the giant nerve fiber of the squid. In: *Neuron-glia interrelations during phylogeny: II Plasticity and regeneration.* A Vernadakis and B Roots (Eds). Humana Press Inc. Totowa N.J.
72. Villegas R and Villegas G M (1968) Characterization of the membranes in the giant nerve fiber of the squid. *J Gen Physiol* **43** (5, Part 2): 73-103.
73. Villegas R and Villegas GM (1981) Nerve sodium channel incorporation in vesicles. *Ann Rev Biophys Bioeng* **10**: 387-419.
74. Villegas R, Bruzual IB and Villegas GM (1968) Equivalent pore radius of the axolemma of resting and stimulated squid axons. *J Gen Physiol* **51** (5, Part 2), 81-92.
75. Villegas R, Villegas GM, Barnola FV and Racker E (1977) Incorporation of the sodium channel of lobster nerve into artificial liposomes. *Biochem Biophys Res Commun* **79**: 210-217.
76. Villegas R, Villegas GM, Rodriguez JM and Sorais-Landáez F (1988) The sodium channel of excitable and non-excitable cells. *Quart. Rev. Biophys* **21**: 99-128.
77. Virchow R (1846) Über das granulirte Anschen der Wandungen der Gehirnventrikel. *Allgem Z Psychiat* **3**: 424-450.
78. Wei AM, Covarrubias A, Butler K, Baker, Pak M, and Salkoff L (1990) K^+ current diversity is produced by an extended gene family conserved in *Drosophila* and mouse. *Science* **248**: 599-603.
79. Westfall IA (1996) Ultrastructure of synapses in the first-evolved nervous systems *J Neurocytol* **12**: 735-46.
80. White GB, Pfahnl A, Haddock S, Lamers S, Greenberg RM and Anderson PA (1998) Structure of a putative sodium channel from the sea anemone *Aiptasia pallida. Invert Neurosci* **3**: 317-326.
81. Woese CR and Fox GE (1977) Phylogenetic structure of the prokaryotic domain: The primary kingdoms. *Proc Natl Acad Sci USA.* **74**: 5088-5090.
82. Woese CR, Kandler O and Wheelis ML (1990) Towards a natural system of organisms: Proposal for the domains Archaea, Bacteria and Eucarya. *Proc Natl Acad Sci USA.* **87**: 4576-4579.
83. Young JZ (1934) Structure of nerve fibers in sepia. *J Physiol (London)* **83**: 27p-28p.

ORIGIN OF SYNAPSES: A SCIENTIFIC ACCOUNT OR THE STORY OF A HYPOTHESIS

ERNESTO PALACIOS-PRÜ*
Electron Microscopy Center, University of Los Andes. Mérida, Venezuela.
prupal@ula.ve
International Institute of Advanced Studies, Simón Bolívar University, Caracas, Venezuela.

Abstract: Synapses are the structural components developed by nature to allow neuronal communication and consequently they are the fundamental sites at which the integrative functions of the central nervous system, such as intelligence and others activities of integration of less importance, are carried through. In this contribution, from both phylogenetic and ontogenetic points of view, the origin of synapses was analyzed and commented in relation to the evolution of intelligence on Earth, herein interpreted as the capacity of matter to understand matter. This approach could also be useful for the comprehension of extraterrestrial intelligence.

1. INTRODUCTION

1.1. BRIEF HISTORICAL NARRATION

The term synaptogenesis was employed for the first time in 1969 by L. M. H. Larramendi to describe the principal ultrastructural events during development that lead to the formation of diverse synaptic contacts integrating the cerebellar cortical circuit. Synaptogenesis, according to Larramendi, is "the actual formation of synaptic adhesion, viewed as the interaction of synaptic competent cells possessing matching of complementary specificities; therefore, on theoretical grounds, it is conceivable that cells may develop their synaptic competence and yet fail to form synapses, if they never come in contact with other cells having matching specificities". This definition, however, only referred to the description of how synapses are formed depending on existing specific membrane sites between neurons that enter into contact.

In 1971, Joseph Altman proposed a possible mechanism in relation to the origin and formation of synapses, which stated that the coated vesicles, originating from the Golgi apparatus, were involved in the initiation of the process of synaptogenesis and were the ones responsible in the formation of the postsynaptic densities (PSDs). Two years later, D. J. Stelzner and his collaborators (1973) upheld Altman's hypothesis,

* Centro de Microscopía Electrónica, Universidad de Los Andes, P. O. Box 163, Mérida 5101-A, Venezuela.

J. Chela-Flores et al. (eds.), Astrobiology, 213–224.

however, they expressed their doubts whether the coated vesicles were endocytic or exocytic. In 1976, Rosemary Rees, Mary and Richard Bunge corroborated the hypothesis of Altman with experiments done on in vitro grown neurons. Maryellen Eckenhoff and Joseph Pysh in 1979 described the internalization of plasma membrane segments by coated vesicles, both in the granular and molecular layers, during cerebellar development, and according to these authors, this process represents a mechanism of neuronal modelling that occurs during late postnatal development. This interpretation opened another controversy about endocytic vesicles since they were thought of as modelling agents instead of synaptogenetic elements.

In 1980, Richard Burry proposed that the formation of synaptic junction is brought about by electrostatic differences between the growing axon and the postsynaptic element. The growth cone is supposed to carry acidic or negative charges and that the postsynaptic membrane has positive charges. The electrostatic attraction between the axon and the dendrite brings about the formation of a synapse. After some years of research, we published our results in 1981 with regard to the origin and formation of synapses between parallel fibers and the Purkinje cells' (PCs) dendritic spines. These studies led us to know the progressive stages of synaptogenesis. During the early stages of cerebellar neurogenesis, before the descent of the granule cells from the external to their final location in the internal granule cell layer, the postmitotic PCs, still showing primitive characteristics, are already localized in their definitive sites. We consider the initial period of sprouting or growth of the PC dendritic tree as the Critical Period (Palacios-Prü et al. 1976a, b, 1978a, b) because it is during this time when the first synapses are formed. The critical period varies according to the animal species: in chick, it coincides around embryonic day 16; in mouse, at postnatal day 7; and in rat, at postnatal day 9.

The first stage of synaptogenesis occurs before the critical period and which we denominated as the Period of Adhesion Plaques Formation. At this stage, the Golgi apparatus produces numerous exocytic clathrin-coated vesicles (Pearse 1980), which expose glycoproteins to the extracellular space. This phenomenon occurs both in the axons and dendrites. We were able to identify their exocytic nature because they did not incorporate the injected electron-dense tracers. The exocytic vesicles form highly electronegative adherence plaques, as shown by the accumulation of cationized ferritin or ferrous fumarate on their surface, that serve as attachment sites between parallel fiber terminals and PC dendritic spines (Figures 1 and 2).

The second stage corresponds to the Cellular Recognition Period. This period starts shortly after the critical period and is characterized by the formation of endocytic vesicles at the sites of adherence between the parallel fiber and PC dendrite. With the use of cationized ferritin and ferrous fumarate, the fate of endocytic vesicles, of which many were seen containing membrane fragments, was traced within the PC cytoplasm (Figures 1and 2).

The third stage corresponds to the Period of Postsynaptic Density (PSD) Formation by the subsynaptic ribosomes. This period is characterized by the appearance of polyribosomes at the adherence sites, where endocytosis of part of the adhesion plaque membrane was produced. As electron-dense granular and fibrous materials were synthesized by the subsynaptic polyribosomes at the postsynaptic side, a progressive increment in the number of synaptic vesicles could be observed at the presynaptic side

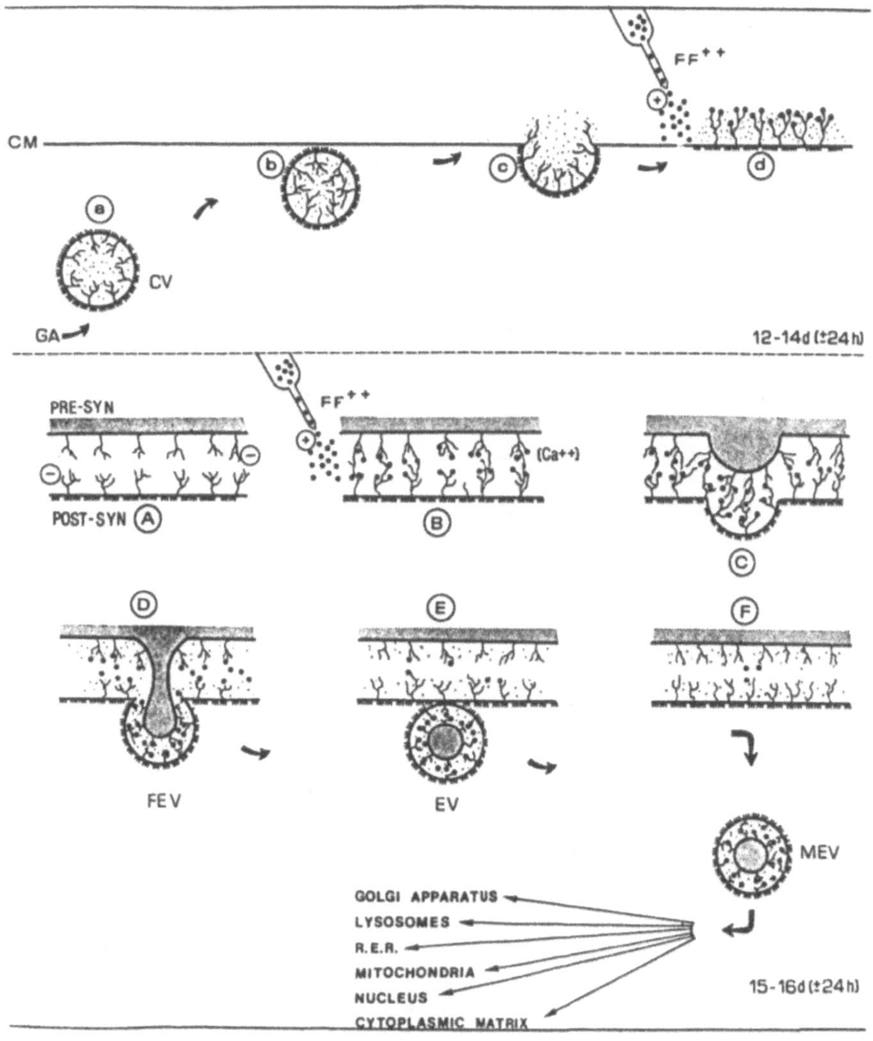

Figure 1. Between days 12 to 14, the Golgi apparatus (GA) produces (a) coated vesicles (CV) that migrate (b) to the surface (c) where they form plaques (d); there the glycoproteic inner coat of CV is exposed after fusion to the cellular membrane (CM). Since the glycoprotein of the adhesion plaques are negatively charged, the cationic tracers attach to the membrane (d). Between days 15 to 16, (**A**) the incoming axon fibers (pre-syn) become attracted by the plaques and attach their own plaques to the postsynaptic counterparts (post-syn), (**B**) the cationic tracer label the anionic glycoproteins of both pre- and postsynaptic sites, (**C**–**E**) once this contact is established, endocytic vesicles are formed (FEV and EV) at this site incorporating part of the axonic membrane, the axoplasm and the injected tracers, (**F**) loaded migrating endocytic vesicles (MEV) deliver their contents into several cellular organelles.

(Palacios-Prü et al. 1981). This stage ends when the synapse has acquired its phenotypic characteristics followed by a decrease in the number of subsynaptic ribosomes (Figures 2 and 3).

With regard the role played by subsynaptic polyribosomes in the synthesis of PSDs, it is neccesary to mention that in 1980, M. Anthony Verity and collaborators showed the capacity of protein synthesis by ribosomes present in synaptosomal preparations from 10-14-day-old rat cerebral cortices. Oswald Steward (1986) demonstrated the reappearance of subsynaptic ribosomes during synaptic reinnervation after experimental deafferentation.

In relation to our experiments, we designed a novel procedure of subcellular fractionation using a mild cellular dissociation procedure, by which we were able to separate the dendritic and axonal processes from neuronal cell bodies. With this method, we were able to determine the amount of RNA and proteins contained in the dendrites, free from contamination with those present in the cell bodies (Figure 4). Our results showed that the dendritic RNA and protein contents increased by about 6-10-fold during the critical period of synaptogenesis in chick cerebellar cortex, compared with their values before the critical period of synaptogenesis (Palacios-Prü et al. 1988). On the other hand, the ultrastructural studies demonstrated the presence of postsynaptic polyribosomes during the critical period which disappeared in the young adult stage. These results reinforced our hypothesis on the polyribosomal origin of PSDs thereby consolidating our proposal in 1981 with regard to the general theory of the origin of synapses (Palacios-Prü et al. 1981).

In a review article of James Vaughn in 1989, he accepted the first stages of our hypothesis, however, he did not put importance on the role played by subsynaptic polyribosomes on PSD formation. On the other hand, Marcus Jacobson, in the third edition of his classic book on Developmental Neurobiology (1991), described Altman's hypothesis on synaptogenesis, including that of Rees and his collaborators (1976). He mentioned the hypothesis of the ribosomal origin of PSD but cited only the publication of Steward and coworkers in 1988. However, from the accumulated evidence with regard the PSD protein composition (Matus 1981, Walsh & Kuruc 1992, Kennedy 1997, 1998) and the molecular nature of clathrin, the coat protein of the coated vesicles (Pearse 1980), the hypothesis of Altman and supported by Rees, Bunge and Bunge lost validity. The findings of Burry in 1980 contributed to the failure of Altman's hypothesis when he demonstrated that the mechanism of cellular recognition should be more specific and complex and not just brought about by merely membrane surface complementarity, as shown by the capacity of axons to adhere on sepharose beads coated with cationized ferritin. Finally, recent reports (Steward & Banker 1992, Steward 1995, Torre & Steward, 1996) have provided increasing evidence of local synthesis and processing of proteins at postsynaptic sites thereby corroborating the hypothesis that we have formulated more than a decade ago on the ribosomal origin of the PSDs.

1.2. CURRENT KNOWLEDGE ON SYNAPTOGENESIS

In the study of synaptogenesis, several investigators have focussed their attention not only on the development of the presynaptic but also on the postsynaptic components of

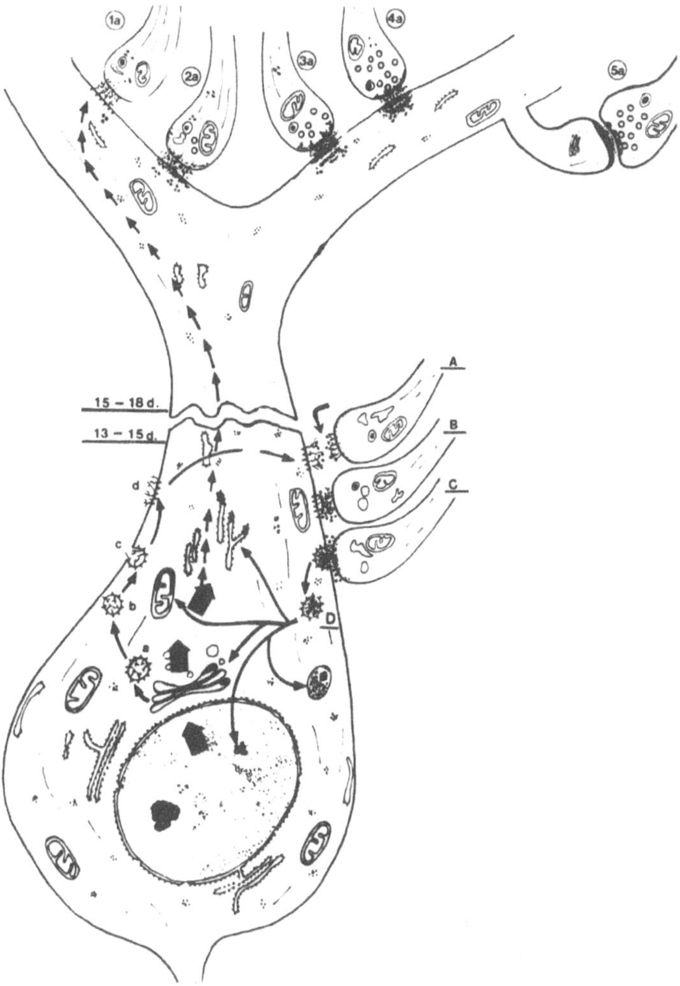

Figure 2. This semi-schematic drawing summarizes the main morphological events that guide cerebellar synaptogenesis. Between days 12 to 14, the Purkinje cell's Golgi apparatus produces numerous coated vesicles that migrate to the surface where they form plaques increasing the adherence to the incoming axons by means of their glycoproteic inner coat which is exposed when the vesicles fuse to the cellular membrane; this part is illustrated in a, b, c, d. The incoming parallel fibers become attracted by the plaques and attach their own plaques to the dendritic counterparts to form symmetrical contacts. Once these contacts are established, endocytic vesicles are formed at these sites incorporating part of the axoplasm, the axonic membrane and the tracers injected (curved arrow), as shown in A, B, C, D. This sequence of events is part of the Period of Adhesion Plaque Formation. Loaded pinocytic vesicles deliver their contents directly by fusion or indirectly at the cellular organelles (continuous arrows). From this point, the Purkinje cell seems to enter a process of cytoplasmic recognition which implies the decodification of the protein of the incoming axon and the subsequent formation of appropriate RNA to produce specific postsynaptic proteins (heavy arrows). This is the Period of Cellular Recognition. After this period specific ribosomes are placed at postsynaptic sites to produce in situ electron-dense postsynaptic proteins between days 15 to 18 (1a, 2a, 3a, 4a) (discontinuous arrow line). This way of making up postsynaptic densities correspond to the Period of Postsynaptic Density Formation (5a) (modified from Palacios-Prü et al. 1981).

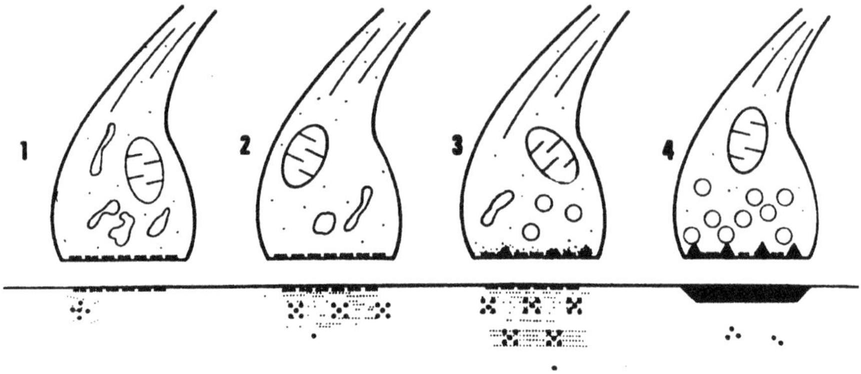

Figure 3. Formation of the postsynaptic densities of parallel fiber-Purkinje cell dendrite synapses by subsynaptic ribosomes during cerebellar development (modified from Palacios-Prü et al. 1988).

the synapse (Siekevitz 1985, Walsh & Kuruc 1992, Kennedy 1997, 1998). Much is already known in relation to the presynaptic element and numerous studies have been reported with regard to the receptor machinery of the postsynaptic membrane. However, in relation to the PSDs, which are disc-shaped proteinaceous structures tightly apposed to the postsynaptic membrane and associated with the receptor apparatus, in spite of the enormous amount of published reports, still much information are required to elucidate the molecular nature of its structure and function.

Subcellular fractions enriched with PSDs were described several years ago by various research groups (Cohen et al. 1977, Kelly & Cotman 1981, Kennedy et al. 1983). Some of the major protein components of the PSD fraction included: cytoskeletal proteins as actin, tubulin, spectrin, myosin, and neurofilament proteins (Cohen et al. 1977, Sahyoun et al. 1986, Adam & Matus 1996); protein kinases as the α and β subunits of calcium/calmodulin-dependent protein kinase II (CaM kinase II) (Levine III & Sahyoun 1988, Burgin et al. 1990); glycoproteins as gp 130, gp 150 (Beesly 1989); receptor-associated proteins as PSD-93 (Roche et al. 1999) and PSD-95 (Hunt et al. 1996); and a new class of synaptic adhesion molecules as densin 180 (Apperson et al. 1996).

As a result of our studies on specific PSD protein components in cerebellar cortex synapses, we reported a putative specific PSD protein of 140 kDa, that was produced during PC spine maturation (Miranda-Contreras & Palacios-Prü 1995). Amino acid analysis of the 140 kDa protein revealed a high content of nonpolar amino acids, such as leucine, isoleucine, glycine, valine and phenylalanine. A hypothetical model relative to the participation of the 140 kDa protein in the molecular organization of the PSD was suggested which may contribute to the understanding of the role played by this structure in synaptic neurotransmission.

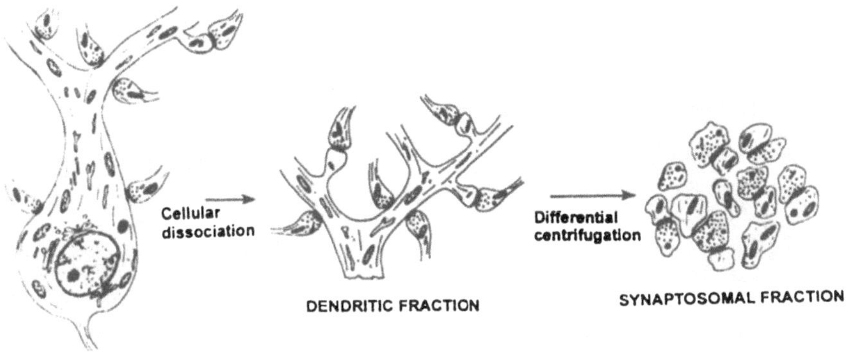

Figure 4. Summary of the procedure for the preparation of synaptosomes (modified from Palacios-Prü et al. 1988).

2. SYNAPTOPHYLOGENESIS: ANOTHER STORY ON THE SAME SUBJECT

The natural history of synapses, which are the principal structural sites of interaction between neuronal collectivities enabling them to perform the automatic and complex functions characteristic of the nervous tissue, has been poorly understood in spite of the extraordinary efforts of Bullock & Horridge (1965) and Lentz (1967).

The evolutionary origin of the nerve cells, neurophylogenesis, is not well-known and the scarce information available is incoherent, disperse and, in general, not up to date. In recent times, interest on this important area of Neuroscience has decreased considerably and it is difficult to find literature in relation to the topic in question. Therefore, we decided to study metazoans of increasing tissue complexity, not strictly taxonomically but following the general sequence of evolutionary taxonomy (Figure 5). We focussed our study on representative samples of the following phyla: sponges, cnidarians, platyhelminthes, molluscs and onycophora. In that sequence of increasing order of complexity, we systematically analyzed at the ultrastructural level the presence of cells with morphological and biological markers that would characterize them as nerve cells. We also focussed our attention on the different intercellular contacts, from simple adhesion plaques to intercellular junctions indicative of the ultrastructural pattern of the synaptic junction.

2.1. SPONGES

In our study of the freshwater sponges, the presence in these metazoans of a complex framework of siliceous material hindered the reliable identification of the cells due to the fact that the needle-like siliceous material caused damages to the cells during sample sectioning. For this reason, these results were not included in this study.

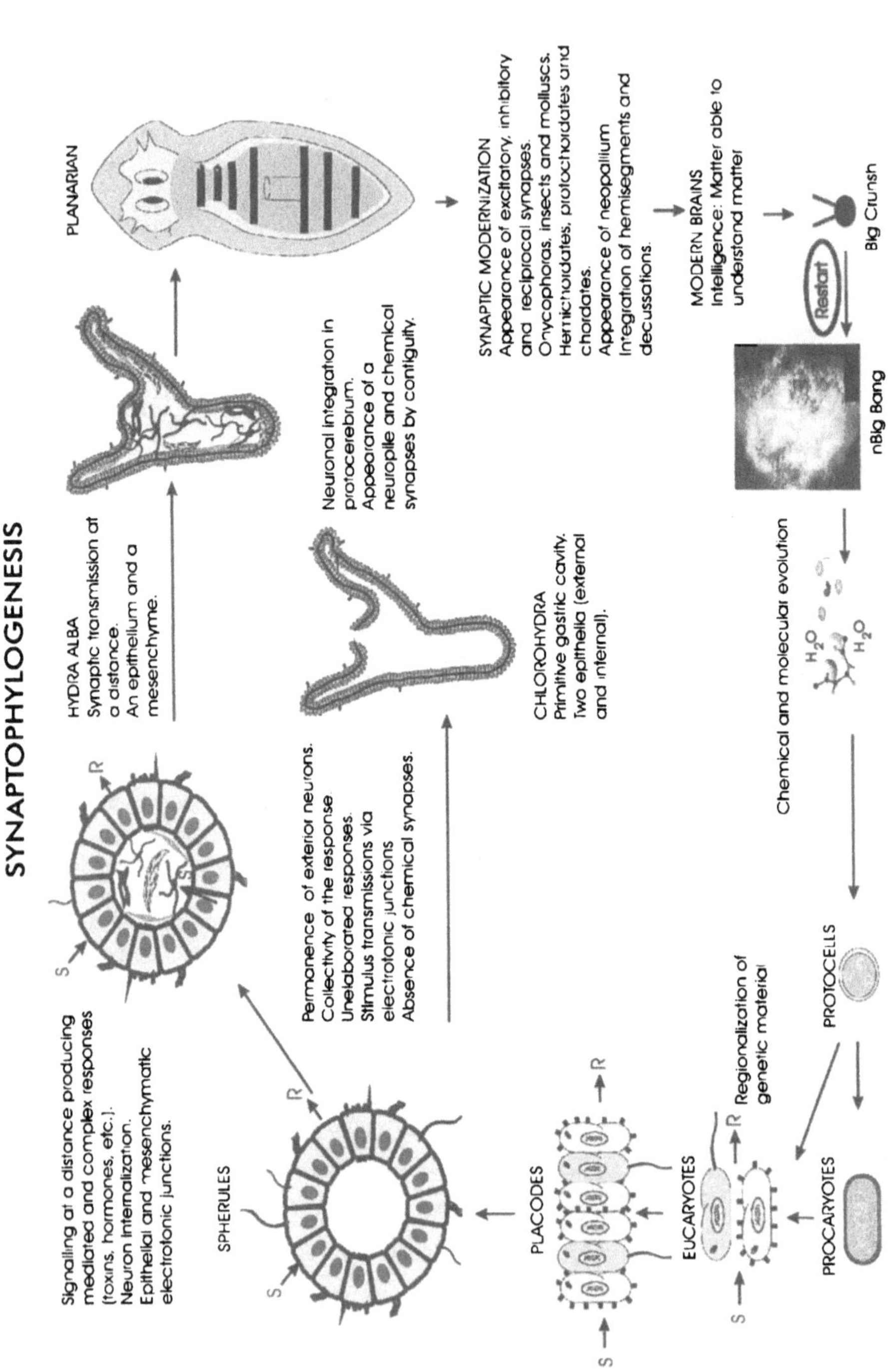
SYNAPTOPHYLOGENESIS
Signalling at a distance producing mediated and complex responses (toxins, hormones, etc.).
Neuron internalization.
Epithelial and mesenchymatic electrotonic junctions.
SPHERULES
HYDRA ALBA
Synaptic transmission at a distance.
An epithelium and a mesenchyme.
PLANARIAN
Permanence of exterior neurons.
Collectivity of the response.
Unelaborated responses.
Stimulus transmissions via electrotonic junctions
Absence of chemical synapses.
CHLOROHYDRA
Primitive gastric cavity.
Two epithelia (external and internal).
Neuronal integration in protocerebrum.
Appearance of a neuropile and chemical synapses by contiguity.
SYNAPTIC MODERNIZATION
Appearance of excitatory, inhibitory and reciprocal synapses.
Onycophoras, insects and molluscs.
Hemichordates, protochordates and chordates.
Appearance of neopallium
Integration of hemisegments and decussations.
MODERN BRAINS
Intelligence: Matter able to understand matter
Big Crunsh
Restart
nBig Bang
Chemical and molecular evolution
H_2O
H_2O
PLACODES
EUCARYOTES
Regionalization of genetic material
PROTOCELLS
PROCARYOTES
S
R

2.2. CNIDARIANS

Within this phylum, we studied two species of freshwater hydra. We denominated chlorohydra (*Chlorohydra viridissima*) to one of them for containing numerous algae in its interior and hydra alba (*Hydra attenuata*) to the other specie because of its white color. The chlorohydra was captured in a lagoon located at an altitude of about 1200 m in the Andes mountains of Mérida, Venezuela, and the hydra alba was obtained from Carolina Biological Supply Company, USA. The chlorohydra is an organism composed of two epithelia, an external one and the other, internal, which limits the gastric cavity. Between both epithelia there is a lamina having the aspects of a basal lamina constituting the mesoglea. According to the literature (Lentz 1967, Wesfall 1980), in both external and internal epithelia are encountered sensory cells or neurons and cnidocytes serving as support and muscle cells. In the gastric cavity, no cells were observed.

In hydra alba, we were able to observe in the gastric cavity a trabecular system composed of cells with fine cellular processes that are adhered to the mesoglea giving to this cavity a spongy aspect. In between these cells, which we denominated mesenchymatic, three types of cells can be distinguished: cells with abundant secretory granules and cilia that form canalicular spaces with their numerous microvilli; cells with dense cytoplasm filled with microfilaments similar to the vertebrate smooth muscle cells; and cells with less dense cytoplasm containing small, clear vesicles similar to synaptic vesicles. Between the mesenchymatic and muscle cells, cellular contacts of the electrotonic type can be seen. Adherence junctions or desmosomes and even septate junctions were observed only between mesenchymatic cells. In all the examined samples, we were not able to see cellular junctions with the ultrastructural characteristics of a synapse.

2.3. PLATYHELMINTHES

The samples of locally captured planarians showed an important advancement in terms of neuronal organization. In planarian, there exists an internalization and aggregation of neurons which has been denominated as the cephalic ganglion. In the latter, a neuropil is formed in which true synapses can be observed. Within the elements of the neuropil, electrotonic junctions were also seen. These chemical synapses show clear neurotransmitter-containing vesicles and large neurosecretory dense-cored vesicles. The majority of these synapses showed postsynaptic densities which are commonly found in the excitatory synapses of the vertebrates. In planarian, cells equivalent to glial cells were not observed.

2.4. MOLLUSCS AND ONYCOPHORA

In these two phyla, the nervous tissue is organized in encapsulated ganglia, and between

---

Figure 5. General hypothesis relative to the natural history of the synapses, the communication machinery of nerve cells. S, stimulus; R, responses.

the nerve cells, glial cells are seen forming perisynaptic lamellae. In onycophora, both asymmetric and symmetric chemical synapses are found, including electrotonic junctions. The ganglia are divided into a cellular peripheral one and a central zone composed of neuropil.

In summary, according to the results obtained in this study, we may conclude that the first identifiable neurons, different from the sensory epithelial cells, are found in the interior of the hydra alba. In chlorohydra, the interior cavity contains algae that are associated with the internal epithelium but do not contain differentiated cells within the cavity. No true chemical synapses, even primitive ones, were found in hydra alba, although electrotonic junctions were seen between neurons, mesenchymatic, muscular and secretory cells.

The first contiguous chemical synapses were seen in the encapsulated ganglia of planarians. This fact allows us to assume that the synapses found in hydras are of the distant type of synapses and not of the contiguous ones, although electrotonic type of transmission were observed. Inhibitory or symmetric chemical synapses are first observed in onycophora. Glial cells start to be clearly distinguished in the ganglia of molluscs and onycophora.

ACKNOWLEDGEMENTS

This study was partially supported by CDCHT-ULA grant M-533-85-03-AA. The author would like to thank José Ramírez and Francisco Durán for their technical assistance.

3. REFERENCES

Adam, G., Matus, A. (1996) Role of actin in the organization of brain postsynaptic densities. *Mol. Brain Res.* **43**: 246-250.

Altman, J. (1971) Coated vesicles and synaptogenesis. A developmental study in the cerebellar cortex of the rat. *Brain Res.* **30**: 311-322.

Apperson, M. L., Moon, L. S. and Kennedy, M. B. (1996) Characterization of densin-180, a new brain-specific synaptic protein of the O-sialoglycoprotein family. *J. Neurosci.* **16**: 6839-6852.

Beesly, W. P. (1989) Mini Review: Immunological approaches to the study of synaptic glycoproteins. *Comp. Biochem. Physiol.* **93**[a]: 255-266.

Bullock, T. H. and Horridge, G. A. (1965) *Structure and Function in the Nervous Systems of Invertebrates.* Vol. I and II. W.H. Freeman and Company. Netherlands.

Burgin, K. E., Waxham, M. N., Rickling, S., Westgate, S. A., Mobley, W. C. and Kelly, P. T. (1990) In situ hybridization histochemistry of Ca^{++}/calmodulin-dependent protein kinase in developing rat brain. *J. Neurosci.* **10**: 1788-1798.

Burry, R. W. (1980) Formation of apparent presynaptic elements in response to polybasic compounds. *Brain Res.* **184**: 85-98.

Cohen, R. S., Blomberg, F., Berzins, K. and Siekevitz, P. (1977) The structure of postsynaptic densities isolated from dog cerebral cortex. I. Overall morphology and protein composition. *J. Cell Biol.* **74**: 181-203.

Eckenhoff, M. P. and Pysh, J. J. (1979) Double-walled coated vesicle formation: evidence for massive and transient conjugate internalization of plasma membranes during cerebellar development. *J. Neurocytol.* **8**: 623-638.

Hunt, C. A., Schenker, L. J. and Kennedy, M. B. (1996) PSD-95 is associated with the postsynaptic density and not with the presynaptic membrane at forebrain synapses. *J. Neurosci.* **16**: 1380-1388.
Jacobson, M. (1991) *Developmental Neurobiology*. Plenum Press, New York.
Kelly, P. T. and Cotman, C. W. (1981) Developmental changes in morphology and molecular composition of isolated synaptic junctional structures. *Brain Res.* **206**: 251-271.
Kennedy, M. B., Bennett, M. K. and Erondu, N. G. (1983) Biochemical and immunological evidence that the "major postsynaptic density protein" is a subunit of a calmodulin-dependent protein kinase. *Proc. Natl. Acad. Sci. USA* **80**: 7357-7361.
Kennedy, M. B. (1997) The postsynaptic density at glutamatergic synapses. *Trends Neurosci.* **20**: 264-268.
Kennedy, M. B. (1998) Signal transduction molecules at the glutamatergic postsynaptic membrane. *Brain Res. Rev.* **26**: 243-257.
Larramendi, L. M. H. (1969) Analysis of synaptogenesis in the cerebellum of the mouse In *Neurobiology of Cerebellar Evolution and Development*, ed. R. Llinás (Am. Med. Assoc. Educ. Res. Found., Chicago), pp. 803-843.
Lentz, T. L. (1967) *Primitive Nervous System*. Yale University Press, U.S.A.
Levine III, H. and Sahyoun, N. E. (1988) Two types of brain calmodulin-dependent protein kinase II: Morphological, biochemical and immunocytochemical properties. *Brain Res.* **439**: 47-55.
Matus, A. (1981) The postsynaptic density. *Trends Neurosci.* **4**: 51-53.
Miranda-Contreras, L. and Palacios-Prü, E. L. (1995) Existence of a putative specific postsynaptic density protein produced during Purkinje cell spine maturation. *Int. J. Dev. Neurosci.* **13**: 403-416.
Palacios-Prü, E. L., Palacios, L. and Mendoza, R. V. (1976a) In vitro vs in situ development of Purkinje cell. *J. Neurosci. Res.* **2**: 357-362.
Palacios-Prü, E. L., Palacios, L. and Mendoza, R. V. (1976b) Formación de circuitos neuronales in vitro. I. Análisis con el método de Golgi de cultivos rotatorios íntegros de cerebelo. *Acta Cient. Venezolana* **27**: 301-308.
Palacios-Prü, E. L., Palacios, L. and Mendoza, R. V. (1978a) Formación de circuitos neuronales in vitro. III. Análisis ultraestructural de cultivos histotípicos de cerebelo y médula espinal. *Acta Cient. Venezolana* **28**: 19-29.
Palacios-Prü, E. L., Palacios, L. and Mendoza, R. V. (1978b) Cultivos neuronales histotípicos: significación y perspectivas. *Acta Cient. Venezolana* **29**: 295-308.
Palacios-Prü, E. L., Palacios, L. and Mendoza, R. V. (1981) Synaptogenetic mechanisms during chick cerebellar cortex development. *J. Submicrosc. Cytol.* **13**: 145-167.
Palacios-Prü, E. L., Miranda-Contreras, L., Mendoza, R. V. and Zambrano, E. (1988) Dendritic RNA and postsynaptic density formation in chick cerebellar synaptogenesis. *Neuroscience* **24**: 111-118.
Pearse, B. (1980) Coated vesicles. *Trends Biochem. Sci.* **5**: 131-134.
Rees, R. P., Bunge, M. B. and Bunge, R. P. (1976) Morphological changes in the neuritic growth cone and target neuron during synaptic junction development in culture. *J. Cell Biol.* **68**: 240-263.
Roche, K. W., Dune Ly. C., Petralia, R. S., Wang, Y. X., McGee, A. W., Bredt, D. S. and Wenthold, R. J. (1999) Postsynaptic density-93 interacts with the δ2 glutamate receptor submit at parallel fiber synapses. *J. Neurosci.* **19**: 3926-3934.
Sahyoun, N., Levine III, H., McDonald, O. B. and Cuatrecasas, P. (1986) Specific postsynaptic density proteins bind tubulin and calmodulin-dependent protein kinase type II. *J. Biol. Chem.* **261**: 12339-12344.
Siekevitz, P. (1985) The postsynaptic density: A possible role in long-lasting effects in the central nervous system. *Proc. Natl. Acad. Sci. USA* **82**: 3494-3498.
Stelzner, D. J., Martin, A. H. and Scott, G. L. (1973) Early stages of synaptogenesis in the cervical spinal cord of the chick embryo. *Z. Zellforsch.* **138**: 475-488.
Steward, O. (1986) Protein synthesis under dendritic spine synapses during lesion-induced synaptogenesis. Evidence for regulation of reinnervation by target cell. *Exp. Brain Res.* **13**: 173-188.
Steward, O., Davis, L., Dotti, C., Phillips, L. I., Rao, A. and Banker, G. (1988) Protein synthesis and processing in cytoplasmic microdomains beneath postsynaptic sites on CNS neurons. *Mol. Neurobiol.* **2**: 227-226.
Steward, O and Banker, A. (1992) Getting the message from the gene to the synapse: sorting and intracellular transport of RNA in neurons. *Trends Neurosci.* **15**: 180-186.
Steward, O. (1995) Targeting of mRNAs to subsynaptic microdomains in dendrites. *Current Opinion in Neurobiology* **5**: 55-61.
Torre, E, and Steward, O. (1996) Protein synthesis within dendrites: Glycosylation of newly synthesized proteins in dendrites of hippocampal neurons in culture. *J. Neurosci.* **16**: 5967-5978.

Vaughn, J. E. (1989) Review: Fine structure of synaptogenesis in the vertebrate central nervous system. *Synapse* **1**: 255-285.

Verity, M. A., Brown, W. J. and Cheung, M. (1980) Isolation of ribosome containing synaptosomes subpopulation with active in vitro protein synthesis. *J. Neurosci. Res.* **5**: 143-153.

Walsh, M. J. and Kuruc, N. (1992) The postsynaptic density: Constituent and associated proteins characterized by electrophoresis, immunoblotting, and sequencing. *J. Neurochem.* **59**: 667-678.

Westfall, J. A., Kinnamon, J. C. and Sims, D. C. (1980) Neuro-epitheliomuscular cell and neuro-neuronal gap junctions in *Hydra*. *J. Neurocytology* **9**: 725-732.

ORIGINS OF LANGUAGE
The evolution of human speech

M.E. MEDINA-CALLAROTTI
Departamento de Idiomas and Instituto de Estudios del Conocimiento (INESCO), Universidad Simón Bolívar
Apartado Postal 89000, Caracas 108, Venezuela

Abstract: We propose that human language is a biological adaptation whose function is communicating information. As such, it has been subject to the mechanisms of evolution and natural selection. The essential characteristics of language as a complex system, basic aspects of Universal Grammar, and the processes whereby language is acquired are presented as evidence of this affirmation.

1. The essence of human language systems

If viewed simply as a system of communication, language is not a phenomenon limited to humans. Nevertheless, certain characteristics of human language are not found in the communicative systems of any other species. Wolves, dolphins, bees and eagles interact by means of mechanisms that, though quite expressive, appear to be fixed, limited, and stimulus-bound. In recent experiments, primates have demonstrated a limited capacity to acquire some complex rules, but this output does not exhibit characteristics essential to human communication.

Basically, human language is creative, systematic and generative: from a finite set of discreet units, humans systematically produce an infinite set of novel utterances. "The organizational principles characteristic of human language are extremely complex and abstract, and the vast majority of persons never consciously attain knowledge of them... [Yet]...this ignorance of the rules that underlie the linguistic system to which we have been exposed places no limits on either our linguistic or communicative competence." (Medina de Callarotti, 1993, p.11) Thus, communication – speaking and listening – is a complex, specialized, and skillful activity, which generally occurs below the level of consciousness, on the basis of a tacit awareness of the rules underlying the system.

Language – henceforth, we refer to human language – is a set of arbitrary symbols with conventionalized meanings used to communicate in a speech community or culture. As a species-specific behavior, language is uniformly distributed and qualitatively similar in all human groups, i.e., it exhibits universal characteristics. It develops spontaneously in early childhood, without conscious effort or formal instruction; i.e., the process whereby language is acquired also exhibits universal characteristics. It is distinguishable from other, more general abilities of information processing or intelligent behavior.

J. Chela-Flores et al. (eds.), Astrobiology, 225–232.

As we will explore in this presentation, language is not an acquired cultural artifact, but rather a specific biological component of the human brain—in a very real sense, a mental organ (Chomsky, 1975). We intend to show that as a biological adaptation whose function is communicating information, language has been subject to the mechanisms of evolution and natural selection (Pinker, 1994). We will proceed within the framework of cognitive science, an integrative discipline which combines tools of psychology, computer sciences, linguistics, philosophy and neurobiology to explain the workings of human intelligence.

2. Chomsky and Universal Grammar

Chomsky (1965, 1975), protagonist of the modern revolution in linguistics and cognitive science, was the first to point out two fundamental facts: first, almost any sentence a person utters, hears and understands, is a new combination of words. Thus, a language cannot be a repertory of conditioned responses, a habit, as behaviorist theory proposed (Skinner, 1957). The brain must contain a recipe or a program – a mental grammar— which can construct an unlimited set of sentences from a finite list of elements. Second, children develop these complex grammars quickly and without formal instruction. As they mature, children are increasingly and consistently capable of producing and interpreting new constructions, which they have never encountered before. "Thus, it is clear that the language each person acquires is a rich and complex construction hopelessly underdetermined by the fragmentary evidence available." (Chomsky, 1975, p.10) Children, Chomsky asserted, must be innately equipped with a plan common to the grammars of all languages, a Universal Grammar, that tells them how to distill the syntactic patterns out of the speech of their parents.

2.1 LANGUAGES ARE INFINITE; CHILDHOOD IS FINITE

Just how "hopelessly underdetermined" language is can be illustrated by facing what linguists have called "the logical problem of language acquisition." The sheer vastness of language induces arithmetic impossibilities when behaviorist paradigms of stimulus-response are applied to language acquisition. For example, acquiring the transitions among grammatical categories necessary to process an English sentence by behaviorist mechanisms would imply that we learned, "...as Miller and Chomsky point out...the value of 10^9 parameters in a childhood lasting only 10^8 seconds." (Brown, 1994, p.24) If a speaker is interrupted at any randomly selected point in a 20-word sentence, an average of 10 different words could be inserted that would be a grammatical and meaningful continuation of the sentence. At some points, only one word is possible, but at others, literally thousands could be inserted; the average is 10. At a rate of 5 seconds per sentence, a person would need a childhood of 100 trillion years, with no time for eating or sleeping, to memorize all of them. (Pinker, 1994) Yet 3- and 4- year-olds produce and understand novel sentences much longer than 20 words. Demonstrating that children know things that they have not been taught has been one of the most interesting areas of linguistic experimentation since the 1960s. The "argument from the poverty of the input," is, as Pinker states, "...the primary justification for saying that the basic design of language is innate." (1994, p.42)

2.2 DEVELOPMENT OF COGNITIVE SYSTEMS

In a much-quoted passage, Chomsky has pointed out the curious fact that during the past few centuries, science has approached physical and mental development in different ways. All accept that the human organism does not learn through experience to have arms rather than wings, nor that the basic structure of the heart is the result of accidental experience. The physical structure of the organism is accepted to be genetically determined, with some variation due in part to external factors. Such genetically determined schemes of growth and development are "...characteristic of the species and...[give] rise to structures of marvelous intricacy. The species' characteristics themselves have evolved over long stretches of time, and evidently the environment provides conditions for differential reproduction, hence evolution of the species." (Chomsky, 1975, pp. 9-10) Conversely, empiricist tradition has approached the study of the development of cognitive structures in higher organisms as if the dominant factor were the social environment. According to this view, the development of human cognitive structures can be explained –though no satisfactory explanations are in evidence—by certain general principles of learning. But, Chomsky argues, "...human cognitive systems, when seriously investigated, prove to be no less marvelous and intricate than the physical structures that develop in the life of the organism. Why, then, should we not study the acquisition of a cognitive structure such as language more or less as we study some complex bodily organ?" (1975, p.10)

The great variety of human languages seems to cast doubt on this proposal, but upon closer consideration, "...even knowing very little of substance about linguistic universals, we can be quite sure that the possible variety of language is sharply limited...." The fact that persons in a speech community develop the same language "...can be explained only on the assumption that...individuals employ highly restrictive principles that guide the construction of grammar." Since humans are obviously not designed to learn one particular language rather than another, "...the system of restrictive principles must be a species property. Powerful constraints must be operative restricting the variety of languages." (Chomsky, 1975, pp. 10-11) Chomsky has claimed that a Martian visitor would conclude that all humans speak a single language. "[This] is based on the discovery that the same symbol-manipulating machinery, without exception, underlies the world's languages." (Pinker, 1994, p.237)

2.3 ASPECTS OF UNIVERSAL GRAMMAR

Even before Chomsky's ideas arrived on the scene, linguists had noted many universal characteristics of languages. Among these are the utilization of the mouth-to-ear channel in hearing persons, substituted by facial and hand gestures in the deaf; a common grammatical code used for production and comprehension; stable meanings for words, established arbitrarily by convention; the processing of sound signals; common patterns and stages in children's linguistic development. Undoubtedly, however, one of the most fascinating discoveries of modern linguistics is Chomsky's: the fact that sentences in all of the world's languages have a common anatomy.

Though detailed analyses of syntactic structures and mechanisms are beyond the scope of this presentation, it is important to clarify that Universal Grammar (UG) does not make "...vague or unverifiable suggestions about properties of the mind, but precise statements based on specific evidence. The general concepts of the theory are inextricably connected with the specific details; the importance of UG theory is its attempt to integrate grammar, mind, and language at every moment." (Cook & Newson 1996, p.3) What follows is an outline of some essential aspects of Universal Grammar (UG) and some simple examples as substantiation.

Defined by Chomsky (1975) as the system of principles, conditions and rules that are elements or properties of all human languages, it is important to remember that UG is a theory of knowledge, not of behavior. UG does not refer to the rules of a particular grammar, but rather to linguistic properties inherent to the human mind. In other words, UG is an innate faculty which sets the limits within which languages can vary.

According to UG theory, "...a speaker knows a set of **principles** that apply to all languages, and **parameters** that vary within clearly defined limits from one language to another. Acquiring language means learning how these principles apply to a particular language and which value is appropriate for each parameter." (Cook & Newson 1996, p.3) Certain principles, or "super-rules" are a general guide for all languages. The principle of **structure dependency** is a simple example (Cook, 1991). In order to form questions in English of sentences such as

'Sam is the cat that is black',

one could propose a simple linear instruction, such as, "Move the second word 'is' to the beginning." This works, and we obtain

'Is Sam the cat that is black?"

However, applying this rule for forming questions to the sentence

'The big cat is the one that is black',

produces

'Big the cat is the one that is black?',

a sentence that would be judged ungrammatical by any English speaker. One could then suggest another rule: "Move the copula 'is' to the beginning of the sentence." But since there are two copulas, one could produce

'Is big the cat is the one that black?'

Thus, which 'is' one moves is quite important. The only instruction that works properly is, "Move the copula 'is' in the main clause to the beginning of the sentence." (Cook, 1991, p.23) This instruction relies on the speaker's knowing enough sentence structure to be able to distinguish the main clause from the relative clause.

Yet very young children, as well as adult speakers who have never been instructed in the tedious meta-language of school-taught prescriptive "grammar", when they produce such inversion questions in English – and in all other languages – exhibit a knowledge of structure, not just of the order of the words. And this knowledge of a highly abstract rule is generally tacit, unconscious, and not easily accessible to speakers (Medina de Callarotti, 1993). Structure dependency, then, is one of the language principles that is built-in to the human mind (Cook 1991). No human language has been found that does not depend on structure.

A piece of information that makes one language different from another is a parameter. To illustrate, let us use X-bar theory, shorthand for a general scheme devised by Chomsky to point out that phrases in all languages consist of a head word X, followed by any number of non-subject arguments, or role-players, <u>in either order</u>. English puts "head" words first, as in 'Kenji ate sushi'. Head-first languages are head-first across all the kinds of phrases, e.g., 'Kenji went to school', 'Kenji is taller than...', 'Did Kenji eat?' Languages that follow this subject-verb-object (SVO) order all have "prepositions." Japanese is a head-last language, with SOV order, and this is demonstrated in all its phrases, e.g. (roughly) 'Kenji sushi ate', 'Kenji than taller...', 'Kenji eat did?' Head-last languages all have SOV order and "postpositions." This remarkable consistency has been found in scores of languages.

UG theory suggests that the principles, the super-rules, are universal and innate. Children do not have to learn a long list of rules when they learn a particular language, because they were born knowing the super-rules. They only need to learn, for example, whether the particular language they are exposed to has the value of the parameter "head-first" or "head-last." By simply "noting" (unconsciously) whether the verb comes before or after the object of any phrase uttered by their parents (e.g. Drink your milk! versus Your milk drink!), the child arrives at certain conclusions that allow him to obtain huge portions of the grammar, all at once. It is as if the child is setting the stops in a complex church organ before playing the music. Once activated, of course, a parameter can make many changes in the surface characteristics of a language. Nonetheless, it appears from incontrovertible evidence that a common plan of principles and syntactic, morphological and phonological rules exists for all languages, with a small set of variable parameters, somewhat like a list of options. "If this theory of language is true, it would help solve the mystery of how children's grammar explodes into adult-like complexity in so short a time. They are not acquiring dozens or hundreds of rules; they are just setting a few mental switches." (Pinker 1994, p.112)

3. Language as an evolutionary adaptation

That language is a species-specific adaptation, not simply a cultural artifact, is evident: although the sophistication of cultural inventions (e.g. counting systems, technology) vary greatly, there is no such thing as a "primitive" language. The universality of complex language is a constant amazement to linguists. Witness the fourteen tenses of verbs in Kivunjo, a Bantu language spoken by a non-industrialized people in Tanzania (Bresnan 1988, in Pinker 1994) or the complicated coordination and subordination mechanisms in Kari'ña,, a language of the Ge-Pano Caribe family spoken by groups in Guyana, Surinam and Venezuela (Beria, 2000).

3.1 THE BIOLOGY OF GRAMMARS

Scores of acquisition studies indisputably demonstrate that children re-invent language, generation after generation, and appear to be "bio-programmed" to proceed from stage to stage in common patterns of linguistic and cognitive development, all within a critical window of opportunity in childhood (Lenneberg 1967). Further evidence comes from creolization processes. Children exposed during the critical period for language acquisition to one of the very limited, grammatically reduced contact languages known as "pidgins", regularly transform them into languages which exhibit the grammatical complexity, expressive richness and abstraction common to all natural languages (Bickerton 1981). Congenitally deaf children, exposed during the critical period to the limited, error-laden signing of hearing parents, convert this faulty input into richly expressive, grammatically complex sign languages. Finally, bountiful data demonstrate that children know more than they have been exposed to or "taught". For example, they happily apply abstract linguistic rules of pluralization and tense marking when experimenters ask them to interact with puppets or drawings of fanciful creatures with invented names. Thus, 3-year-olds blurt out that a picture has one "wug", whereas the other has two "wugs" and that "wugging" is what these creatures do! Over-generalization of verb endings is an endless source of evidence of children's unconscious, ongoing analysis and manipulation of highly abstract rules, e.g. My teacher holded the baby rabbits. I finded Renée. Children are obviously relying on knowledge that cannot be attributed to the input to which they have been exposed.

Evidence that language is a biological adaptation with definite localization in the brain is available from aphasics, patients whose language has been affected due to damage in lower areas of the frontal lobes of the left hemisphere. Sufferers of Broca's or Wernicke's aphasia exhibit specific problems of a grammatical and lexical nature, although neither motor control of muscles involved in speech nor general non-verbal intelligence are affected. The general intelligence of sufferers of Specific Language Impairment (SLI) does not appear to be affected, but their dramatic difficulties with language have been linked to a defective dominant gene.

Studies of persons whose language faculty is intact although their general cognition has been affected further corroborate the biological nature of the language faculty. Some hydrocephalic, spina biphida patients with significant retardation exhibit linguistic hyperdevelopment. Verbal fluidity and grammaticality have also been observed in patients with grave intellectual deficits, e.g., some sufferers of Alzheimer's disease, schizophrenia, and autism. The very rare "chatterbox" or Williams' syndrome, associated to a defective gene, affects development and results in severely retarded persons who, nevertheless, are highly competent in speech production and comprehension.

As Eimas and collaborators (1971) have demonstrated in ingenious experiments, human babies as young as 3 weeks can discriminate sounds significant to all the world's languages, although by 10 months they are no longer universal phoneticists and, like their parents, can only discriminate sounds significant in the language spoken around them.

Scovel (1988) points to the initial high position of the larynx, which permits human infants to suckle milk and breathe through the nose simultaneously, as can all mammals. The human infant loses this capacity gradually, as the larynx descends down the throat until, at about the age of two, it reaches its adult position. Thus a true pharynx is formed, which becomes a common passage for ingestion of food and air. No other animal exhibits this anatomical trait. Articulatory limitations in infants are attributable to the undescended larynx. The lowered resonance cavity, however, greatly flexibilizes articulation, since the tongue is liberated and a third resonating chamber, the pharynx, is formed. This, added to the nasal and oral cavities, forms a resonating, multi-component vocal tract. Significantly, humans pay a terrible price for this adaptation: the characteristics that allow us to speak are directly responsible for our being, of all animals, the most susceptible to death by choking. Advantages attributable to an increased articulatory capacity must have outweighed such a frightening limitation.

3.2 SPEECH AND THE BRAIN

Monod (1972, in Lightfoot 1982) asserts that once language—however primitive—made its appearance, "...it could not fail to increase the survival value of intelligence, and so ... create a formidable and oriented selective pressure in favor of the development of the brain....As soon as a system of symbolic communication came into being, the individuals, or rather the groups best able to use it, acquired an advantage over others incomparably greater than any that a similar superiority of intelligence would have conferred on a species without language...the selective pressure engendered by speech was bound to steer the evolution of the central nervous system in the direction of a special kind or intelligence: the kind most able to exploit this particular, specific performance with its immense possibilities." (p.10)

There are clear examples of evolution prizing one specialized faculty and producing enormous development of the corresponding part of the brain; for example, 7/8ths of a bat's brain is devoted to hearing. "The workings of evolution are channeled in a certain direction because some new feature arising by chance mutation makes available a higher level of performance and adaptation." (Lightfoot 1982, pp.10-11) Luria (1973) affirms that once certain mutations started to produce a more powerful brain system, any new combination of genes that perfected it were "powerfully favored". Indeed, he adds, in recent human evolution, everything else was neglected in favor of increased brain power. Humans lost many adaptations useful to other animals (early sexual maturity, fur). "In exchange, [they] won the brain and with it the faculty of language, speech, thought, and consciousness. The central role of speech and language in the development of thought-power and in the success of man as a species suggests that a major part of the evolution of the human brain...must have been a continuous perfecting of the speech centers, which are located on the left side of the brain." (pp. 138-9)

A controversy of sorts must be noted. In some texts, Chomsky appears to have based his theory of the "language organ" on firm bases of evolutionary theory; elsewhere, however, he has expressed a certain skepticism about the capacity of mechanisms of natural selection to explain the development of innate mental structures. Arguing from the mainstream of modern evolutionary biology, Pinker cogently refutes his colleague, stressing the importance ascribed to natural selection. Since the fundamental problem for biology resides in explaining "complex design," Darwin is arguably history's most important biologist, Pinker affirms, precisely because he demonstrated how such "organs of extreme perfection and complication" could arise from the purely physical process of natural selection. This underscores "...the adaptive complexity of the language instinct... All the evidence suggests that it is the precise wiring of the brain's microcircuitry that makes language happen..." (Pinker, 1994, p. 362) And in terms of Darwinian fitness, language awards a powerful advantage.

A final comment is in order. Whereas the majority of complex systems (sound, climate, light, geology) are "blending systems", language, as we have said, is a discrete combinatory system. A finite number of discreet elements (words) are arranged to produce an infinite set of larger structures (sentences) with properties that are very different from those of its elements. The only other example of such a system in the natural world is the genetic code of DNA. Is it mere coincidence that life and mind, the two systems in the universe that astound us with their open-ended complex design are both based on discrete combinatorial systems? "Many biologists believe that if inheritance were not discrete, evolution as we know it could not have taken place." (Pinker, 1994, p. 85) Dawkins (1986) proposes, furthermore, that natural selection is not only the correct explanation for life on earth, but is certainly the correct explanation for anything we would be willing to call "life" anywhere in the universe.

4. References

Beria, J. 2000. Estudio de las estructuras coordinadas y subordinadas en la sintaxis de la lengua Kari'ña. Unpublished Master's Thesis. Universidad Simón Bolívar, Caracas, Venezuela.

Bickerton, D. 1981. *Roots of Language*. Ann Arbor, MI: Karoma Publishers.

Brown, H.D. 1994. *Principles of Language Teaching and Learning*. 3rd Edition. Englewood Cliffs, NJ: Prentice Hall Regents.

Chomsky, N. 1965. *Aspects of the Theory of Syntax*. Cambridge, MA: The MIT Press-

Chomsky, N. 1975. *Reflections on Language*. New York: Pantheon Books.

Chomsky, N. 1988. *Language and Problems of Knowledge. The Managua Lectures*. Cambridge, MA: The MIT Press.

Cook, V. 1991. *Second language learning and language teaching*. London: Edward Arnold.

Cook, V. and Newson, M. 1996. *Chomsky's Universal Grammar: An Introduction*. 2nd Edition. Oxford, UK: Blackwell Publishers Ltd.

Dawkins, R. 1986. *The Blind Watchmaker*. New York: Norton.

Eimas, P.D., Siqueland, E.R., Jusczyk, P. & Vigorito, J., 1971. Speech perception in infants. *Science*, **171**, 303-306.

Lenneberg, E. 1967. *The Biological Foundations of Language*. New York: John Wiley & Sons.

Lightfoot, D. 1982. *The Language Lottery. Toward a Biology of Grammars*. Cambridge, MA: The MIT Press.

Luria, S. 1973. *Life: The Unfinished Experiment*. New York: Scribner.

Medina de Callarotti, M.E. 1993. *Capacidad Metalingüística. Un estudio de la detección y explicación de anomalía*. Caracas, Venezuela: Equinoccio.

Pinker, S. 1994. *The Language Instinct*. London: Allen Lane-The Penguin Press.

Scovel, T. 1988. *A Time to Speak. A psycholinguistic inquiry into the critical period for human speech*. New York: Newbury House Publishers.

Skinner, B.F. 1957. *Verbal Behavior*. New york: Appleton-Century-Crofts.

Section 6.
Philosophical implications of the search for extraterrestrial life

ASTROPHYSICS AND META-TECHNICS

ERNESTO MAYZ VALLENILLA
Unidad de Filosofía. Instituto de Estudios Avanzados (IDEA)
Apartado 17606, Parque Central, Caracas 1015[a], Venezuela.

Abstract: This paper wants to provide a stimulus for reflecting on the ways *meta-technics* are having fundamental effects on Contemporary Science, including Astrophysics and Astrobiology. According to the author, traditional human technology is being replaced and superseded by "*meta-technics*". *Meta-technics* seeks to overcome the traditional *anthropomorphic*, *anthropocentric*, and *geocentric* aspects of all previous technology. It also goes beyond a primary reliance on the human sense of sight as the highest form of perception, –what the author refers to as an "optic-luminic" ordering of reality. Contemporary Science and Technology is captured on an optic-luminic *logos* and its corresponding language; now: what *meta-technics* reveals is a trans-human ordering of reality or a trans-optic *logos* that has important implications on the foundations and conceptual basis of all scientific disciplines. It is simply with the goal of raising such issues, and of beginning to ask such questions in the fields of Astrobiology and Astrophysics, that Mayz Vallenilla gave the following lecture.

1. Introduction

Let me begin by expressing my firm conviction regarding the legitimacy and validity of contemporary scientific knowledge, both in respect to its methodology and its results. Therefore, it would be a temerity, more than plain senselessness, to doubt scientific theories and the great achievements obtained through its application in Astrophysics and Astrobiology.

What I will explain today –making use of the premises that I developed in my book *Foundations of Meta-Technics*– will then be, not a refusal or denial of the founding basis that inform the current cosmographic and cosmological image of the *Universe*, but rather a critical effort to try to analyze the *techno-rational foundations* of its epistemic horizons... producing, as a consequence, the necessary broadening and concomitant variation that underlies them... without yet achieving an adequate expression.

2. Perplexities and problems

Let us question on the first place, what motive or reason is there to connect *Astrophysics* and/or *Astrobiology* to *Meta-technics*? Perhaps, before an audience such as this one, primarily composed of scientists familiarized as much with *Astrophysics* as with *Astrobiology*, the right procedure is to explain what *Meta-technics* is, and why it can have (as, in fact, it has had) such a radical and decisive influence on those reputable and attracting Disciplines.

At the beginnings of the 80s, when I created the term "*meta-technics*" –and used it for the first time before an international audience at the World Philosophy Congress held in Brighton–, my aim was to highlight two main aspects implicit in its very denomination: a) that such term designated a new stage or mode for "*technics*", the purpose of which was to overcome the *anthropomorphic*, *anthropocentric* and *geocentric* characteristics that prevailed up to that moment in *traditional technology*, as

J. Chela-Flores et al. (eds.), Astrobiology, 235–240.

much in relation to its *ars operandi*, as in relation to the *nature* and *functions* of its own instruments; and b) that such overcoming of the *traditional technology* implied, *eo ipso*, a radical transformation and transmutation, as much of *Reason* as of *Rationality* which, until then, had served as basis for technology as an expression of the *will to power* that guides it... and the impetus of which, acting as the root and engine of that *technology*, activated the epistemic effort through which the human being sought to dominate otherness or reality (including in such otherness or reality, *Nature as such* and *man* himself as a *natural being*).

But what does this explosive affirmation mean, which states that *Meta-technics* implies and presumes overcoming *Reason* and *rationality* that inform and sustain *technology* and/or *traditional techno-scientific knowledge*?

Plainly stated, from a strictly philosophical point of view, it means the following: that *Reason* and *rationality*, from its beginnings with the Pre-Socratics and Parmenides, up to Husserl, Heidegger and all of their current successors, have been identified (etymologically and philosophically) with a "*seeing*" (νοεῖν)... such "*seeing*" being of a sensitive and empirical, or eidetic and intelligible nature.

Therefore, the *rational-truth*, as such, is considered the expression of an *evidence* (term, as it may be easily noted, derived from the Latin *videre*, which means "*to see*")... be it that such *rational-truth* embodies an *empirical evidence* and/or an *intellectual evidence*... in accordance with its corresponding optical mode as well as with the corresponding mode of being of its objective correlates.

Now: being *Reason* synonymous with a "*seeing*", such "*seeing*" ... (in order to fulfill its role) must be nurtured and must make innate use of the *visible light* (from the Greek φώς, the latter from φάος)... taking possession, depicting and processing such type of waves from the electromagnetic spectrum that, due to their longitude and frequency, adapt themselves to the apprehensive and receptive capabilities of the human eye as such.

The above mentioned is indeed one of the principal aspects that *Meta-technics* tries to overcome before *traditional technology*... and one that it has convincingly accomplished in our days, through the construction of new instruments capable of apprehending otherness or reality and its manifestations in a manner incomparably greater, richer and radically different from the traditional one... strictly and exclusively human-bounded, or –may we say– reduced to *optic* and *luminic* characteristics on its limits as well as on its foundations.

In fact: otherness or reality, as such, is nowadays apprehended or grasped, for instance:

- through *sonic* or *ultrasonic waves*, be it with the *radar* and the *sonar* or through the assistance of other more complex and sophisticated devises that provide echo-sonograms of such otherness or reality;

- through *thermic stimuli*... in the sensors used to guide missiles or ballistic projectiles;

- through *radio waves* in radio-telescopes;

- through *infra-red rays* in some satellites-telescopes specifically designed to this end;

- through *nuclear magnetic resonance*, where it should suffice to mention instruments such as the SQUID (Superconducting Quantum Interference Devise), the action of which allows to depict some magnetic fields utilized to register the functioning of neurons; the MRI (Magnetic Resonance Imaging), which opens access to millimetric structures in brain tissue; or the MRS (Magnetic Resonance Spectroscopy), which provides data on the functioning of cerebral chemistry in activities such as learning, memory, emotions and vision itself;

also mentioning the complex sensors which, using the *Doppler-Fizeau effect* and the different longitudes of *neutral* and *ionized hydrogen waves*, have amplified and renewed, up to unimaginable boundaries, the most surprising research in Physics, Biology and Astrophysics... as well as, on the other hand, *radioactive carbon*, which applied to chronography and/or chronology, has allowed the calculation of a time, undetectable to any kind of human measurement under chronometric and/or chronological parameters of an optical and luminic nature.

Now: thanks to this *meta-technical logos*, today it is possible to order and make space and time intelligible (as fundamental *basis* or *organizational supports* for otherness or reality) through codes or precepts totally different to the traditional ones.

This radical transformation, in its basis as in its parallel spatial-temporal orderings of otherness or reality, is directly projected not only on all the epistemological and ontological concepts and notions created by philosophical tradition (as for instance, the notions and concepts of Being and of Nothingness, of Affirmation and Denial, of Conscience and Knowledge, etc.), but also, in parallel, on the *basis* and *foundations* themselves, that act as implied assumptions in Disciplines as fundamental as Logic, Geometry and Mathematics... additionally affecting, as an inevitable consequence, the idea or concept of *Nature* (Physis), as well as the structure and purposes of the *human institutions*, which are founded over such surmounted basis and notions that have been left behind (for example, those of Language, Ethics, Politics or State Science)... encompassing, of course, even the basis and foundations of Biology, Anthropology and Antropogony themselves... which are, without any doubt, also subject to the transformation propitiated by the *meta-technic revolution*... under the scope of which, as we are seeing it today, the human being himself may be subject to be transformed and/or transmuted in his innate *physis* and, therefore, in his somatic and psycho-physic congenital limitations, with sequels and consequences that are not hard to imagine... in every sense... particularly the axiological one... as we currently hear and observe in our daily morning news.

But this radical *meta-technical trans-formation* and *trans-mutation*, of course, also affects the notions and concepts employed by Astrophysics and Astrobiology... the scope of which spreads out from the notion or concept itself of *Cosmos* or *Universe* (structured and ordered as a *whole* or *totality*, based on the traditional representation of a spatial-temporal binomial notion that is optic-luminic in nature and origin) as well as on the determinations or notions equally optic-luminic such as those of "here and now" ("*hic et nunc*") or those of "before and after" (*ante* or *prius... post* or *deinde*), the natures of which are, ostensibly and undeniably, of the same genealogy.

It was based on these notions and corresponding etymologies (optic-luminic and optic-spatial) that Greeks and Latins, from assumptions and traditions of Phythagorean origin, conceived and represented the *Cosmos* (κόσμος) and/or the Universe (*universus*), as eidos, species or figures.

In fact: in Greek, κόσμος means *order*... and such *order* was the one that governed the *World* as much as *Heaven*. Also, *Universus-a-um*, means (in Latin) "*what, with one single impulse, is driven, oriented, and directed toward*"... and translated precisely the Greek word, κόσμος... which referred to a *whole* or to the *totality*.

Now: there is no other notion more ostensibly optic-spatial, than that of the *whole* or the *totality*... still present and acting in our own conception of the *Cosmos* and of the *Universe* as such.

Also, on these basis and syntax –optic-spatial and optic-luminic– were and are still conceived such notions as a *beginning* or as an *end* and/or an *ending*... likewise (and not by mere coincidence) that of an *expansion* and/or that of a *contraction*; and also, even beyond those, the optic-spatial representations and notions of *particles* and *waves*... all of them created, without exception, based on meanings of identical origin ("optic-spatial" and "optic-temporal")... and in consequence, strictly *anthropomorphic*, *anthropocentric* and *geocentric*.

Can we now understand the explosive or detonating effect that the *meta-technics thesis* contain in themselves? They mean –no more, no less– the radical questioning of the *optic-luminic logos* prevailing up to our own time, on which *cosmo-logy* and *cosmo-graphy* are founded.

In fact: it is not only the concept or notion of a *Cosmos* (which in Greek expressly means, as we have indicated, *order* or *ordering*) what is being questioned... but, indeed, what is at stake is the nature and efficiency of the *ordering* itself... which is rooted and originated from the *optic-luminic order* (represented by the *Cosmos*).

Or going straight to the matter: If the *Cosmos*, as such, is a concept or a construct developed and sustained only through the notion of an optic-luminic conception of *space* and *time* (anthropomorphic, anthropocentric and geocentric)... can man nowadays continue to assert the validity of the concept of *Cosmos* as an hermeneutic instrument, ordering otherness or reality in an emblematic manner? Is such otherness or reality reduced in its parameters, dimensions, ingredients, potentials and energetic modalities, to those which have been detected, described and dominated by men through the use of their anthropomorphic, anthropocentric and geocentric instruments and devises?

Or, prying even deeper: What are *Space* and *Time* themselves, and what ordering do they have... are they apprehended, and organized through a syntax, no longer through the limited windows of the *human eye* and of *visible light*... but in a *trans-optical*, *trans-luminic* and *trans-human* manner through *meta-technical* instruments and devices as those now normally used?

Would optic-luminic characteristics, as those represented by the *line* (γραμμή), the *figure* (σχῆμα), the *limit* or *frontier* (ὅρος, πέρας)... of a clear Euclidian-Phythagorean genealogy, subsist in Space, as such?

And if such notions or characters disappear from *Space* as such... can, and should *Time* continue to be ordered and interpreted using images and metaphors of a *spatial origin*... such as those of a *flowing* or *elapsing* that is endowed with a *distance* and a *direction* (as well as with a *before* and an *after*) of a clear *optic-luminic* genesis, proceeding and meaning?

Are there "arrows" and "trajectories" in *Time*... or are these also just metaphors... as are dates, cycles and periods... in which we continue to divide *Time* into segments, according to its presumed *spatial extension*?

And would it not be valid to assume the same about *spatial notions* –likewise applied to *Time*– such as those of an *origin* or a *beginning*, and/or those of an *end* or an *ending*? Does the first one not come from the Latin word *origo*... and the latter from *orior*, *oriris*, *oriri*, *ortus*... from which also derives the substantive *ortu-us*... which designated sunrise and the first rays of light?

And its opposite –may we say, the notion of an *end* or an *ending* (from the Latin word *finis*... which translated the Greek ὅρος)– did it not designate, as such, the *edge* or *limit* of a *field* or *territory* obviously *spatial* and *optic*?

Or now, asking provokingly: is not the very hypothesis of the *"Big Bang"* inscribed, perhaps, within this same notion of an *origin*... without observing that it assumes (at least etymologically) the prior existence of *light* and the *sun*? And, isn't it likewise with the genealogy and with the ontic-epistemic texture that the optic-luminic models have of an assumed "*open-ended universe*", "*closed*" or "*plane*"?

As one may notice and understand, these are just a few questions... stimulating and unsettling... that in their most simple and general manner, without going into further details, or making in-depth reference to their truly philosophical and epistemological background, I have deemed convenient to disseminate at this important meeting with specialists in *Astrophysics* and *Astrobiology*... since many of the basic concepts of such Disciplines –such as *physis* and *bios*– are not, in the very least, foreign or immune to the fundamental criticism that the questioning of their *anthropomorphic, anthropocentric and geocentric basis of support* projects on *contemporary science.*

This *basis of support* –may we say, the very *fundamentals* on which both *Astrophysics* and *Astrobiology* are built– are, in our opinion, constructed and developed over the questionable preeminence or supremacy of an *optic-luminic logos*... already overcome, *de facto*, by the advances and achievements of *meta-technical reason* and *rationality*... thanks to the instruments built by that very same *reason* and *rationality.*

3. Conjectures and new horizons

Although today I have abused of your patience in excess –adducing the multiple philological and philosophical annotations inserted throughout my presentation– please allow me to explain the reason and need for these.

Meta-technics being addressed, as one of its main objectives, toward the discovery, critique and overcoming of the *optic-spatial* and *optic-luminic fundamentals* that support *human thinking* and *instituting...*, *language*, as such, is the *proto-institution*, fundamental and basal, to that *thinking* and *instituting*... and, accordingly, in that very *language* are gathered all the limitations of anthropomorphic, anthropocentric and geocentric interpretations of otherness or reality –of optic-luminic and optic-spatial genealogy–, that we have questioned in their *fundamentals* or *supporting basis* throughout this presentation.

Now: we are prisoners of this *language* (may we say, limited users, conditioned by the limits, syntax and belief assumptions that nurture it) –as the fundamental human proto-institution– without being aware that it is the one that surreptitiously holds and organizes our *ratio*, as well as the consolidation and construction of the *otherness* or *reality* in which we live, nurtured and interwoven by its etymologies and syntax, tacit and hidden, as ingredients of our own life and existence... and intangible ground of our world.

Astrophysics and *Astrobiology* could in this manner be prisoners of the same prison or cavern in which we all live... given our inborn somatic-psychic constitution, and the supremacy that the *seeing* or νοεῖν have over our *reason* and *rationality* as building and ordering agents of the *otherness* or *reality* that we inhabit.

Such captivity –ancestrally larval and tacitly dominating human race from its very genesis– is expressed through *word* and *language*... and without historical, cultural or racial limitations, is necessarily imposed by the very *factum* to which we have made reference.

Therefore, to trans-mute and overcome the enslaving *optic-luminic limits* of *language* (and as such *limits*, all those of concepts, notions, meaning, etc., that we use as repertoire of our knowledge)... is an urgent and decisive imperative of our times... one which we ought to feel incited and obliged to address in view of the very same challenges, perplexities and defiances that are ostensibly set forth by the discoveries and revelations, made by the instruments created by a *meta-technical reason* and/or *rationality*... designed and construed by man himself, in an agonizing and creative struggle, to conquer and overcome the anthropomorphic, anthropocentric and geocentric limitations that until now oppressed and suffocated him.

How can we apprehend and decipher data, configurations, and a new syntax –without translating them into optic-luminic images or codes– that provide us with revolutionary *meta-technical sensorium* available to man today? Is this possible? Will we be able to articulate some day the syntax and the fortuitous *metaxys* of *abyss* and *chaos*... without equating them (from within our linguistic prisons) to a *Non-Being*, a *Dis-Order*, or a simple and negative *An-Archy*?

These questions are left unanswered... but I humbly believe that the path outlined (and barely glimpsed at) opens up some perspectives that lead to urgent and necessary reflections.

4. Acknowledgements

The author is sincerely grateful to Ms. Morella Lazzari G. and Ms. Gladys Arellano for their efforts and dedication in the translation of this manuscript, as well as to professor Dr. Fabio Morales for the final revision of this article.

DEEPER QUESTIONS
the Search for Darwinian Evolution in our Solar System

JULIAN CHELA-FLORES

The Abdus Salam International Centre for Theoretical Physics (ICTP), Office 276, P.O.Box 586; Strada Costiera 11; 34136 Trieste, Italy,
and
Instituto de Estudios Avanzados, Apartado 17606, Parque Central, Caracas 1015A, Venezuela.

1. Position of humans in the totality of all earthly species

It was characteristic of the Enlightenment-the movement of ideas current during the 18th century- to distrust tradition in cultural matters. Truth was to be approached through reason. At the end of that period Auguste Comte founded a movement which advocated that intellectual activities should be confined to observable facts. The reason why this movement was called "positivism" is that observable facts were called 'positive' by Comte.

This point of view was developed much later by a group of philosophers working in Vienna at the beginning of the 20th century. They were known as the "Vienna Circle". They maintained that scientific knowledge is the only kind of factual knowledge. We discuss the main cultural implications of Astrobiology against the background of such philosophical and theological constraints.

The Vienna Circle maintained that all traditional doctrines are to be rejected as meaningless. They went beyond positivism in maintaining that the ultimate basis of knowledge rests on experiment. Since they were also considering the unification of science and were using mathematical logic in their version of extreme positivism came to be known as 'logical positivism'.

Although some scientists have adopted this philosophy, either consciously or unconsciously, the fact remains that modern science begins with Galileo, who initiated the tradition of formulating theories based on observation and experimentation. No underlying philosophy was adopted then, or need to be adopted now beyond the dialogue theory/experiment. On the other hand, there is a large number of issues that science cannot handle, or even formulate. Bertrand Russell in his *History of Western Philosophy* makes this point even in much stronger terms that I have just done (Russell, 1991): *"Almost all the questions of most interest to speculative minds are such as science cannot answer"*

Positivism avoided all consideration of ultimate issues, including those of metaphysics and religion. However, as anticipated by Russell, the reduction of all knowledge to science is a matter that debate has not yet settled at all. Paramount amongst all the issues of first causes and ultimate ends is precisely the main topic we have discussed in this book. To our 19th-century colleagues the problems of the origin and distribution of life in the universe were issues that were to be excluded from the scientific discourse, even by those that had adopted the mildest form of positivism.

J. Chela-Flores et al. (eds.), Astrobiology, 241-246

These problems are approachable by scientific methods. Both subjects have a remarkable history of valuable efforts by some of the best scientific minds of the 20th century. The long list of scientists began with Alexander Oparin, John Haldane and included others. Because of the progress achieved, extreme aspects of first causes and ultimate ends are naturally inserted in the science of astrobiology; yet, neither of the two problems (origins and distribution of life) is solidly set on scientific bases: it has been impossible to synthesise a living organism, and no signal form an extraterrestrial civilisation has yet made contact with highly sensitive detectors. In view of this unsettled state of affairs, it seems unavoidable that a reasonable collective approach to the deepest questions in astrobiology should be encouraged by all sectors of culture. As an illustration of the limits of science, Russell provides us with one example: *Is man what he seems to the astronomer, a tiny lump of impure carbon and water impotently crawling on a small and unimportant planet?*

To address the Russellian question, we must first decide on the place of man amongst the Earth biota. From the perspective of biology, human beings represent only a single species among four thousand mammals. Yet, this is a small number when compared with the 30 million species that are expected to constitute the whole of the Earth biota. One aspect of this bewildering abundance of species of which humans are only one, has led to the metaphor (Gould, 1991): *If the history of evolution were to be repeated, such an alternative world would teem with myriad forms of life, but certainly not with humans.*

Our main concern is not the origin and evolution of our own species. Our main concern is rather the likelihood that the main attributes of Man would rise again, if the history of evolution starts all over again elsewhere, not in a hypothetical Earth that would be miraculously reconstructed.

We are mainly concerned with the repetition of biological evolution in an extrasolar planet, or satellite, that may have had all the environmental conditions appropriate for life. Such attributes are, for example, a large brain and consciousness. These features of Man evolved from primates over the last 5 to 6 million years. This is in plain contrast with molluscs. Members of this large phylum of invertebrates to which snails, mussels and squids belong, have survived since the Lower Cambrian. Their first appearance occurred 500 million years ago, about 100 times earlier than the first appearance of Man.

In an extraterrestrial environment the evolutionary steps that led to human beings would probably never repeat themselves. However, the possibility remains that a human level of intelligence may be favoured when a combined effect of natural selection and cultural evolution are taken together. This is independent of the particular details of the phylogenetic tree that may lead to the intelligent (non human) organism. Conway-Morris in *The Crucible of Creation* states it briefly: "The role of contingency in evolution has little bearing on the emergence of a particular biological property" (Conway-Morris, 1998). I would like to illustrate the inevitability of the emergence of particular biological properties with examples of convergent evolution, a phenomenon that has been recognised by students of evolution for a long time (Tucker Abbott, 1989).

In the phylum of molluscs the shells of both the camaenid snail from the Philippines, or a helminthoglyptid snail from Central America, resemble the members of European helcid snails. These distant species (they are grouped in different Families), in spite of having quite different internal anatomies, have grown to resemble each other outwardly over generations of response to their environment. In spite of considerable anatomical diversity, molluscs from these distant families have tended to resemble in a particular biological property, namely, their external calcareous shell. A second example is provided by swallows (Passeriformes) a group which is often confused with swifts (Apodiformes), but are not related to them. In fact, the taxons are orders, rather than families, as in the first example. Members of these two orders differ widely in anatomy and their similarities are the result of convergent evolution on different stocks that have become adapted to the same ways of living in ecosystems that are similar to both species. In the light of these examples the question of whether our intelligence is unrepeatable goes beyond biology and the geological factors mentioned in the metaphor on the repetition of the history of evolution.

Indeed, the question is rather one in the domain of the space sciences; in which the radio astronomers have led the way with the SETI project (Drake and Sobel, 1992). The question of whether we are alone in the cosmos concerns astrometry measurements for the search for extrasolar planets. This activity has led to the current revolutionary view that planets of our solar system are not unique environments that may be conducive to the origin, evolution and distribution of life in the universe (Mayor *et al.*, 1997) . The presence of a dozen planets in the cosmic neighbourhood of the Sun, argues in favour of the ubiquity of life in the universe. It seems plausible that if the right environments exist elsewhere, some of the biological attributes of Man may have repeated themselves.

2. The search for microbial life in our own solar system

A separate question, much closer to our capability to perform practical experiments concerns the search for microbial life in our own solar system. So we believe it is appropriate to shift our attention away from 'attempting a full and coherent account of the Phenomenon of Man' (Teilhard de Chardin, 1965). Instead we should focus attention at the level of a single cell. Indeed, we feel that the progress of molecular biology forces upon us a search for a full and coherent account of eukaryogenesis, the first transcendental transition in terrestrial evolution at the cellular level which led to intelligence The cosmic search for extraterrestrial intelligence ought, in our opinion, begin with a single step, namely the search for the first cellular transition on the pathway to multicellularity, and inevitably to brains (due to their selective advantage). This emphasis on the eukaryotic cell as a 'cosmic imperative', has been referred to as the phenomenon of the eukaryotic[•] cell (Chela-Flores, 1997; 1998). The task of understanding the origins of the eukaryotic cell is not easy (De Duve, 1995). But let me dwell on clarifying the terms being used.

We have already reviewed some arguments that suggest that the problem of the position of Man in the cosmos depends critically on the evolution of microorganisms up to the level in which eukaryogenesis occurred. This forces upon us the question of the position of the eukaryotic cell in the cosmos, as the main focus of our attention. I feel that such a radical break with the past has some implications in our understanding of the origin and destiny of Man. Nevertheless, none of these arguments lie outside the scope of the question raised in the Papal Message to the Pontifical Academy of Science ((John Paul II, 1997): *New knowledge has led to the recognition that the theory of evolution is no longer a mere hypothesis.*

In spite of this important step, the acceptance of evolution has not led to a consensus amongst scientists, either on its mechanism, or on its implications. Nevertheless, in spite of this shortcoming, we shall base our subsequent arguments on Darwin's theory of evolution. This leads us to a discussion of some of the implications that such a search might imply for the dialogue between science and natural theology.

Constraints imposed by philosophy and theology on our view of life mostly favours a special place of man in the universe. In the case of philosophy there is a continued quest for the impact of technological progress on the future of mankind. We have already encountered the perspective of cultural evolution on the breakdown of a straight coupling between chance and necessity, namely the continued accumulation of mutations that may favour the adaptation to changing environmental conditions. A separate question concerns the changes in our theological outlook that may follow the incorporation of knowledge of the place of Earthly biota in the cosmos. Would there be problems in the traditional Judeo-Christian-Muslim view of Deity as being confined with the affairs of Man? Jastrow argues that the Deity is omnipotent and can be concerned with the affairs of as many intelligent species as there may exist in the universe. Yet the question remains whether the original image of God as portrayed in the Scriptures would be acceptable to its current interpretation, if SETI or a more restricted search in our own solar system were to confront us with parallel evolution that would point towards the existence of life in innumerable worlds? A first contact with extraterrestrial life would confront us with new problems to be solved in biology. For example, a more extensive view of taxonomy would be needed. We would have to learn to classify new organisms. This would be within the domain of scientific enquiry.

A different problem, beyond the limits of science, would concern the subject of Divine Action. The monotheistic tradition may be traced back to the New Kingdom of Ancient Egypt (1379-1362 BC). Although it is debatable whether the system initiated during the 18th Dynasty may be called monotheism, in *"The Hymn to the Aten"*, close parallels to the verses of *Psalm 104,* have been pointed out, not only in words, but also in thought and sequence, anticipating this part of the Bible by several centuries (Simpson, 1972). Closer to our own experience, in the Judeo-Christian-Muslim tradition the ancient concept of Divine Action and its implication in natural theology has been extensively reviewed in the literature (Russell, 1996; Barbour, 1995). In parallel to the Christian tradition a position that has been discussed since the Enlightenment regards the confrontation between science and religion. It maintains that God acts only in the beginning, creating the universe and the laws of nature. This thesis is called deism, usually taken to imply that God leaves universal evolution to its own laws, without intervening, once the process of creation has taken place. On the other hand, within the Christian tradition an approach towards integration has been advocated in the relationship between science and theology. It concerns the problem of biological evolution.

Like Darwin, John Paul II, while referring to the living world, for good reasons has not put the main emphasis on first causes and ultimate ends. At the beginning of this chapter we emphasised that for the first time within science there is a branch, namely astrobiology, which makes first causes and ultimate ends its own subject matter. We identify as a first cause the origin of life in the universe; the distribution of life in the universe may be identified as an ultimate end. With respect to human beings there is much ground to cover yet in the road of convergence between science and religion. Once again, the Papal Message refers to remaining points still to be discussed.

One example is provided by one aspect of the human being that is exclusively discussed in the context of natural theology, namely, "the moment of transition to the spiritual which cannot be the object of [scientific] observation".

3. Questions for a dialogue rather than a debate

We have already touched upon the concept of evolutionary convergence. We do not consider the above-mentioned 'rewinding of the evolutionary clock' as a thought-experiment, but as a real possibility that may have occurred ubiquitiously in extrasolar planets. Evolution may not produce Man again, but within the scope of science we can discuss the possible convergence of some of the attributes that are characteristic of human beings. For instance, language and intelligence are two attributes which are of extreme importance for the search of extraterrestrial life.

What questions would a first contact with extraterrestrial life imply for both science and religion? We have endeavoured to demonstrate that contact need not come only at the level of a fully intelligent message; contact could come first in the form of detecting the first cellular steps towards intelligence; in other words, through eukaryogenesis. There remains a difficulty of addressing those attributes of human beings that are raised in theology, but not in science (spiritual dimension). The question has been formulated a little more precisely (Coyne, 1998): *Is a creationist theory required to explain the origins of the spiritual dimension of the human being?* While we are still not in a position to answer this question, we have endeavoured to gather a number of efforts within science that suggest that contact with extraterrestrial life cannot be excluded in the future. Such an experience would give us a unique opportunity; it would provide us with a solid point of reference on which to base original discussions of the implications of all the attributes of human beings. In such discussions the participants should be scientists and natural theologians. Facing the discussion of this possibility now is neither premature nor idle:

Exploration on Earth in the 15th century led to the difficulty of widening the horizons of the accepted attributes of Man. The confrontation of Europeans with the native Americans proved to be traumatic. In retrospect, the dialogue that took place in Valladolid, Spain, between Bartolomé de las Casas, ex-Bishop of Chiapas (Guatemala), and the learned Juan Ginés de Sepúlveda is still of considerable interest. The question of the attributes that characterise Man was raised on that occasion.

We are still not ready to decide on what makes us human till we reach consensus on what is our position, first on the tree of life and, subsequently, when we understand what is the position of our tree of life in a universal context.

References

Barbour, I. G. (1995) Ways of relating science and theology, in R.J. Russell, W.R. Stoeger, SJ and G.V. Coyne (eds.), *Physics, Philosophy and Theology. A common quest for understanding* (2nd. ed.), Vatican Observatory Foundation, Vatican City State, pp. 21-48.

Chela-Flores, Julian (1997) Cosmological models and appearance of intelligent life on Earth: The phenomenon of the eukaryotic cell, in Padre Eligio, G. Giorello, G. Rigamonti and E. Sindoni (eds.), *Reflections on the birth of the Universe: Science, Philosophy and Theology,* Edizioni New Press, Como, pp. 337-373.

Chela-Flores, Julian (1998) The Phenomenon of the Eukaryotic Cell, in Robert John Russell, William R. Stoeger, and Francisco J. Ayala (eds.), *Evolutionary and Molecular Biology: Scientific Perspectives on Divine Action,* Vatican City State/Berkeley, California, Vatican Observatory and the Center for Theology and the Natural Sciences, pp. 79-99.

Conway-Morris, Simon (1998) *The Curcible of Creation. The Burgess Shale and the Rise of Animals* Oxford University Press, New York, pp. 9-14.

Coyne SJ, G. (1998) The concept of matter and materialism in the origin and evolution of life, in J. Chela-Flores and F. Raulin (eds.), *Chemical Evolution: Exobiology. Matter, Energy, and Information in the Origin and Evolution of Life in the Universe,* Kluwer Academic Publishers, Dordrecht, The Netherlands, pp. 71-80.

De Duve, Christian (1995) *Vital dust: Life as a cosmic imperative,* Basic Books, New York, pp. 160-1 68.

Drake, Frank and Sobel, Dana (1992) *Is there anyone out there? The scientific search for Extraterrestrial Intelligence* Delacorte Press, New York, pp. 45-64.

Gould, Stephen J. (1991) *Wonderful life. The Burgess Shale and the Nature of History,* Penguin Books, London, pp. 48-52.

John Paul II (1997) Papal Message to the Pontifical Academy, *Commentarii* **4**, N. 3. Vatican City, pp. 15-20. [cf., also *La traduzione in italiano del Messaggio del Santo Padre alla Pontificia Accademia delle Scienze,* L'Osservatore Romano, 24 October 1996, p. 7.]

Mayor, Michel, Queloz, Didier Udry, Stephane and Halbwachs, Jean-Lois (1997) in C.B. Cosmovici, S. Bowyer and D. Werthimer (eds.) *From Brown Dwarfs to planets, in Astronomical and Biochemical Origins and the Search for Life in the Universe,* Editrice Compositore, Bologna, pp. 313-330.

Russell, B. (1991) *History of Western Philosophy and its Connection with Political and Social Circumstances from the Earliest Times to the Present Day,* Routledge: London, p.13.

Russell, J.R. (1996) Introduction. In: Quantum Cosmology and the Laws of Nature. Scientific Perspectives on Divine Action, in R.J. Russell, N. Murphy and C.J. Isham (eds.) (2nd ed.), Vatican Observatory Foundation, Vatican City State, pp. 1-31.

Simpson, William Kelly, (ed.), (1972) *The Literature of Ancient Egypt. An Anthology of Stories, Instructions and Poetry,* Yale University Press, New Haven, pp. 7-9 and pp. 289 295.

Teilhard de Chardin, Pierre (1965) *The phenomenon of man,* Fontana Books, London, p. 33.

Tucker Abbott, R. (1989) *Compendium of landshells.* American Malacologists, Melbourne, Florida, USA, pp. 7-8; Austin, Jr., A.L. (1961) *Birds of the World* Paul Hamlyn, London, p. 216.

Section 7.
Round-table

REPORT ON A ROUND TABLE: "MUSIC OF THE SPHERES"

JUAN G. ROEDERER
Geophysical Institute
University of Alaska-Fairbanks
Fairbanks, AK 99775, USA

A Round Table was held with the participation of three distinguished panelists from Venezuela, artist *Jacobo Borges,* composer *Diana Arismendi* and art historian *Irene McKinstry de Guinand,* specially invited for that purpose. Also participating were three representatives from the School on Astrobiology. The purpose of the discussion was to attempt to link the arts with science and to address the question: Would other intelligence also exhibit artistic creativity?

The Round Table was opened by *Guillermo Lemarchand,* who reminded the participants that one of the very first attempts to link science with art was undertaken by Pythagoras in the 5h Century B.C. in his study of the vibrations of a string and the generation of the tones of a musical scale. Even much longer before that, the archeological find of a 42,000-plus year old flute, made of bone, convincingly shows from the position of the finger-holes that a musical scale based on integer number ratios was already in use at that time, Dr. Lemarchand concluded his remarks with an account of the history of planning this particular Round Table for the School of Astrobiology.

After their introduction by Julián Chela Flores, who pointed out the notable accomplishments and renown of the guest speakers in their country and world-wide, the three artists made their presentations. This was followed by an address by *Chela Flores* describing the aims of the discipline of Astrobiology and the reasons for including the arts in the deliberations. At the conclusion of the extensive discussions, *Juan Roederer* gave a succinct summary of the main arguments presented by the artists.

All participants were duly impressed with the detail and depth of the thoughts on the subject presented by the three artists. It would do injustice to the panelists to pretend to summarize their presentations here because of the great variety of themes that were addressed. All one can do is try to identify some common theme, or Leitmotiv, that threads through these presentations.

Jacobo Borges talked about the relativity and mutability of artistic values and the evolution of art in general. Art is interactive with nature; it is not only the product of humans but also of that which surrounds humans. Beauty is found when one relates to things, but also demands that these things relate back to oneself. Referring to statements by Picasso and Poincaré, Borges said that the goal both in art and science should be not to search but to

J. Chela-Flores et al. (eds.), Astrobiology, 249–250.

find. Diana Arismendi demonstrated that the creative process is governed by templates and rules, but pointed out that external elements also contribute in substantial ways. The brain is the "great integrator". Different epochs are characterized by differences in the relative importance of rules vs. external influences. She pointed out the great abstraction that can be found in music, and illustrated this by playing the tape of one of her compositions masterfully performed by a local clarinetist. Irene McKinstry de Guinand elaborated on the idea that art was indeed the very first manifestation of truly human capacity or intelligence. The course of art through the ages marked the course of history; the great constructions and artworks indeed represent the best testimony of the passage of time. Creativity always contains subjective expressions of the moment wherever it is found-whether on Earth or, maybe soon in the future, elsewhere in the Universe.

Chela Flores closed the Round Table discussing the close connection between the evolution of the brain in humans and the appearance of works of art. He concluded that given the strong selective advantage of brain evolution, this aspect of humans should appear again in other evolutionary lines. This in turn implies, according to him, that artistic creativity should be linked inexorably with other intelligence.

Section 8.
Contributions from participants

ULTIMATE PARADOXES OF TIME TRAVEL

GUSTAVO E. ROMERO
Instituto Argentino de Radioastronomía,
C.C.5, 1894, Villa Elisa, Buenos Aires, Argentina
email: romero@irma.iar.unlp.edu.ar

AND

DIEGO F. TORRES
Departamento de Física, UNLP,
C.C. 67, 1900, La Plata, Buenos Aires, Argentina
&
Instituto Argentino de Radioastronomía,
C.C.5, 1894, Villa Elisa, Buenos Aires, Argentina
email: dtorres@venus.fisica.unlp.edu.ar

Abstract. We briefly present some paradoxes of time travel, and discuss their possible solutions. We also analyze the epistemological status of the Principle of Self-consistency.

1. Introduction

One of the most perplexing aspects of General Relativity is that its field equations have solutions that allow the existence of closed time-like curves (CTCs). CTCs are space-time trajectories such that a particle can traverse them, always moving towards its own future, and intersect itself in the past. Such a situation corresponds to what science-fiction writers have called "time travel". A time machine is a device that can generate or use a special energy-momentum distribution to distort the geometric structure of space-time to the level at which CTCs are formed (strong time machine) or allowed (weak time machines) to some physical systems ("time travelers").

The most studied solutions of Einstein field equations that allow CTC formation are wormhole space-times. [1] If a wormhole exists, then it can

[1]For a discussion on wormhole features and observational bounds see the work by Torres and Romero (2000) in this volume.

J. Chela-Flores et al. (eds.), Astrobiology, 253–257.

be converted into a time machine through the relativistic motion of one of the mouths, or using strong gravitational fields in order to induce large redshifts (e.g. Morris et al. 1988, Novikov 1989).

2. Against time machines

The opponents of time machines are legion. To many people time machines appear to be hideous devices that unlock the door to all kind of ontological bizarrerie. However, since CTCs are naturally expected in General Relativity, if we are going to take seriously this theory we should also take seriously the possibility that CTCs can actually occur in the Universe, and then explore their consequences as far as our theoretical knowledge allows it.

The opposition to time machines has adopted two main forms. First, through the chronology protection conjecture (Hawking 1992): Nature abhors time machines and they are not physically feasible: always that a CTC is near to be formed, some laws of nature produce a situation that destroys the machine. The main mechanism to enforce chronology protection is the back-reaction of vacuum polarization fluctuations (see Earman 1995 for a discussion). However, it has been recently shown, and admitted by Hawking and followers (Cassidy & Hawking 1998), that this mechanism can fail in creating an effective protection.

In second place, chronology protection has been sought in the apparent logical paradoxes produced by time travel. The best known case is the "grandfather paradox": I go to the past, and in a bloody impulse I kill my own grandfather before he had the opportunity to meet my grandmother. Where am I from then?[2] This kind of paradoxes have been used to claim that time travel is illogical.

At this point a semantic note on the word "paradox" is required. A paradox can be:

1. An apparent logical situation that actually is not.
2. An apparent illogical situation that actually is perfectly logic.

It has been argued (e.g. Friedman et al. 1990) that the paradoxes just point out the existence of "consistency constraints" in the laws of Nature. With such constraints all time travel paradoxes would be of class 2. More specifically, it has been proposed the following

PRINCIPLE OF SELF-CONSISTENCY (PSC): The only solutions of the laws of physics that can occur locally in the real Universe are those which

[2]The reader interested in ALL kind of time paradoxes should run to see the fascinating book by Nahin 1998, by far the most comprehensive work on the subject.

are globally self-consistent.

This mean that I cannot kill my grandfather (a local action) because in the far future this would generate an inconsistency with my birth. Other way to state this is that I cannot kill my grandfather because nobody killed him in the past. Just consistent histories can develop in the Universe. In a series of papers (e.g. Echeverría et al. 1991), the application of the Principle of Self-Consistency has been used to shown that in different physical (simple) situations no paradoxes arise at all.

3. What is the epistemological status of the PSC?

What is this principle of consistency? Is it a law, like Einstein field equations or the Newton's laws? Is it tautological, i.e. it has not a factual content? Does it refer to our capacity to understand a logical world?.

We think that the PSC is not a physical law, in the sense that it says nothing on the kind of entities that populate the Universe. Its reference class is not a class of individuals, the ontological furniture of the world, like it is the case for the usual physical laws. We propose that the PSC is a metanomological statement (Bunge 1961), like, for instance, the principle of covariance. This means that the reference class of the principle is not formed by things, but by laws. The usual laws are restrictions to the state space of a physical system. Metanomological statements are laws of laws, i.e., restrictions on the global network of laws that thread the Universe.

Earman (1995) has suggested that the requirement of self-consistency must be considered as a new law. This can be objected through the simple determination of the reference class of the principle. In order to accomplish this, the principle should be translated to logical notation and the dominion of the bound variables in the logical quantifiers must be determined (for details see Romero & Torres 2000). When this is done it becomes clear that the principle deals just with laws. The requirement of consistency constraints is then pointing out the existence of deeper level super-laws, which enforce the harmony between local and global affairs in space-time.

4. Self-existing objects trapped in CTCs?

Although the PSC eliminates grandfather-like paradoxes, other highly perplexing situations remain. The main of these situations is the possibility of an ontology with self-existing objects. Here there is a graphic description of one of such objects: I have a book and I give it to a friend of mine saying: "Would you be so kind as to keep safe this book a whole year and return it to me at this same day and hour, next year". "Sure!" says my friend, and a year later gives me back the book. Then, I take the book, run into my time machine, appear a year before in front of my friend and say: "Would you be so kind as to keep safe this book a whole year and return it to me at this same day and hour, next year".

Who made the book?. It seems to be trapped in a time loop. There are not inconsistencies to be avoided: every local action is consistent with the global loop. There is no causality violation, no lawless trajectories, nothing but a book never created, never printed, but, somehow, existing in space-time. It has been suggested that if time travel is unavoidable, then we should accept an ontology of self-existing objects (Nerlich 1981). They just are out there, trapped in space-time. There is no sense in asking where they come from.

The acceptance of such a bizarre ontology proceeds, we believe, from an incorrect application of the PSC. It is usually discussed within the framework of General Relativity, but actually it encompasses all physical laws. What should be demanded is total consistency, and not just consistency in the solutions of Einstein or other field equations.

In order to show this, we have taken into account thermodynamic effects when considering CTCs in wormhole space-times. We have specifically shown that any real fluid increases its entropy when traveling in a casual loop threading two wormhole mouths (see, again, Romero & Torres 2000 for details). This means that the state of the system changes along the loop, and then that the loop cannot be consistently closed, because initial and final states do not match. The PSC therefore rules out these trajectories in space-time. There cannot be self-existing objects by the same reason that there is no grandfather paradox: global consistency must be fulfilled in the Universe.

5. Conclusions

Time travel is not only a justification to let fly our imagination. It can be a powerful tool to probe the deep levels at which the laws of nature are led to their most extreme manifestations. It is a conceptual instrument that can allow us to clarify not only the very foundations of General Relativity, but also the complex relations among the different theories we use to describe the Universe in which we live.

We have described here how certain apparent paradoxes originated by time travel can be conveniently dissolved. In particular, we are not committed to accept an ontology of self-existing objects if we take seriously the possibility of formation of traversable CTCs. There remain, however, some important epistemological issues arisen from information flux to the past. But this is a different story, that will be treated elsewhere (Romero & Torres 2000).

Acknowledgements

The authors participation in this School has been possible thanks the kind support provided by Fundación Antorchas and the ICTP. They also thanks the CONICET and the agency ANPCT (PICT 03-04881).

References

Bunge, M. (1961) Laws of physical laws, *Am. J. Phys.* **29**, 518-529.

Cassidy, M.J. and Hawking, S.W. (1998) Models for chronology selection, *Phys. Rev* **D57**, 2372-2380.

Earman J. (1995) *Bangs, Crunches, Whimpers, and Shrieks: Singularities and Acausalities in Relativistic Spacetimes*, Oxford University Press, New York.

Echeverría, F., Klinkhammer, G., and Thorne, K. S. (1991) Billard balls in wormhole spacetimes, *Phys. Rev.* **D44**, 1077-1099.

Hawking, S.W. (1992) Chronology protection conjecture, *Phys. Rev.* **D 46**, 603-611.

Morris, M.S. and Thorne, K. S. (1988) Wormholes in spacetime and their use for interstellar travel: a tool for teaching general relativity, *Am. J. Phys.* **56**, 395-412.

Nahin, P.J. (1998) *Time Machines: Time Travel in Physics, Metaphysics and Science Fiction*, Springer-Verlag and AIP Press, New York.

Nerlich, G. (1981) Can time be finite?, *Pacific Philosophical Quaterly* **62**, 227-239.

Novikov, I. D. (1989) An analysis of the operation of a time machine, *Soviet Physics JETP* **68**, 439-443.

Romero, G. E. and Torres, D. F. (2000) Self-existing objects and auto-generated information in chronology-violating spacetimes, in preparation.

Torres, D. F. and Romero, G. E. (2000) Do wormholes exist?, these *Proceedings*.

DO WORMHOLES EXIST?

DIEGO F. TORRES
Departamento de Física, UNLP,
C.C. 67, 1900, La Plata, Buenos Aires, Argentina
&
Instituto Argentino de Radioastronomía,
C.C.5, 1894, Villa Elisa, Buenos Aires, Argentina
email: dtorres@venus.fisica.unlp.edu.ar

AND

GUSTAVO E. ROMERO
Instituto Argentino de Radioastronomía,
C.C.5, 1894, Villa Elisa, Buenos Aires, Argentina
email: romero@irma.iar.unlp.edu.ar

Abstract. In this paper, we assess the possible existence of natural wormholes in extragalactic space.

1. Introduction

Mankind has always shown a fascination to break speed limits, from blood-driven carriages to inter-planetary probes. The state of the art in ultra-fast vehicles is ion-propelled engines, like NASA's Deep Space I, whose maximum velocity is about 99 200 km/h. Even with this high velocity, and disregarding obvious technical problems, a trip to the nearest star would require 40 000 years. Are we confined to our own corner of the galaxy, within the Solar System? Is the constraint imposed by the velocity of light necessary telling us that we shall never be able to reach other stars? A possible answer to these questions depends on the topology of the space-time.

Imagine that space-time is represented by a sheet of paper. Between two arbitrary points, we can always draw a straight line: this stands for a geodesic. Is this line the shortest path between these two points? Generally,

J. Chela-Flores et al. (eds.), Astrobiology, 259–262.

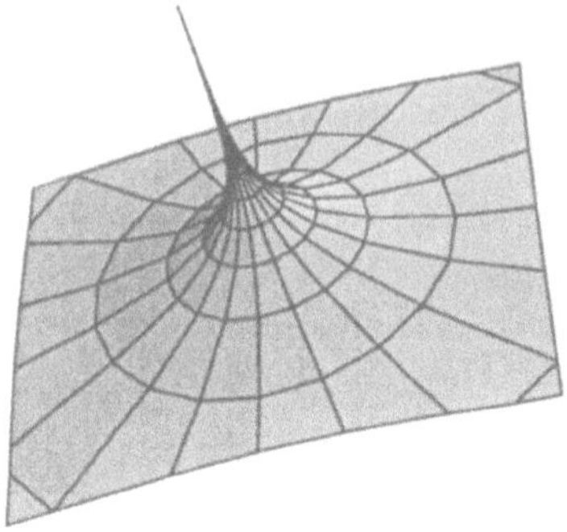

Figure 1. Black hole embedding diagram.

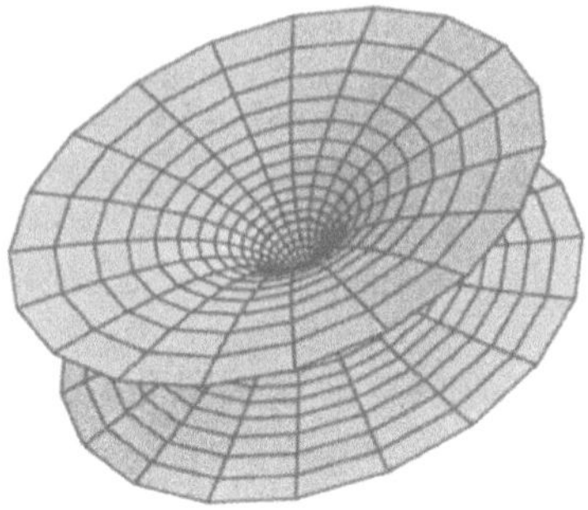

Figure 2. Wormhole embedding diagram.

the answer would be yes; but suppose that the paper (the space-time) is folded in such a way that both points coincide. If we could open a tunnel from one side of the paper to the other, the length of the shortest path would be zero. This is the intuitive idea of the wormhole concept.

Technically speaking, a wormhole is a region of space-time with non-trivial topology. It has two mouths connected by a throat. The mouths are not hidden by event horizons, as in the case of black holes, and, in addition, there is no singularity that could avoid the passage of particles, or travelers, from one side to the other. Embedding diagrams of both, black holes and wormholes are shown in Figs. 1 and 2.

In this paper we shall review the possible observational consequences of the existence of wormhole-like objects in the universe.

2. What wormholes are made of?

The usual approach to solve Einstein's field equations is to locally assume an energy and momentum distribution for matter and fields, and then to derive the corresponding metric. For studying wormhole properties, this method is inverted: we impose a tunnel-like geometry, and we then determine the characteristics of the matter that threads it. The crucial point is to ensure that the throat does not collapse under its own gravitational attraction: this requires very peculiar constraints in the stress-energy tensor that make it violate the so-called energy conditions.

This is equivalent to say that, at some points of the throat, a gravitational repulsion should be exerted. Graphically, a bundle of light rays entering in the upper mouth, which firstly is radially converging, should get defocused, leaving the lower mouth with diverging trajectories. For an external observer, the only wormhole mouth which is visible to her, would appear as a diverging lens when acting upon the light of distant point sources. This fact can be used to trace observational signatures of natural wormholes.

3. On the detection of natural wormholes

In a wormhole lensing configuration, light rays coming from a distant galaxy are deflected in such a way that two intensity enhancements occur (for a picture of the process see the article by Parsons in New Scientist N^o2127, March 28, 1998, p14). These intensity peaks are called caustics. Between them there is an umbra region, where the observer receives no light at all.

The temporal profile is specular in character. Firstly, the intensity rises in a longer time than that it takes to decay, later, after the umbra is traversed, vice versa. If the distant point-like source is the nucleus of an active galaxy, which emits gamma rays, both caustics are similar to observed gamma ray bursts, and repetition measurements could constrain the total number of wormholes that may actually exist in the universe: if wormholes exist, we expect two gamma ray bursts of specular temporal profile, separated in time, coming from the same position of the sky.

There are more than 2000 observed gamma ray bursts, mainly by BATSE experiment on-board NASA's Compton Satellite. Of them, 5% at most, could be physically associated with repetition. 4% have a peculiar temporal profile: a flux which rises in a shorter time than what it takes to decay. These data can be used to extract the first upper bound on the possible existence of macroscopic amounts of matter that violates the energy conditions (Torres et al. 1998a,b). It turns out to be bounded by $|\rho_w| < 10^{-36}$ g cm^{-3}, which should be compared with the critical density $\rho_c \sim 10^{-29}$ g cm^{-3}. Although cosmologically insignificant, ρ_w still permits that millions of wormholes of stellar size could populate the universe.

4. Discussion

We have analyzed a subsample of 631 gamma ray bursts from BATSE data, and extracted from them the best candidates for wormhole lensing (Romero et al. 1999, Anchordoqui et al. 1999). However, the positional error boxes of BATSE detections are about several degrees, and it makes necessary an statistical analysis. Evidence for wormhole existence is, by now, highly unconvincing, and we can expect to tighten the bound even more with the forthcoming generation of gamma ray satellites.

But even if natural wormholes would not exist, an advanced enough civilization could, perhaps, construct one. In order to do so, they will have to change the topology of space-time. Classical general relativity ensures us that by doing so, bizarre consequences will appear. Particularly, it can be proved that it implies the appearence of closed time-like curves (see the work by Romero and Torres in this volume). It should be noticed, however, that general relativity breaks down at a quantum level, and that new effects, not taken here into account, could then arise.

Wormholes, or any other kind of warp-propulsion devices (all of them involving matter that violates the energy conditions) seem to be the only known way that could help our far descendants to break their galactic isolation, and look for their fate, in the stars.

Acknowledgements

The authors participation in this School has been possible thanks the kind support provided by Fundación Antorchas and the ICTP. They also thanks CONICET, and the agency ANPCT (PICT 03-04881).

References

Anchordoqui L. A., Romero G. E., Torres D. F. and Andruchow I. (1999) *In search of natural wormholes*, Mod. Phys. Lett. **A14**, 791-798.

Romero G. E., Torres D. F., Andruchow I., Anchordoqui L. A. and Link B. (1999) *Gamma ray bursts with peculiar temporal profiles*, Mon. Not. R. Ast. Soc. **308**, 799-806.

Torres D. F., Romero G. E. and Anchordoqui L. A. (1998a) *Wormholes, gamma ray bursts and the amount of negative mass in the universe* (Honorable Mention in the Gravity Foundation Research Awards, 1998), Mod. Phys. Lett. **A13**, 1575-1582.

Torres D. F., Romero G. E. and Anchordoqui L. A. (1998b) *Might some gamma ray bursts be an observable signature of natural wormholes?*, Phys. Rev. **D58**, 123001.

HETEROGENEOUS RADIOLYSIS OF SUCCINIC ACID IN PRESENCE OF SODIUM- MONTMORILLONITE. IMPLICATIONS TO PREBIOTIC CHEMISTRY

M. COLÍN-GARCÍA, A. NEGRÓN-MENDOZA AND S. RAMOS-BERNAL
Instituto de Ciencias Nucleares, U.N.A.M. Circuito Exterior C. U. 04510, México D.F., México.

1. Introduction

Solid surfaces may be of a great importance in prebiotic chemistry, as site of concentration, adsorption and catalysis of many compounds. Among the solid surfaces geologically relevant for this purpose are clay minerals. Clays have an enormous distribution in time and space. They have high affinity to organic compounds (Ponnamperuma, *et al.*, 1982, Rao, *et al.*, 1980). In this geological framework, it is relevant to consider the contribution of solid surfaces. On the other hand, succinic acid molecule may play an important role in chemical evolution studies. This is due that this compound, readily formed from many different reactions of prebiotic significance, acts as raw material for the synthesis of more complex compounds like porphyrins. Succinic acid also acts as substrate of many organisms, and it has a high stability toward a high radiation field. The aim of this work is to establish how a clay mineral (sodium-Montmorillonite) affects the radiolysis of succinic acid yielding more complex acids.

2. Experimental Procedures

2.1. PREPARATION OF SAMPLES

Aqueous solutions of succinic acid 0.1 M, at natural pH, were prepared, employing triple distilled water. All chemicals used were of the highest purity available. The glassware was cleaned with a sulfonitric solution according to the procedures recommended in Radiation Chemistry (Donnell & Sangster, 1970). For this study we used Na-Montmorillonite of Wyoming bentonite. Different quantities of clay were used (0, 0.5, 0.7 and 1 g). Five milliliters of succinic acid solutions were mixture with the Na-Montmorillonite. This procedure was in a glass tube, evacuated and saturated with argon. In the case of the system without clay the samples were prepared in glass syringes. The samples were deareated with argon.

J. Chela-Flores et al. (eds.), Astrobiology, 263–266.

2.2. IRRADIATION

The irradiations were carried out in two gamma sources of ^{60}Co. One of high intensity (Gammabeam 651 PT). In this source the radiation doses were from 46.29 kGy to 277.78 kGy. To study the formation of gaseous products, a second source was used. This is a low intensity gamma source (Gammmacell-200). The radiation doses were from 3.7 Gy to 6.2 Gy.

2.3. ANALYSIS OF NON-VOLATILE PRODUCTS

For the analysis of non-volatile products, the mixture was centrifuged. A measured amount of the supernatant was evaporated until dryness. Then methyl esters were prepared according to Negrón-Mendoza *et al.*, 1983. The analysis was done in a GC chromatograph, using a glass column (1.82 m in length and an internal diameter of 4 mm) packed with Silar 7C. For the GC-MS analysis a capillar column of methyl silicon (12 m in length and 0.33 μm of inner diameter) was used. This chromatograph was coupled to a mass spectrometer.

2.4. GAS AND CLAY ANALYSIS

The production of CO_2 was followed as function of the dose. Previously 0.5 g Na-Montmorillonite were evacuated in a vacuum line for three hours. After 5 ml of the solution of succinic acid were added to the clay. The system was evacuated for 15 minutes more. The gaseous products formed by the irradiation were extracted employing a Toepler gauge connected to a Gas Chromatograph. The products were identified by their GC retention time.

The clay was recovered from solutions by centrifugation, washed with water, and finally dried at 50 °C. The clay was analyzed by infrared spectroscopy (Perkin Elmer model 500-FT-IR) using bromide potassium disks.

3. Results

3.1. SAMPLES WITHOUT CLAY

The irradiation of aqueous succinic acid produced many compounds. The main product obtained was the 1,2,3,4-butanetetracarboxylic acid (dimer of succinic acid). Other carboxylic acids were also identified, like oxalic, malonic, malic, carboxysuccinic, 1,2,3-butanetricarboxylic, tricarballylic, aconitic and citric acids (Figure 1).

The principal feature of these series of experiments was the production of the dimer, as the principal way of decomposition of succinic acid. The decomposition increases as a function of the dose.

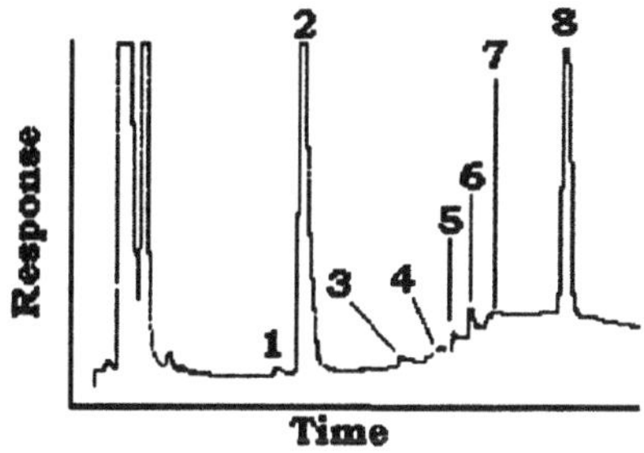

Figure 1. Gas chromatogram of methyl esters of carboxylic acids formed by the irradiation of succinic acid in the absence of clay. (1) Malonic, (2) succinic, (3) malic, (4) carboxysuccinic, (5) 1,2,3-butanetricarboxylic, (6) tricarballilic+aconitic, (7) citric, (8) dimer.

3.2. SAMPLES WITH CLAY

In presence of clay the number of products identified decreased considerably. At the maximum concentration of Na-Montmorillonite (1 g), aconitic, tricarballilyc and the dimer of succinic acid were the only products observed (peaks 6 and 8 in Figure 1). The production of the dimer as a function of the doses shows that when the amount of clay increases, the production of dimer decreases.

Gaseous products were also detected and identify, as CO_2 and H_2 .The production of CO_2 was greater in samples with clay. This formation increases as function of the dose. The source of H_2 is from the radiolysis of water and from the abstraction reactions produced during the radiolysis.

3.3. CLAY ANALYSIS

Various adsorption bands characterize the typical IR-spectrum of Na-Montmorillonite. The band in 3630 cm^{-1} corresponds to Si-OH. Clays usually have water associated, this correspond to the band in 1636 cm^{-1}. The bands in 918 cm^{-1}, 880 cm^{-1} and 798 cm^{-1} are referred as the bands of Si-O-Al, Si-O-Mg and Si-O-Fe (Wilson, 1994). The analysis by IR of the Na-Montmorillonite treated e irradiated with aqueous succinic acid shows the presence of the acid and the IR spectrum is modified in several bands. Two bands increase in magnitude, the 3630 cm^{-1} and 3428 cm^{-1} bands, also the last one shows a small displacement. The appearance of a new band in 1700 cm^{-1} is evident and is due to the carbonyl group of the succinic acid. These results show that there is an interaction between clay and the succinic acid.

4. Discussion and General Remarks

In an aqueous system the radiation interacts with water molecules. Very reactive species are formed due to this interaction (H, OH, $e^{\cdot}_{aq}$, H_2 and H_2O_2). They attack in a secondary way to the succinic acid molecules, yielding the observed products.

$HOOC\text{-}CH_2\text{-}CH_2\text{-}COOH + {\bullet}H \text{ or } {\bullet}OH \longrightarrow HOOC\text{-}CH_2\dot{C}H\text{-}COOH + H_2 \text{ or } H_2O$

succinic acid — radical 1

$HOOC\text{-}CH_2\dot{C}H\text{-}COOH \longrightarrow HOOC\text{-}CH_2\text{-}(CH\text{-}CO_2H)_2\text{-}CH_2COOH$

radical 1 — dimer of succinic acid

In the system without clay the main reactions induced by radiation takes place via free radicals. The principal reaction is the dimerization. However, in presence of clay, the results showed that there are changes in the mechanism. First the number of products diminished and the generation of CO_2 increased lineally with the radiation dose. Thus, there is a preferential way of decomposition. In this system is, the decarboxylation reaction. Thus, the main products obtained are the CO_2 and the propionic acid, but the last acid was not identified because its high reactivity in the conditions studied. In this case, the clay may play a role as an energy moderator following the next pathway:

$\text{Clay} + \gamma \text{ radiation} \longrightarrow \text{clay}^{+} + e^{-} \text{ and } [\text{clay}]^{*}$

$[\text{clay}]^{*} + \text{succinic acid} \longrightarrow \text{clay} + [\text{succinic acid}]^{*}$

$[\text{ succinic }]^{*} \longrightarrow {\bullet}CH_2\,CH_2 - COOH + {\bullet}COOH$

${\bullet}COOH \longrightarrow CO_2 + {\bullet}H$

The present study was a further attempt to gain more insight into the role played by radiation-induced reactions in solid surfaces in chemical evolution studies. As in previous studies (Negrón-Mendoza and Ramos Bernal, 1998), the results obtained suggest that the clay alter the reaction mechanism in a preferential way for some reactions, acting as moderator in energy transfers process. In the case of carboxylic acids this way is the decarboxylation versus other reactions that take place in the system without clay.

Acknowledgement

This work was partially supported by a CONACYT grant.

5. References

Donnell, J.H. and Sangster, D.f. (1970). *Principles of Radiation Chemistry.* American Elsevier Publishing Company, United States. 176 pp.

Negrón-Mendoza, A., Albarrán, G. and Ramos-Bernal, S. (1996). Clays as natural catalyst in Prebiotic Processes. In: *Chemical Evolution: Physics of the Origin and Evolution of Life.* (Chela-Flores, J. & Raulin, F. Eds.) Kluwer Academic Publishers, Netherlands. pp. 97-106

Negrón-Mendoza, A. and Ramos-Bernal, S. (1998). Radiolysis of carboxylic acids adsorbed in clay minerals. *Radiat. Phys. Chem.* **52**(1-6): 395-397

Ponnamperuma, C., Shimoyama, A. and Friebele, E. (1982). Clay and the Origin of Life. *Origins of Life.* **12**:9-40

Rao, M., Odom, D.G. and Oró, J. (1980). Clays in Prebiological Chemistry. *J. Mol. Evol.* **15**: 317-331

Wilson, M.J. (1994). *Clay Mineralogy: Spectroscopic and Chemical Determinative Methods.* Chapman & Hall, Great Britain. 367 pp.

CONDENSED MATTER SURFACES IN PREBIOTIC CHEMISTRY

S. Ramos-Bernal and A. Negrón-Mendoza
Instituto de Ciencias Nucleares, UNAM
México, D.F., 04510 México

Abstract

Chemical reactions among organic compounds adsorbed in condensed matter are significantly enhanced when irradiated with gamma rays. Such combination of interactions is considered to have been served for the prebiotic evolution of molecules. The interaction of minerals as alumino-silicates with gamma rays and its relationship to surface chemistry is discussed. The importance of the above radiation-solid interaction is supported, within the framework of the capabilities of these solid minerals, to serve, with its behavior, to the evolution of organic molecules. The characterization of the solids, as well as identification of defects, gave some clues about the storage and transfer of energy and they are connected with the luminescent properties of these solids. Therefore, the interaction of these condensed phases with gamma rays, conduces to some commonly accepted concepts of heterogeneous catalysis by which surface activity might be enhanced with the presence of this energy inputs and may have relevance in the early Earth

1. Introduction

To considerate multiphase systems would be much more realistic for the simulation of the primitive Earth's environment, than the monophase system generally used. The contribution of solid phases is of great relevance to the experimental simulation of the prebiotic Earth. The prebiotic concentration of organic molecules in solid state has also been thought of as one of the most important contributions to the evolution of such molecules, without this concentration factor by solids the reactions among organic compounds would be difficult in many cases. Besides the concentration capacity, also the solids have a potential catalytic activity for very important reactions that may occurred during the prebiotic Earth. In other words, after the solids adsorbed organic molecules, would then catalyze the synthesis of small molecules and oligomers into bio-polymers. If in addition, an energy input was introduced then many reactions may have enhanced.

J. Chela-Flores et al. (eds.), Astrobiology, 267–270.

2. Interaction of Solids with Ionizing Energy

Very few types of potential sources of energy in the primitive Earth are able to penetrate the condensed matter; ionizing radiation is the one of these few. Therefore is proposed as relevant to this scenario. The investigation of the ionizing radiation effects on solids led to the radiation physics and chemistry to contribute to a greater understanding of condensed matter. Henceforth both subjects are of great relevance in the study of the prebiotic Earth.

3. Main Features about Radiation Catalytic Effect.

The interaction of ionizing radiation with solids produces: excitation, degradation, as well as storage, and transfer of energy within the solid. Hence, the role of solids as substrates for prebiological chemistry and inorganic prototype of life should be considered along with this high energy radiation. On the other hand, organic chemical reactions within solids can be improved via heterogeneous catalysis with the presence of ionizing radiation. Several examples of these types of enhancement are clay-mediated decomposition of carboxylic acids acid exposed to ionizing radiation (Ramos-Bernal and Negrón-Mendoza, 1992; Negrón-Mendoza et al., 1993; 1995).

The trapped non-equilibrium charge carries within the solid, play a very important role within this radiation-induced catalysis. A correlation between the changes of optical, and crystallographic properties of solids under irradiation, and the possible organic reactions inside solids may give some important clues in the synthesis of complex molecules during the prebiotic period of the Earth. Chemical reactions within irradiated solids are known as topochemical reactions and are the surface-chemical equivalent of the light emission in radiation induced thermoluminiscence.

4. Charge Transfer

The interaction of radiation with condensed matter can produce charge transfer to the organic compounds absorbed on its surface. In multi-phase systems there are modes of energy transfer across interface that do not exist in homogeneous phases. In other words, solids can store energy from the gamma radiation and release it in various forms. The energy transfer from solids to adsorbate is possible. Therefore, it is likely that energy deposited in the bulk solid by penetrating energy sources, such as ionizing radiation, should be transduced to minor energies form before transfer to a distant interfacially adsorbed reactant.

The distribution of energy, in the damaged material, is distributed between the excited states and changes made in the energy-level structure. The absorption of energy such as ionizing radiation is considerable less well understood than optical transitions, since the energy absorbed may be fractionated into: (a) heat; (b) luminescent decay; (c) chemical changes in the solid; (d) transference to an adsorbed reactant; or (e) energy stored as separated charge pairs. Therefore, the absorption and storage of electronic energy within a crystalline lattice substrate will have a significant impact on the chemistry of adsorbed organics. If the effect of bulk lattice excitation is to alter the surface electronic distribution. Then there must exist modes of energy transfer between the lattice substrate and the reacting medium. The role of electronic energy transfer in heterogeneous catalysis is not very well characterized. Then further development in this topic is needed to understand its relevance in prebiotic chemistry. The luminescence phenomena are indicative of energy storage and charge mobility in the solid and they are produced by solvatation / desolvatation and mechanical stress. The trapped separated charge pairs produced by high-energy radiation frequently are long lived, manifestations of these metastable excited states are EPR signals and predicted optical absorption properties.

4.1 CHANGE OF PROPERTIES

Properties of the solid that are sensitive to radiation damage are to be correlated with the chemical change of the organic compounds adsorbed. For the substances of most interest to chemists, the properties of principal concern are thermal, electromagnetic, physical, and electrical properties. Mechanical properties are, of course, definitely affected in irradiated solids, but these are of minor importance in inorganic and organic crystals. A significant mechanical effect in complex inorganic crystals is a general destruction of the crystal forces producing changes in density. Another case is when significant chemical change has occurred, pulverization. In some simple salts such as LiF hardening may occur and the yield stress may increase, in some cases by as much as a factor of 2 (Ramos-Bernal et al., 1983).

5. Radiation effects in organic compounds

It is very important to elucidate, how much the solid is involved in the chemical transformation that is taking place at its surface. In other to have some idea of this, some properties of the solid must be monitored and correlated with the adsorption of organic compounds into its surface. The role of minerals as selective adsorbed, concentrator and catalyst for polymerization have already been demonstrated (Negrón-Mendoza et al., 1993,1995). A review of the data on adsorption shows that for most compounds of biological relevance, the adsorption is larger in acidic pH. Sometimes the binding of these types of compounds drops to almost null adsorption at pH=8 . Since the pH of the primitive ocean was almost 8 this may disqualify some results. Nevertheless there are several natural conditions that attain higher acidities, and thus improve adsorption binding supporting in this way the possible role of microenvironments in the primitive Earth.

Irradiation of organic systems will produce radicals, new stable molecular species, trapped electrons, various excited states, holes, interstitials, and so forth. These are detected by the usual techniques of EPR, NMR, optical spectroscopy, X-ray, infrared, photo conductivity, electron microscopy and conductivity measurements. It must be remembered however, that detection of the various primary species will depend on the temperatures of irradiation.

Organic chemical reactions within solids can be gotten better via heterogeneous catalysis with the presence of ionizing radiation. The enhancement of these reactions can be accounted for: transfer processes, redox reactions, stabilization of intermediates, and finally by acidity behavior, that can be brought from either Bronsted or Lewis type sites.

Since irradiation forms pair of opposite charges, the formation of trapped non-equilibrium charge carries occur within the solid. These carriers play a very important role within this radiation-induced catalysis.

6. Final Remarks.

Changes into the solid will affect the organic compounds sorbed, then a correlation between the changes of, for example, optical, mechanical and crystallographic properties of solids under irradiation, and the possible organic reactions inside solids may give some important clues in the synthesis of complex molecules during the prebiotic period of the Earth.

It was considered here the potential for altering, via irradiation, the capabilities of solids to serve as substrates for prebiological chemistry. To support for this broadly defined idea about the role of the interaction of high energy radiation with solids, our group has provided some results using clay minerals (Negrón-Mendoza, et al.,1993; Negrón-Mendoza and Ramos-Bernal, 1995; Ramos and Negrón-Mendoza ,1992).

This shows potential relationships between minerals, electronic excitation, and surface reactivity, as applied to chemical evolution. The role of electronic energy transfer in heterogeneous catalysis on solid surfaces is fundamental. It is necessary to include spectroscopic properties of catalysts in order to see their reaction-promoting capabilities. In photo-electrochemical systems the mode of energy transfer is electron (hole) transfer, driven by a potential difference between the electronic energy of the light-absorbing catalytic surface and the redox potential of the solution phase reactant. For electronic excitation to produce chemistry effectively, an energy barrier between electron/hole pairs is needed to prevent wasteful electron/hole recombination. This barrier is provided by the space charge potential near the interface. Given these high excitation energies, impurity centers have a much larger impact on surface chemistry.

On the other hand the thermoluminescent properties provide the basis for estimating the potential significance of solid surfaces role for catalyzing reactions relevant to prebiotic chemistry. Yet more specific experiments should be made to test this hypothesis about the role played by solids and ionized energy. This hypothesis is principally based on the coupled transport of electrical charge and electronic energy through the solid, which operates via production and mobility of electron/hole pairs. This is to say the mobility of charge/electronic excitation between defect centers serves as the basis for a primordial inorganic electron transport chain.

7. References

Negrón- Mendoza A., Albarrán G., Ramos-Bernal, S. (1993). Transformation of malonic acid adsorbed on a clay mineral by gamma irradiation.. *Radiation Phys and Chem,* **42**, 1003-1006

Negrón-Mendoza A., Ramos-Bernal S., Albarrán G. (1995) Enhanced decarboxylation reaction of carboxylic acids in clay minerals. *Radiation Physics and Chemistry* **46**, 565-568

Ramos –Bernal, S. and Negrón-Mendoza A.(1992) Radiation heterogeneous processes of ^{14}C-acetic acid adsorbed in Na-montmorillonite. *J. Radioanalytical Nuclear Chemistry,* **160**, 487-492

Ramos-Bernal, S., Murrieta, H., Aguilar, M., and Rubio,J. (1983). Dose rate dependence of first stage coloration of NaCl doped with divalent impurities. *Radiation Effects Letters,* **68**, 176.

IRRADIATION OF ADENINE ADSORBED IN NA-MONTMORILLONITE. *IMPLICATIONS TO CHEMICAL EVOLUTION STUDIES*

A. GUZMAN, S. RAMOS-BERNAL AND A. NEGRON-MENDOZA
Instituto de Ciencias Nucleares, UNAM.
Circuito Exterior 04510, Mexico D.F., Mexico.

1. Introduction

Adenine is an important compound in biological systems, such as genetic and energy utilization processes. Adenine is readily formed in prebiotic conditions. Its synthesis and stability in environmental conditions is of paramount importance in chemical evolution.

There are several routes for the synthesis of adenine and other nitrogen bases simulating the primitive Earth (see Miller and Orgel, 1974 and the references therein). However, it is one thing is to be able to show that an organic compound can be synthesized in prebiotic conditions, and it is quite another to be sure that is sufficiently stable to have been available to the first organism.

To ensure that organic compounds endured in the primitive Earth there are several possibilities: a) they were synthesized continuously and they reach steady state concentration. b) The compounds present a long half-life in the environmental conditions of the primitive Earth. c) Solid surfaces protect the organic compound adsorbed in the clay, as it was postulated by Bernal in 1951.

In spite of the importance of the presence of adenine in the prebiotic environment, there are few papers that deal with the survival of this type of compound. In particular, there are none about the study of the stability of adenine in an aqueous medium, at high temperatures, or in the presence of high radiation fields.

On the other hand, clay minerals might have played an important role on the early Earth. Clays are known to have a high affinity for organic compounds. They are considered the most likely inorganic material to promote organic reactions at the interface of the hydrosphere and lithosphere. Bernal (1951) suggested several roles for the clays. He suggested that clays act as concentrators of biological precursor molecules. In the clays polymerization to macromolecules was possible. Clays may protect these molecules from high-energy radiation.

In previous studies it was determined that the adsorption of the adenine in clay happens readily in an acid pH. For Na-Montmorillonite, the adsorption occurs mainly inside the interlamellar channel (Perezgasga *et al.*, 1993.).

The attempt of this work is to show preliminary results about the studies of the stability of adenine in a high radiation field. The work is divided in two folds:

1) To study the stability of free adenine in an aqueous solution exposed to a high-energy radiation.

J. Chela-Flores et al. (eds.), Astrobiology, 271–274.

2) To study the protector role of clays when they have an adsorbed compound and the clay-organic system is exposed to radiation. For this purpose, we study the radiolysis of adenine adsorbed in Na-Montmorillonite.

2. Experimental

The experimental part is divided in three stages: 1) Radiolysis of aqueous solutions of adenine at two different pHs. 2) Adsorption/ desorption experiments of adenine from a clay mineral. 3) Study of the heterogeneous radiolysis of the water-adenine-clay system.

2.1 RADIOLYSIS OF FREE ADENINE

2.1.1. Preparation of Solutions and Irradiation
Two solutions of adenine 10^{-3} M at pH = 7 and 2 were prepared. The oxygen was removed by passing argon through solutions. Solutions were irradiated in glass ampoules at different doses, using a source of ^{60}Co (Gammabeam 651-PT).

2.1.2 Adsorption/ Desorption Experiments
The adsorption of the adenine at pH 2 in Na-Montmorillonite was done as it is described previously (Perezgazga *et al.*, 1996). Adenine left in the solution was analyzed by HPLC.

The adenine was removed from the clays with the following procedure: After pH adjustment with HCl and an adsorption period of 60 minutes (with continuos shaking), the clay-adenine was centrifuged. The supernatant contained the adenine that was not adsorbed into the clay. The clay pellet was washed with distilled water of the same pH as the original solution. This solution was measured by HPLC. After centrifugation, the clay was extracted with solutions at different pH. Several experiments were made adjusting the pH with NaOH at 7, 10, 12, and 14, and shaking the suspensions for 60 min,

2.1.3 Radiolysis of Adenine-Na-MontmorilloniteComplex
One hundred milligrams of clay were mixed with 3 ml of adenine 10^{-3} M; this amount is below the cation exchange capacity of the clay. The samples were adjusted to pH 2, and they were shaking for 60 minutes. After this time, the samples were centrifuged. The supernatant was removed and analyzed by HPLC. The clay was washed with 3 ml of water and centrifuged again. Another 3 ml of water was added and later the pH was adjusted to 2 and the samples were irradiated at different radiation doses. After the irradiation the supernatant was analyzed again by HPLC.

2.2 HPLC ANALYSIS

The analysis of the samples was done by high-pressure liquid chromatography (HPLC) in a Varian chromatograph model 8055 with a column packed with MCH-10. The detector used was a Varian ultraviolet detector at 260 nm. The mobile phase was a mixture of two solutions. Solution A: a methanol-water mixture (80:10 v/v). Solution B: a pH 2 buffer solution of KH_2PO_4 0.01 M and H_3PO_4 0.035 M. The gradient used was from 94% of solution B up to 60 % of solution B in 12 minutes.

Figure 1. HPLC chromatogram of adenine 1 x 10^{-3} M irradiated at 1.75 kGy.
A) pH 7, B) pH 2.

3. Results

Radiolysis of Free Adenine. The destruction of adenine in terms of molecules destroyed for each 100 eV given to the system was 1.1 and 0.8 molecules at pH 2 and 7 respectively. These values represent the radiochemical yield and they are low values. These results showed that adenine is resistant when it is exposed to a high radiation field of 1.2 kGy/h (1 Gy= 1J/Kg).

Figure 1 shows the chromatograms of the adenine 10^{-3}M irradiated at 1.75 kGy at pH 2 and 7. Five compounds are formed from the irradiation at pH 7 and their yield increases lineally with the doses. Still they remain unidentified. There was a difference in the formation of products at pH 2 in which only three products were detected.

Desorption of Adenine from the Clay. To remove the adenine from the Na-Montmorillonite, several experiments were made, varying the pH of the extracting solution. The best results were obtained with a NaOH solution of 10^{-14} M. With this solution the extraction of adenine was almost quantitative. For more concentrated solutions (1x10^{-2}M) two extractions were necessary for removing the adenine from the clay, and the recovery was almost quantitative.

Irradiation ofAdenine-Na-Montmorillonite System. The results obtained with the adenine-clay system shows that there was not a decomposition of adenine. The role of clay as protector of the organic molecules was observed in these experiments.

4. Discussion

Adsorption of nucleic acid bases and their derivatives on clay minerals may have played an important role in primordial organic chemistry. Since it was first proposed by Bernal in 1951, it has been frequently argued that organic compounds are adsorbed, concentrated and protected by clays against radiation. In this article, we consider the irradiation of adenine adsorbed in a clay mineral in order to study its stability in such conditions.

The rate of hydrolysis of adenine to hypoxanthine seems so be slow enough at neutral pH that the instability of these substances does not represent a serious problem.

There are many works in the literatures that deal with the irradiation of aqueous, oxygen-free solution of purines. Still, the radiolysis of this kind of compounds is not well established, and it is poorly understood despite the considerable effort that as been made in this area (Mosqueira, *et al*, 1996).

The complexity of the reactions is considerable and the sites of the reaction with the purine ring are speculative (Cadet and Berger, 1985). From quantum chemical calculation it is suggested that the radicals of the water radiolysis attack the adenine molecule mainly at N-7, C-8 double bond (Pullman and Pullman, 1961). After the attack to the C-8 position with the hydrogen radical (from the water molecule), the imidazole ring of the adenine at the 7-8 bond can be opened. The steady state radiolysis of adenine showed a low yield of destruction, regardless of the high reactivity of adenine toward the attack of water decomposition products. This suggested that adenine goes through an overall reconstruction pathway.

Ionizing radiation on solids may be viewed as a mechanism that produces defects in the solid itself. These defects are responsible for possible chemical reactions of molecules adsorbed in an irradiated solid. Energy transfer in clay is an important phenomenon as it may induce chemical reactions. Thus, the energy of the radiation may be partially deposited into the clay leading to excitation and ionization. The energy taken up by the solid may be transferred to the adsorbed substance, both the water and the substrate. Hence, a radiolytic degradation of adsorbed nucleic bases is expected for a long exposition time. It has been calculated that the energy delivered in the prebiotic epoch 3.8 Ga ago from ^{40}K in clays was 1.75 x 10^{-2} Gy/year (1 Gy= 1J/Kg).

5. Concluding Remarks

The results obtained show that the gamma irradiation of aqueous adenine is pH dependent. The products formed at pH 2 are different from those formed at pH 7. Also the rate of decomposition of adenine is different. It increases slightly at pH 2.

Purine bases, like adenine, are also able to resist radiation. This is a distinct advantage since the molecules that were formed by ultraviolet light, ionizing radiation, or electric discharges had to survive in order to interact with each other to form more complexes molecules. In these series of experiments the protection role of the clays toward ionizing radiation was observed. However, these are very preliminary results and they still needs to investigate more deeply.

Acknowledgement. The authors are very gratefully to the Organizing Committee for the financial support to attend to The Ibero-American School of Bioastronomy.

6. References

Bernal, J.D. (1951) *The Physical Basis of Life*, Routledge and Kegan Paul, London

Cadet, J., and Berger, M.: (1985) *Int. J. Radiat. Biol.* **47**, 127-143.

Miller, S.L. and Orgel L. (1974) *The Origins of Life on Earth*, Prentice-Hall, Inc. New Jersey.

Mosqueira, G., Albarrán, G. and Negrón-Mendoza, A. (1996) A Review of conditions affecting the radiolysis due to ^{40}K of nucleic acid bases and their derivatives adsorbed on clay minerals, *Orig. of Life and Evol. Bios.*, **26**, 75–94.

Perezgasga, L., Negrón-Mendoza, A., De Pablo-Galán, L., Mosqueira, G. (1993) Site of adsorption of purines, pyrimidines and their corresponding derivatives on sodium montmorillonite. *Orig. of Life and Evol. Bios.*, 4,47

Pullman, B., and Pullman, A.: 1961, *Nature* **189**, 725.

ACCUMULATION OF ALKANES $\geq$ n-C_{18} ON THE EARLY EARTH

VICENTE MARCANO[1], PEDRO BENITEZ[1] AND ERNESTO PALACIOS-PRÜ[1,2*]
[1] *Electron Microscopy Center, University of the Andes, P. O. Box 163, Mérida, Venezuela. prupal@ula.ve*
2 International Institute of Advanced Studies, Simon Bolívar University, Caracas, Venezuela.

There is evidence that, in the early Earth, a considerable diversity of hydrocarbons produced endogenously or delivered by meteorites, comets and IDPs, could be presents on its surface, in the atmosphere and hydrosphere (Kvenvolden et al. 1970, Deamer et al. 1989, Chyba & Sagan 1992, McCollom et al. 1999). The formation of such hydrocarbons may be explained by the occurrence of Fischer-Tropsch-type (FTT) reactions (Ferris 1992, McCollom et al. 1999), mainly during the condensation of the solar nebula near 10 AU (Anders et al. 1973, Fegley & Prinn 1989, Gaffey 1997). Other pathways are thermal decomposition of iron oxalate (McCollom & Simoneit 1999) and structurally non-selective Miller-Urey (free radical) reactions (Lasaga et al. 1971, Sagan & Chyba 1997). Particularly Miller-Urey reactions yield tholin-like compounds having mainly saturated, aliphatic and polycyclic aromatic hydrocarbons (Sagan & Khare 1971, Clarke & Ferris 1997, Cleaves & Miller 1998). This paper shows the probable amounts of the heavy *n*-alkanes accumulated on the Earth at the time (t) from calculation based in the contribution of organic material by endogenous and exogenous sources.

It has been estimated a minimum value of 0.05% of alkanes $\geq$ *n*-C_{18} corresponding to the total mass for carbonaceous chondrites from the Murchison meteorite (Kvenvolden et al. 1970), which constitute ~ 17% of the total hydrocarbons and ~ 50% of the total amount of alkanes. Assuming from Mukhin et al. (1989) data that the hydrocarbon/carbon ratio is roughly constant, the total mean concentration of the hydrocarbons trapped in the delivered matter on the early Earth may be estimated depending from the carbon total content ($M\gamma$) present in it. Therefore, the amounts of trapped alkanes $\geq$ *n*-C_{18} (M_{HA}) from several exogenous sources are $M_{HA} = 0.17[M\gamma(\eta / \chi)]$, where η and χ are constant values and constitute the mean hydrocarbon amount (η) produced from a determined amount of total carbon (χ); 0.17 constitute the fraction of alkanes $\geq$ *n*-C_{18} presents in the delivered matter and eliminate the net amount of low molecular weight hydrocarbons presents in it. Results calculated on the basis of hydrocarbon/carbon ratio taking $\eta = 0.11\%$ and $\chi = 0.61$ wt% from experimental data (Mukhin et al. 1989) and $\eta = 0.35\%$ and $\chi = 2.0$ wt% from Murchison data are very

* To whom correspondence should be mailed.

J. Chela-Flores et al. (eds.), Astrobiology, 275–278.

TABLE I. Total amounts of carbon in exogenous sources. Amounts of alkanes $\geq n\text{-}C_{18}$ were calculated from evaporating-impact simulation experiment (1) and Murchison data (2).

Exogenous sources	Total carbon*	Alkanes $\geq n\text{-}C_{18}$	
		1	2
IDPs	~ 10%	~ 0.30%	~ 0.29%
Cometary matter	~ 14-17%	~ 0.47%	~ 0.46%
Asteroidal matter	~ 1.3%	~ 0.04%	~ 0.038%
Carbonaceous chondrites	~ 2-5%	~ 0.1%	~ 0.1%
Non-carbonaceous chondrites	~ 0.1-0.2%	~ 0.004%	~ 0.004%
Micrometeorites	~ 8%	~ 0.24%	~ 0.238%

* Nagy (1975), Anders & Owen (1977), Delsemme (1988), Cronin et al. (1988), Anders (1989), Chyba & Sagan (1992), Engrand et al. (1994) and Clemett et al. (1998).

similar (Table I) demonstrating its usefulness to calculate the total amounts of heavy *n*-alkanes delivered on the Earth.

Assuming that interplanetary DPs has accumulated linearly with geologic time without suffering changes during its entry (Chyba & Sagan 1992, Gaffey 1997), then the total amount of alkanes $\geq n\text{-}C_{18}$ delivered from these particles at a particular time (t) would be $m_{IDP\text{-}HA}(t) = \zeta\,[3.2 \times 10^6\,(t)]$, where $m_{IDP\text{-}HA}(t)$ is given in kg yr^{-1}, $\zeta = 2.95 \times 10^{-3}$ is a factor that constitute the mean fraction of alkanes $\geq n\text{-}C_{18}$ in IDPs (Table I), and 3.2×10^6 kg yr^{-1} is the chosen minimum accreted mass of IDPs (Kyte & Wasson 1986, Anders 1989) (Table II).

An additional source of exogenous heavy *n*-alkanes on early Earth, could be the accretion of dust as the Solar System passed through interstellar clouds (INDPs). According to Greenberg (1981), we assume that during each period of $\sim 1.63 \times 10^8$ years the Earth could have passed through one cloud, accreting organic carbon during each passage at a rate (α) ~ 1.5×10^9 kg . Then, the mass of carbon ($M\gamma$) accreted during the passage of a number of clouds $n = (t/1.63 \times 10^8)$ at the time (t) is $M\gamma_{INDPs}(t) = \alpha(t/1.63 \times 10^8)$. Resolving the equation $M_{HA} = 0.17[M\gamma(\eta/\chi)]$, the total mass of heavy *n*-alkanes delivered on the Earth at the time (t) is obtained (Table II).

The total mass m(*t*) from large impacted bodies (LB) on the Earth after some time (t) having masses in the range m_{min} ($= 4.39 \times 10^{15}$ kg = 14 km in diameter) to m_{max} ($= 1.36 \times 10^{20}$ kg = 440 km in diameter) (Sleep et al. 1989) is given by:

$$m_{LB}(t) = \xi\{3.7 \times 10^8[t + 2.3 \times 10^{-11}(e^{t/\tau} - 1)](m_{max}^{\;1-b} - m_{min}^{\;1-b})\} \quad (1)$$

where b = 0.54 is a power law exponent, $\xi \approx 23.54$, $\tau = 0.144$ Gyr, corresponding to a 0.1-Gyr decay half-life for the impactor population, and m is expressed in kg (Chyba 1990, Chyba & Sagan 1992). Because of the uncertainties in the amounts of organic molecules surviving the airburst, bodies < 14 km in diameter were neglected in the equation. There are several opinions about the type of impactors and their frequency of entry to the Earth's orbit (Shoemaker et al. 1979, Grieve 1982, Hartmann 1987). We take 10% of the impactor mass to be cometary and the 90% asteroidal, divided in 5.5%

TABLE II. Comparison of total mass (kg) of alkanes $\geq n\text{-}C_{18}$ deposited on the surface of the Earth between 4000-3600 Myr ago for several exogenous and endogenous sources.

Sources	Contribution of *n*-alkanes at the time (Gyr)		
	4.0	3.8	3.6
Large impactors	7.68×10^{18}	7.73×10^{18}	7.77×10^{18}
Post-impact recombination	7.05×10^{13}	9.87×10^{13}	1.26×10^{14}
Interplanetary DPs	4.72×10^{12}	6.6×10^{12}	8.49×10^{12}
Interstellar DPs	1.41×10^{8}	1.96×10^{8}	2.52×10^{8}

carbonaceous and 84.5% non-carbonaceous objects (Anders 1989, Chyba & Sagan 1992). Therefore, the net amount of those impactors from the total amount, $m_{LB}(t)$, is obtained from the factors: 10^{-1} for comets; 5.5×10^{-2} for carbonaceous, and 8.45×10^{-1} for non-carbonaceous, and the total amount of alkanes $\geq n\text{-}C_{18}$ delivered by exogenous contribution, $M_{LBA}(t)$, is

$$M_{LBA}(t) = \varphi_1[m_{LB}(t)0.1]+\varphi_2[m_{LB}(t)0.055]+\varphi_3[m_{LB}(t)0.845] \quad (2)$$

where φ is the fraction of heavy *n*-alkanes presents in cometary matter ($\varphi_1 \approx 4.75\text{x}10^{-3}$), carbonaceous asteroidal matter ($\varphi_2 \approx 10^{-3}$), and non-carbonaceous asteroidal matter ($\varphi_3 \approx 4 \times 10^{-5}$) (Table II).

The amounts of *n*-alkanes produced by post-impact recombination are obtained by $M_{HA} = 0.17[M\gamma(\eta/\chi)]$, taking the data for η and χ from Mukhin et al. (1989) and Kvenvolden et al. (1970) and the rate of organic carbon incorporated into organics by post-impact recombination, $M\gamma = (4.6 \times 10^6 \text{ kg yr}^{-1})\, f(t)$ in the early Earth given by Chyba & Sagan (1992), which yields $M_{EHA} = 0.17[4.6 \times 10^6(t)(\eta/\chi)]$ (Table II). Because of the uncertainties in the estimation of the amounts of hydrocarbons produced from the hydrothermal systems and in the oceanic lithosphere, they were neglected in the calculation.

In summary, the amounts of alkanes $\geq n\text{-}C_{18}$ delivered by interstellar DPs are lower than those delivered by interplanetary DPs, indicating that this is a better contributor. However, the production of *n*-alkanes by reactions from post-impact recombination shows higher yields than those obtained from interstellar and interplanetary DPs. Neglecting the decomposition rate and the contribution by impactors, it may be estimated that a layer of ~ 1.1 m thickness 3600 Myr ago and ~ 0.9 m thickness 3800 Myr ago might cover the Earth's surface. Large impacts up to ~ 100-km in diameter could occurred still 3600 Myr ago and generated excessives temperatures on Earth able to melt most of the planet's mass. Therefore, it is difficult to suppose that an accumulation of *n*-alkanes before of this age could be possible except during the intervals among impactor fall. All the sucesses described here may be commons to other planetary bodies (e.g. Venus, Mars) or satellites (e.g. Titan) belonging to our solar system and probably in analogs of the early solar system (Oró et al. 1992, Koerner 1997).

REFERENCES

Anders, E.: 1989, *Nature* **342**, 255.

Anders, E. and Owen, T.: 1977, *Science* **198**, 453.

Anders, E., Hayatsu, R, and Studier, M.: 1973, *Science* **182**, 781.

Chyba, C. F.: 1990, *Nature* **343**, 129.

Chyba, C. F. and Sagan, C.: 1992, *Nature* **355**, 125.

Clarke, D. W. and Ferris, J. P.: 1997, *Origins of Life Evol. Biosphere* **27**, 225.

Cleaves, H. J. and Miller, S. L.: 1998, *Proc. Natl. Acad. Sci.* **95**, 7260.

Clemett, S. J., Chillier, X. D. F., Gillette, S., Zare, R. N., Maurette, M., Engrand, C. and Kurat, G.: 1998, *Origins of Life Evol. Biosphere* **28**, 425.

Cronin, J. R., Pizzarello, S. and Cruikshand, D. P.: 1988, in *Meteorites and the Early Solar System*, eds. J. F. Kerridge and M. S. Mathews, University of Arizona Press, Tucson, pp. 268-288.

Deamer, D. W., Harang, E. A. and Seleznev, S. A.: 1989, *Origins of Life Evol. Biosphere* **19**, 291.

Delsemme, A.H.: 1988, *Phil. Trans. R. Soc. Lond.* **A 325**, 509.

Engrand, C., Michel-Levy, M. C., Jouet, C., Kurat, G., Maurette, M. and Perreau, M.: 1994, *Meteoritics* **29**, 464.

Fegley, B. and Prinn, R. G.: 1989, in *The Formation and Evolution of Planetary Systems*, eds. H. A. Weaver and L. Danly (Cambridge University Press), pp. 171-393.

Ferris, J. P.: 1992, *Origins of Life Evol. Biosphere* **22**, 109.

Gaffey, M. J.: 1997, *Origins of Life Evol. Biosphere* **27**, 185.

Greenberg, J. M.: 1981, in *Comets and the Origin of Life*, ed. C. Ponnaperuma (Reidel, Dordrecht), pp. 111-127.

Grieve, R. A. F.: 1982, *Geol. Soc. Am. Spec. Pap.* **190**, 25.

Hartmann, W. K.: 1987, *Icarus* **71**, 57.

Koerner, D. W.: 1997, *Origins of Life Evol. Biosphere* **27**, 157.

Kvenvolden, K Lawles, J., Pering, K., Peterson, E., Flores, J., Ponnaperuma, C., Kaplan, I. R. and Moore, C.: 1970, *Nature* **228**, 923.

Kyte, F. T. and Wasson, J. T.: 1986, *Nature* **232**, 1225.

Lasaga, A. C., Holland, H. D. Aand Dwyer, M. J.: 1971, *Science* **174**, 53.

McCollom and Simoneit, B. R. T.: 1999, *Origins of Life Evol. Biosphere* **29**, 167.

McCollom, T. M., Ritter, G. and Simoneit, B. R. T.: 1999, *Origins of Life Evol. Biosphere* **29**, 153.

Mukhin, L. M:, Gerasimov, M. V. and Safonova, E. N.: 1989, *Nature* **340**, 46.

Nagy, B.; 1975, *The Carbonaceous Meteorites*, Elsevier, New York.

Oró, J., Mills, T. and Lazcano, A.: 1992, *Origins of Life Evol. Biosphere* **21**, 123.

Sagan, C. and Khare, N.: 1971, *Science* **173**, 417.

Sagan, C. and Chyba, C.: 1997, *Science* **276**, 1217.

Shoemaker, E. M., Williams, J. G., Helin, E. F. and Wolfe, R. F.: 1979, in *Asteroids*, ed. T. Gehrels (University of Arizona Press), pp. 253-282.

Sleep, N. H., Zahnle, K. J., Kasting, J. F. and Morowitz, H. J.: 1989, *Nature* **342**, 139.

ADVANTAGES OF THE ALKANES ≥ n-C_{18} AS PROTECTORS FOR THE SYNTHESIS AND SURVIVAL OF CRITICAL BIOMOLECULES IN THE EARLY EARTH

VICENTE MARCANO [1], PEDRO BENITEZ[1] AND ERNESTO PALACIOS-PRÜ[1,2] -

[1] *Electron Microscopy Center, University of the Andes, P. O. Box 163, Mérida, Venezuela.* prupal@ula.ve

[2] *Internacional Institute of Advanced Studies, USB, Caracas, Venezuela.*

The origin of life is hardly understandable in the adverse conditions of the early Earth (Maher & Stevenson 1988, Zahnle & Sleep 1997) unless the formation of different lengths of carbon compounds, as well as amino acids and primitive nucleobases aggregates, are conceived under the effect of a protective mechanism (Miller & Bada 1988, Bada et al. 1995, Levy & Miller 1998). However, temperatures ≤ 350°C were possible 3600 Myr ago due to the multiple impacts of ~ 14 km bolides (Lyons and Vasavada 1999) and the CO_2 greenhouse effects (Kasting & Ackerman 1986). During this age, hydrophobic coats of alkanes ≥ *n*-C_{18} could served as protective mechanism and allowed the yield of long molecules avoiding its hydrolysis (Marcano et al. 1999).

However, synthesis of biomolecules had have other factors inhibiting its formation and survival during the early Earth, such as decomposition by UV-radiation and electric discharges. Based on experimental data, it is shown how the synthesis of biomolecules could be protected by *n*-alkane environments against high temperatures (140-240°C), spark discharges and UV-radiation (Cleaves & Miller 1998). Amino acids synthesized in the atmosphere (Schlesinger & Miller 1983, Kobayashi et al. 1998) or delivered by exogenous sources (Chyba & Sagan 1992) and deposited on *n*-alkane environments could be condensed, decomposed and involved in other reactions yielding various nitrogenous compounds by thermochemical reactions produced from CO_2 greenhouse effect and/or by small impacts shock waves.

A mixture of six α-amino acids in prebiotic proportions (Gly, DL-Ala, DL-Glu, DL-Asp, DL-Val and DL-Pro, 8:4:4:2:2:1) was utilized. The molecules obtained were analyzed by a Hitachi 557 spectrophotometer and by IR spectra utilizing a Fourier transform IR spectrometer Perkin-Elmer, and quantified by modified Lowry protein assay. The molecular weights of the molecules having peptide bonds were estimated by SDS-PAGE, using 20% gel. Products were obtained in a 2-liters glass reactor connected to a vacuum pump, containing 100 ml of mineral oil and the mixture of amino acids. Mineral oil was used (Sigma, d = 0.85 g/cm^3) as a model of primordial alkanes ≥ *n*-C_{18}.

- To whom correspondence should be mailed.

J. Chela-Flores et al. (eds.), Astrobiology, 279–282.

TABLE I. Amounts and molecular weight ranges of peptide bond-containing molecules.

Temperature °C	mg/ml	Yield (%)*	Molecular weight (KD)
120-130	0.10	1.56	-------------
140-150	0.12	1.81	5, 37, 43
170-180	0.16	2.37	5, 10
180-190	0.47	7.06	5, 10, 25
190-200	0.73	10.95	5
210-220	1.10	16.5	5
230-240	0.24	3.63	-------------

* Yield = [mg/ml peptide bond-containing molecules x 150 ml / 10^3 mg amino acids] x 10^2.

Reactor was exposed to UV-radiation (253.7 nm lamp, 22 W) and spark discharges obtained from a high-frequency Tesla coil for 120 minutes at several temperature ranges (120-240°C). When heated, the content of the flask reached 100% of their final temperature within ~ 10 min. This was measured by a thermocouple inserted into of the flask. One type of prebiotic-like atmosphere was selected ($H_2O + CO_2$) in equimolar concentrations and pressures ranging between ~ 0.6-0.8 bars.

Mineral oil subjected to different temperatures containing a prebiotic mixture of amino acids, yielded 1.5-16.5% of Lowry positive compounds, the higher yield of these compounds were obtained at 210-220°C range (Table I). Obtained products, free of chloroform/hexane extracted compounds, showed a brownish pigmentation, which became darker according to temperature increment. The pigment showed an absorption in the range of λ 400 nm, the region of the Soret band of porphyrins. The results showed an important production of organic molecules at several temperatures according to IR spectra (Figure 1). The IR spectral data for the obtained products at temperatures ranging between 180-200°C included bands of alcohols (CH–OH), substituted amides (–CONH–), bands of nitrogenous aromatics and one strong band at 1401 cm^{-1} corresponding to unsubstituted amide. Products obtained between 210-220°C showed bands of alcohols, two bands at 1713 (strong) and 1625 cm^{-1} (weak) for substituted amides (–CONH–), and one band at 1395 cm^{-1} corresponding to unsubstituted amide. Products obtained in higher temperatures (230-240°C) showed bands of alcohols, two bands for substituted amides and another one corresponding to unsubstituted amide. It is important to point out that IR spectra corresponding to mineral oil before and after of the simulation experiments showed no modification of the molecular structure of the *n*-alkanes (Figure 2). SDS-PAGE analysis of crude material showed several bands, according to the Coomassie blue staining method (Table I). Band ~ 5 KD showed higher concentration suggesting the presence of chains having peptide bonds in the heterogeneous molecular product (Table I). IR spectral data corresponding to the products obtained between 180-220°C, revealed the presence of two bands corresponding to substituted amides suggesting the formation of peptide bonds. The peptide bond-containing substance was produced in higher amounts at the 180-220°C range and at temperatures > 220°C decrease abruptly. IR spectra corresponding to the products obtained in temperatures > 220°C showed a weaker absorption at the range

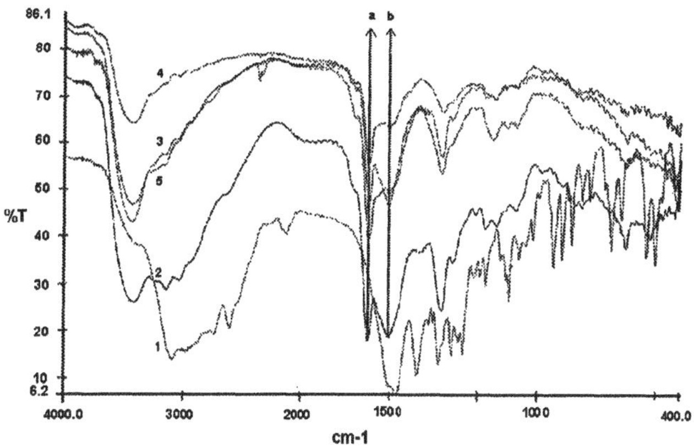

Figure 1. Comparison of IR spectra corresponding to thermal products. Prebiotic mixture of amino acids (1), and products obtained between 180-190°C (2), 190-200°C (3), 210-220°C (4) and 230-240°C (5) are indicated. Transmittance values correspond to the sample 1 (a and b amide peaks).

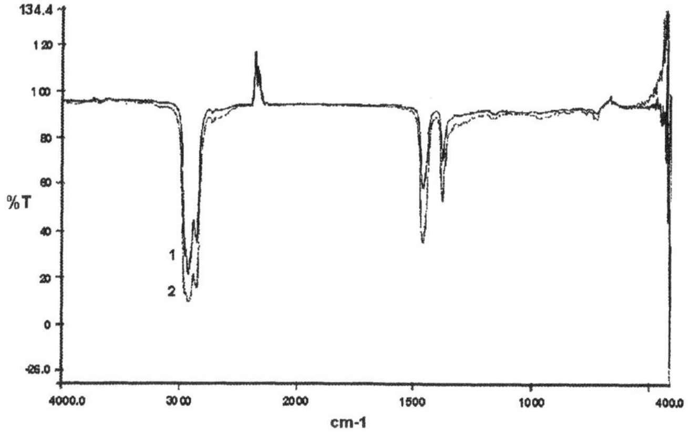

Figure 2. Comparison of IR spectra corresponding to mineral oil before (1) and after (2) of the experiments.

1625-1630 cm^{-1} corresponding to substituted amide (Figure 1). IR bands of nitrogenous aromatics, suggest the presence of porphines, pyrroles or pyridines in the synthesis product. On the other hand, synthesis of peptide bond-containing molecules or other nitrogenous molecules was not possible in the absence of mineral oil.

In summary, the existence of thermostable *n*-alkane environments on the Earth's surface (< 3600 Myr) offer various advantages for the survival of critical biomolecules. The first advantage is the facilitation of protective and anhydrous conditions for the synthesis of complex organic molecules by dehydration of precursors deposited on it; second, the *n*-alkane hydrophobic coats could protect several biomolecules against rapid hydrolysis; third, such coats may avoid or reduce the oxidative decomposition of biomolecules immersed in them caused by strong oxidants: fourth, absorbances of UV light between 200-280 nm allow the protection of amino acids and peptide-bonds against photolysis, and fifth, due to the low dielectric constant (2.0-2.5 x 10^6 cycles) and conductivity (~ 0.5 μs/cm) of these alkanes, these environments could serve as an insulator thereby protecting the synthesis and accumulation of molecules from alterations by electron excitation.

Since the α-amino acids utilized were abundant throughout the early Earth's surface, this work shows that the thermochemistry, photochemistry and electrochemistry of α-amino acids immersed in protective hydrocarbon environments are potential contributors to the richness of terrestrial organic chemistry. These reactions in hydrocarbon environments should modify a portion of the α-amino acids population in important ways. In particular, this process should transform amino acids into a ensemble of organic compounds having amides, alcohols and nitrogenous aromatics including probably peptides, porphines, pyrroles and pyridines.

Connections of those molecules with the carbonaceous fractions of meteorites may also exist (Deamer et al. 1989). Protective hydrocarbon coats and organic molecules having similar features as those proposed, existing in the early Earth are not probably difficult to find in other regions of our solar system (Clarke & Ferris 1997, Irvine 1998), and in extra-solar planets spaced geometrically (Marcy & Butler 1996) or ß Pictoris Vega-like stars (Koerner 1997). Critical biomolecules coexisting there, in the presence of protective heavy hydrocarbon environments, may increase the number of stellar bodies on which the formation and accumulation of complex and long molecules may have occur.

REFERENCES

Bada, J. L., Miller, S. L. and Zhao, M.: 1995, *Origins of Life Evol. Biosphere* **25**, 111.

Clarke, D. W. and Ferris, J. P.: 1997, *Origins of Life Evol. Biosphere* **27**, 225.

Cleaves, H. J. and Miller, S. L.: 1998, *Proc. Natl. Acad. Sci.* **95**, 7260.

Chyba, C. F. and Sagan, C.: 1992, *Nature* **355**, 125

Deamer, D. W., Harang, E. A. and Seleznev, S. A.: 1989, *Origins of Life Evol. Biosphere* **19**, 291.

Irvine, W. M.: 1998, *Origins of Life Evol. Biosphere* **28**, 365.

Kasting, J. F. and Ackerman, T. P.: 1986, *Science* **234**, 1383.

Kobayashi, K., Kaneko, T., Saito, T. and Oshima, T.: 1998, *Origins of Life Evol. Biosphere* **28**, 155.

Koerner, D. W.: 1997, *Origins of Life Evol. Biosphere* **27**, 157.

Levy, M. and Miller, S. L.: 1998, *Proc. Natl. Acad. Sci. USA* **95**, 7933.

Lyons, J. and Vasavada, A. R.: 1999, *Origins of Life Evol. Biosphere* **29**, 123.

Marcy, G. W. and Butler, R. P.: 1996, *Astrophys. J.* **464**, 147.

Maher, K. A. and Stevenson, D. J.: 1988, *Nature* **331**, 612.

Marcano, V, Benitez, P. and Palacios-Prü E.: 1999, in *Abstracts of the 12th International Conference on the Origin of Life, ISSOL'99*, held in San Diego (USA), p. 112.

Miller, S. L. and Bada, J. L.: 1988, *Nature* **334**, 609.

Schlesinger, G. and Miller, S. L.: 1983, *J. Mol. Evol.* **19**, 376.

Zahnle, K. J. and Sleep, N. H.: 1997, in *Comets and the Origin and Evolution of Life*, eds. Thomas, P. J. Chyba, C. F. and McKay, C. P. (Springer, New York), pp. 175-208.

EVIDENCE OF A NITROGEN DEFICIENCY AS A SELECTIVE PRESSURE TOWARDS THE ORIGIN OF BIOLOGICAL NITROGEN FIXATION IN THE EARLY EARTH

LEONEL CALVA-ALEJO*, DELPHINE NNA MVONDO*, CHRISTOPHER P. MCKAY**, AND R. NAVARRO-GONZÁLEZ*

**Laboratorio de Química de Plasmas y Estudios Planetarios*
Instituto de Ciencias Nucleares, Universidad Nacional Autónoma de México, Circuito Exterior, Ciudad Universitaria, Apartado Postal 70-543, México D.F. 04510 México.

***Space Science Division, NASA Ames Research Center*
Moffet Field, CA 94035 USA.

We report an experimental study of the possible effects of lightning discharges on the nitrogen fixation rate during the evolution of the Earth's early atmosphere. The composition of the early atmosphere evolved from predominantly carbon dioxide to predominantly nitrogen. Our results indicate that the energy yield of production of nitric oxide, the main form of fixed nitrogen, drastically decreased from $\sim 1.3\times10^{16}$ molecule J^{-1} at the time of the chemical evolution process when the CO_2 mixing ratios were probably of the order of 0.65($\pm$0.15) to $\sim 1.1\times10^{14}$ molecule J^{-1} at the time of the start of the rise of oxygen in the atmosphere when the CO_2 mixing ratio was ~0.025. Assuming that the lightning dissipation rate has remained constant over time, it is predicted that the annual production rate of NO may have dropped from $\sim 6.5\times10^{11}$ g at ~ 4 Gyr ago to $\sim 5.5\times10^{9}$ g at ~ 2.2 Gyr. This depletion in NO production may have caused catastrophic consequences to the first microbial communities putting on them selective pressures that eventually triggered the development of a highly expensive synthetic pathway, namely biological nitrogen fixation. Sixteen ATP molecules are required to convert atmospheric nitrogen into ammonia by this process.

J. Chela-Flores et al. (eds.), Astrobiology, 283.

RNA-BINDING PEPTIDES AS EARLY MOLECULAR FOSSILS

LUIS DELAYE and ANTONIO LAZCANO
Facultad de Ciencias, UNAM
Apdo. Postal 70-407
Cd. Universitaria, 04510
México D.F., MÉXICO

Abstract

Comparisons of complete cellular genomes indicate that a set of genes whose products synthesize, degrade, or interact with RNA molecules are among the most highly conserved sequences common to all living beings, and therefore may have been present in their last common ancestor, i.e., the cenancestor. In order to obtain insights on the evolution of sequences which may date from an early evolutionary period during which RNA played a genetic role prior to the emergence of DNA genomes, we have analyzed the conserved RNA-binding sites of these highly conserved molecules, since these may be some of the recognizable peptides in our dataset. The characteristics of some of these highly conserved amino acid stretches which are essential in RNA metabolism are discussed.

1. Introduction

The early steps of the evolution of life on Earth can be investigated by analyzing the extant molecular fossil record. A *molecular fossil* as defined by Maizels and Weiner (1994), is *any molecule whose contemporary structure, function (or its phylogenetic distribution) provides a clue to its evolutionary history.* Thus, phylogenetic markers such a rRNA (Woese 1987) are good molecular fossils, but the same is true for other biological molecules.

The discovery of ribozymes has given considerable credibility to prior suggestions on the existence of the RNA world, a hypothetical stage before the evolutionary development of DNA genomes protein-based metabolism. Whether RNA was the first genetic macromolecule or not is a matter of debate, but acceptance of the RNA world can help define the evolutionary *polarity* of a number of molecular traits, i.e., to recognize which character states are ancestral and which are derived. Thus, if RNA preceded DNA as the reservoir of cellular genetic information, and proteins predate DNA, it is likely that proteins which interact with RNA are older than those that do so with DNA.

J. Chela-Flores et al. (eds.), Astrobiology, 285–288.

It is highly unlikely that the first proteins were complex enzymes with exquisitely finely tuned catalytic activity. Although the first peptides that were synthesized biologically (i.e., ribosome-mediated translation), could be positively selected by two properties that are not mutually exclusive, i.e. chaperone-like properties or catalytic activity, in these paper we explore the possibility that the first peptides could have enhanced the catalytic properties or biological functions of ribozymes simply by stabilizing their structures. This chaperone-like property would be the primitive equivalent to the stabilizing effect that the protein subunit of RNAse P plays *in vivo* (Guerrier-Takada *et al.*, 1983), and can in principle be explored by a detailed analysis of extant RNA-binding sites.

Comparisons of completely sequenced cellular genomes (Table 1) from the three primary domains (i.e., Bacteria, Archaea and Eucarya) suggest that a set of genes whose products synthesize, degrade, or interact with RNA molecules are among the most highly conserved sequences common to all living beings (Tekaia *et al.*, 1999). It is thus possible that their corresponding RNA-binding sites, which may be among the oldest motifs in current sequence databases, can provide important insights on the early evolution of ribosome-mediate polypeptide synthesis. Here we report the preliminary outcome of such analysis, and discuss the evolutionary significance of our findings.

2. Material and methods

A list of RNA-binding domains reported in the literature was compiled, which include the those of highly conserved proteins as defined by Tekaia *et al.* (1999). This dataset is now being completed with searches on the following databases: SWISS-PROT (http://www.expasy.ch/sprot/), SRS (http://srs5.hgmp.mrc.ac.uk/), and PROSITE (http://www.expasy.ch/prosite/). The phylogenetic distribution of these RNA-binding sites was analyzed following the three domain taxonomic scheme.

3. Results and discussion

In this first phase of our study we have restricted ourselves to the RNA-binding motifs reported in the literature. Thus, it is likely that the results reported represent a lower limit of the different functional kinds of RNA-binding sequences, i.e., that there are many other as yet undetected cases of such polypeptides.

The distribution of such RNA-binding domains among different highly conserved proteins with different functions (Table 1), suggests that domain recruitment and fusion have taken place in the early evolution of such polypeptides. It also implies that some of these domains are probably older than the proteins in which they are found today, i.e., they are molecular fossils in the sense described above.

RNA-binding domain	Highly conserved proteins	References
O/B fold	RpS17, AspRS, LysRS, PheRS, AsnRS	Arnez and Cavarelli, 1997; Eriani *et al.*, 1990
RNP	EF-G, PheRS	Liljas and Garber, 1995; Mosyak *et al.*, 1995
Left handed βαβ cross over	RpS5, Ef-G	Stams *et al.*, 1998
HU DNA-binding like	RpS7, RpL14	Draper and Reynaldo, 1999
Novel α/β fold	HisRS, ProRS, ThrRS, GlyRS*	Arnez *et al.*, 1995; Cusack *et al.*, 1998; Sankaranarayanan *et al.*, 1999; Logah *et al.*, 1995

Table 1. RNA-binding domains found in some of the highly conserved proteins.
*GlyRS from *Thermus thermphilus*; Rp (ribosomal protein); RS (aminoacyl-tRNA synthetase).

One example of such molecular fossil could be the ribosomal protein S8 (RpS8). This polypeptide is one of the core ribosomal proteins. It binds to the 16/18S rRNA with high affinity, and plays a central role in the assembly of the 30S subunit of the ribosome. The *E. coli* RpS8 also regulates its gene expression by binding to its own mRNA, thereby acting as translational repressor of the *spc* (spectinomycin-resistance) operon. The *spc* operon includes the genes of the ribosomal proteins L14, L24, L5, S14, S8, L6, L18, S5, L30, and L15 (Mattheakis & Nomura, 1988). The RpS8 target site on the *spc* mRNA is similar to the 16S rRNA S8 binding site in both at primary and secondary structure levels (Cerreti *et al.*, 1988; Gregory *et al.*, 1988).

The ribosomal protein S8 resulted from the fusion of two domains. Its N-terminal domain is similar to portions of the DNase I and *Hae*III methyltransferase that bind to DNA (Davies, *et al.*, 1996). The C-terminal domain seems to be an RNA-binding domain found only in this protein. Mutants with only eight residues missing from the C-terminus exhibit significantly lowered RNA-binding properties (Uma *et al.*, 1995). The level of conservation of the primary structure of the RpS8 across the three cellular domains and of the *spc* operon in the two prokaryotic domains (Siefert *et al.*, 1997), together with the fact that almost the same RNA structure is recognized by the protein for its function on the rRNA and for its own regulation in the mRNA, strongly suggest that this mechanism of auto-regulation of gene expression may be among the oldest ones that evolved during the evolution of the protein biosynthesis.

Perhaps not surprisingly, some of the RNA-binding domains resemble DNA-binding domains, such as the ETS DNA-binding motif-like of RpS4, or the helix-hairpin-helix motif of RpS13. It is possible that such domains would correspond to polypeptides with functions already in the RNA-protein world, adapted to the DNA-RNA-protein world.

We are currently working on the construction of a catalog of RNA-binding domains. Analysis of this dataset, is expected to provide insight into the characteristics of which may be some of the oldest polypeptides still recognizable today.

Support from the DGAPA-UNAM/ PAPIIT IN213598 project is gratefully acknowledged.

4. References

Arnez, J.G. and Cavarelli, J. (1997) Structure of RNA-binding proteins, *Quart. Rev. Bioph.* **30**, 195-240

Arnez, J.G., Harris, D.C., Mitschler, A., Rees, B., Francklyn, C.S., and Moras, D. (1995) Crystal structure of histidyl-tRNA synthetase from *Escherichia coli* complexed with histidyl-adenylate, *EMBO J.* **14**, 4143-4155

Cerretti, D.P., Mattheakis, L.C., Kearney, K.R., Vu, L., and Nomura, M. (1988) Translational regulation of the *spc* operon in *Escherichia coli*. Identification and structural analysis of the target site for S8 represor protein, *J. Mol. Biol.* **204**, 309

Cusack, S., Yaremchuk, A., Krikliviy, I., and Tukalo, M. (1998) $tRNA^{Pro}$ anticodon recognition by *Thermus thermophilus* prolyl-tRNA synthetase, *Structure*, **6**, 101-108

Davies, C., Ramakrishnan, V., and White, S.W. (1996) Structural evidence for specific S8-RNA and S8-protein interactions within the 30S ribosomal subunit: ribosomal protein S8 from *Bacillus stearothermophilus* at 1.9 Å resolution, *Structure*, **4**, 1093-104

Draper, D.E., and Reynaldo, L.P. (1999) Survey and summary RNA binding strategies of ribosomal proteins, *Nucl. Acids Res.* **27**, 381-388

Eriani, G., Dirheimer, G. and Gangloff, J. (1990) Aspartyl-tRNA synthetase from *Escherichia coli*; cloning and characterization of the gene, homologies of its translated amino acid sequence with asparaginyl- and lysyl-tRNA syhtetase, *Nucl. Acids Res.* **18**, 7109-7118

Gregory, R.J., Cahill, P.B., Thurlow, D.L., and Zimmermann, R.A. (1988) Interaction of *Escherichia coli* ribosomal protein S8 with its binding sites in ribosomal RNA and messenger RNA, *J. Mol. Biol.* **204**, 295-307

Guerrier-Takada, C., Gardiner, K., Marsh, T., Pace, N., and Altman, S. (1983) The RNA moiety of ribonuclease P is the catalytic subunit of the enzyme, *Cell* **35**, 849-57

Liljas, A. and Garber, M. (1995) Ribosomal proteins and elongation factors, *Curr. Opin. Struct. Biol.* **5**, 721-727

Logah, D.t., Mazauric, M.-H., Kern, D., and Moras, D. (1995) Crystal structure of glycyl-tRNA synthetase from *Thermus thermophilus*, *EMBO J.* **14**, 4156-4167

Maizels, N. and Weiner, A.M. (1994) Phylogeny from function: evidence from the molecular fossil record that tRNA originated in replication, not translation, *Pro. Natl. Acad. Sci. USA* **91**, 6729-34

Mattheakis, L.C. and Nomura, M. (1988) Feedback regulation of the *spc* operon in *Escherichia coli*: Translational coupling and mRNA processing, *J. Bacteriol.* **170**, 4484-92

Mosyak, L., Reshetnikova, L., Goldgur, Y., Delarue, M., and Safro, M.G. (1995) Structure of phenylalanyl-tRNA synthetase from *Thermus thermophilus*, *Nature Struct. Biol.* **2**, 537-547

Sankaranarayanan, R., Dock-Bregeon A.C., Romby, P., Caillet, J., Springer, M., Rees, B., Ehresmann, C., Ehresmann, B., and Moras, D. (1999) The structure of threonyl-tRNA synthetase (Thr) complex enlightens its repressor activity and reveals an essential zinc ion in the active site, *Cell*, **97**, 371-381

Siefert, J.L., Martin, K.A., Abdi, F., Widger, W.R., and Fox, G.E. (1997) Conserved gene clusters in bacterial genomes provide further support for the primacy of RNA, *J. Mol. Evol*, **45**, 467-72

Stams, T., Niranjanakumari, S., Fierke, C.A., and Christianson, D.W. (1998) Ribonuclease P protein structure: evolutionary origins in the translational apparatus, *Science*, **280**, 752-755

Tekaia, F., Dujon, B., and Lazcano, A. (1999) Comparative genomics: products of the most conserved protein-encoded genes synthesize, degrade, or interact with RNA. Abstract of the 12th International Conference on the Origin of Life & 9th ISSOL meeting, ISSOL'99 (San Diego, California U.S.A. July / 11-16 / 1999) pp. 53

Uma, K., Nikonowicz, E.P., Kaluarachchi, K., Wu, H., Wower, I.K., and Zimmermann, R.A. (1995) Structural characterization of *Escherichia coli* ribosomal protein S8 and its binding site in 16S ribosomal RNA, *Nucleic Acids Symp. Series* **33**, 8-10

Woese, C.R. (1987) Bacterial evolution, *Microbiol. Rev.*, **51**, 221-271

ON THE ROLE OF GENOME DUPLICATIONS IN THE EVOLUTION OF PROKARYOTIC CHROMOSOMES

S. ISLAS, A.CASTILLO, H.G. VÁZQUEZ, and A. LAZCANO
Facultad de Ciencias
Apartado Postal 70-407
Cd. Universitaria 04510
México D.F.,México.

1. Introduction

It is generally accepted that primitive cells were endowed with relatively small genetic systems with reduced encoding capacities. How the transition from such small genomes to the more complex ones observed in extant prokaryotes took place is still unknown, but it may have involved duplication of genes and of larger segments, horizontal transfer, cell fusion events, and perhaps even whole genome duplications (Casjens, 1998).

The role of whole genome duplications was first discussed for vertebrate evolution by Ohno (1970). Whether this took place or not is still a controversial issue (Hughes, 1999). On the other hand, compilation of data on mycoplasma DNA content led Wallace and Morowitz (1973) to suggest that the discontinuities in the frecuency distribution of genome sizes of their sample could be explained by succesive genome duplications that were assumed to have taken place also took place in other prokaryotes. Some time ago this idea appeared to be supported by the position between functional-related genes in the circular map of *Streptomyces coelicolor* (Hoopwood 1967) and *Escherichia coli* (Zipkas and Riley 1975). Furthemore, the results of statistical analysis of a sample of 603 prokaryotic genomes was interpreted as evidence of several rounds of entire duplications that had started with a modal value of aproximately 0.8 Mb, and led to genome of 1.6 Mb, 4 Mb, and other minor peaks with higher values (Herdman 1985).

With the development of pulsed-field gel electrophoresis (PFGE), a technique that allows the separation and analysis of large DNA fragments and the direct study of the physical structure of prokaryotic genomes, the accuracy in the determination of genome size has been significantly improved (Shimkets, 1998). Here we report the results of an analysis of a sample of 246 prokaryotic genome sizes obtained by PFGE, that includes both Bacteria and Archaea, available as of June 1999. Although this database is likely to be biased and does not represents the full range of prokaryotic diversity, the different peaks we have observed in the discontinuous distribution of DNA content (Figure1) do not support the idea that several duplications beginning an

J. Chela-Flores et al. (eds.), Astrobiology, 289–292.

hypothetical ancestral minigenome have taken place during evolutionary time.The posibility of a correlation between oxygen response and genome size is also discussed.

2. Material and methods

A genome size database was constructed with the 246 prokaryotic genome sizes determined by PFGE reported in the publications included in the NCBI database PubMed (http://www.ncbi.nlm.mih.gov/PubMed/). This database is available upon request. The information was completed with a description of the organisms' phylogenetic positions and lifestyle. The organisms' response to oxygen was based on the original reports and on data from the *Bergey's Manual of Bacterial Determination* (1994): anaerobe (organisms with negative oxygen response), microaerophilic (organisms that need ≤ 21% of oxygen), facultative anaerobe (organisms with double response to free oxygen), and aerobes (organisms that require ≥ 21% of oxygen). Statistical analyses were performed using the X^2 in Microsoft Excel ^{tm} program.

3. Results

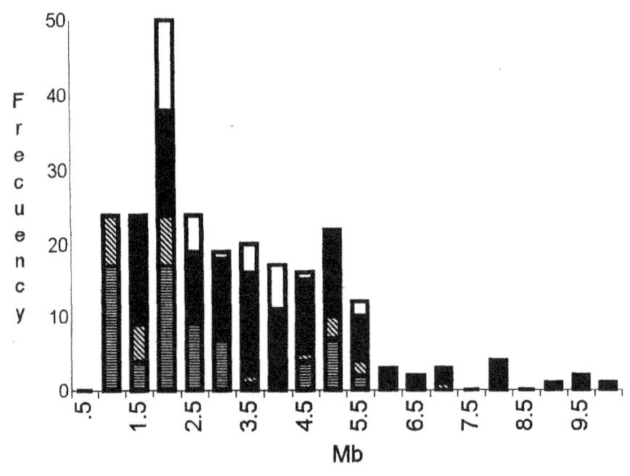

Figure 1. Distribution of prokaryotic genome sizes with respective oxygen response: the white bars correspond to anaerobes, diagonal lines, to microaerophilics; horizontal lines, to facultatives and grey points, to aerobes.

The distribution of prokaryotic genome size and their oxygen response in our sample is shown in Figure 1. The first two bars corresponding to organisms with parasitic lifestyle; in the first bar only low and double response to oxygen are represented. From the third bar onwards, the four types of metabolic response to oxygen are represented, including both parasitic and free-living organisms. The facultative anaerobes have genome sizes ranging from the smallest ones to middle sized (0.57-570 Mb). The aerobic organisms have the largest genome reported, while the strict anaerobes include the genomes with sizes between (1.12- 9.50 Mb), and the negative response include the genomes sizes with DNA content is (1.60-5.40 Mb) are never extreme values include as 9.50 Mb.

4. Discussion and conclusions

Herdman (1985) has argued that the peaks observed in the discontinuos size distribution of his sample of bacterial genomes provided support for the hypothesis that the evolution of prokaryotic DNA content could be explained by whole genome duplications. However, although our sample is larger and based in a more accurate technique for detecting DNA content, we have found no evidence corroborating his conclusions. The range of prokaryotic genome sizes available as of June 1999, ranges from 0.573 Mb (*Mycoplasma genitalium)* to 9.5 Mb (*Stigmatiella erecta).* Since the smallest free-living prokaryotes included in our sample have genome sizes in the range of 1.6 Mb to 1.9 Mb, according to the whole genome duplication hypothesis we would expect to find peaks with modal values of 3.2 Mb, 6.4 Mb, and 12.8 Mb. This is clearly not the case (Figure 1).

The smaller genome sizes (0.57 to 1.5 Mb) in our sample correspond to parasitic organisms of the Mollicutes group. The later include other mycoplasma that have, whose slightly larger genomes such as *Anaeroplasma, Asteroleplasma,* and *Spiroplasma*, oscillate between 1.5 to 1.78 Mb. Other groups with reduced genome sizes are the rickettsia and the spirochaete. Not all these organisms with reduced DNA content are anaerobes,. which can be explained by recognizing that these organisms (genome size $\leq$ 1.78 Mb), are the outcome of a complex series of secondary adaptations that have led to the polyphyletic reduction of their genome dimensions. Thus, neither the mycoplasma or the rickettsia are accurate models of ancestral Archean organisms.

It has been suggested that the first microorganisms were anaerobic heterotrophes, and the later availability of oxygen promoted the apparence of new metabolic capacities (Oparin, 1938). Our results show a correlation between genome size and oxygen response. As seen in Figure 1, the larger genomes are found solely in bacteria, all which are strict aerobes with complex life cycles. These results strongly suggest that such species evolved during late Proterozoic times, once the levels of free-O_2 in the terrestial atmosphere had reached values comparable to the extant ones.

There are no reports available in the literature of Archaea with genome size comparable to those of *Stigmatiella* (Casjens, 1998) (Figure 1). Whether this reflects the evolutionary strategies of the Archaea domain, it is not clear, but our interpretation may be limited by the current descriptions of prokaryotic diversity.

The database analyzed here is biased by the medical and economical significance of the organisms, and does not reflect in an accurate way the actual biodiversity of prokaryotes. Pathogens and parasites are clearly overepresented because of their medical and economic significance in human, animal, and crop plant life. Nonetheless, the large number of microaerophilic and facultative organisms from different phylogenetic groups in our sample probably reflects the succesful adaptation to increasingly higher levels of oxygen in the terrestrial atmosphere.

Finally, we would like underline the fact that in spite of the limitation of our database there are no indications of free-living prokaryotes with genome sizes smaller than 1.53 Mb (*Fervidobacterium islandicum*). This observation casts doubts on the existence of nanobacteria (Kajander and Cificioglu 1998; Aboll, 1999), whose genomes are assumed to be at least one order of magnitude smaller than those of mycoplasma.

Acknowledgments
Support from DGAPA-UNAM/PAPIIT IN213598 project is gratefully acknowledged.

References

Aboll, A. (1999) Battle lines drawn between nanobacteria researchers. Nature **401**: 105
Casjens, S. (1998) The diverse and dynamic structure of bacterial genomes *Annu Rev genet* **32**:339-377
Herdman M (1985) The evolution of bacterial genomes . In: Cavalier Smith T (ed) *The Evolution of genome size*. John Wiley, London
Holt J G *et al* (1994) *Bergey's Manual of Determinative Bacteriology*. Williams and Wilkins. Baltimore, Maryland, USA, p.787
Hopwood D A (1967) Genetic analysis and Genome structure in *Streptomyces coelicolor. Bacteriol Rev* **31**: 373-403
Hughes, A. L. (1999) Phylogenies of developmentally important proteins do not support the hypothesis of two rounds of genome duplication early in vertebrate history *J Mol Evol* **48**:565-576
Kajander,E.O. and Cificioglu, N. (1998) Nanobacteria: an alternative mechanism for pathogenic intra-and extracelluklar calcification and stone formation Proc. Natl. Acad Sci USA **95**: 8274-8279
Ohno S (1970) *Evolution by Gene Duplication*. Springer Verlag, New York
Oparin A. I. (1938) *The Origin of Life*, MacMillan, New York.
Shimkets, L .J. (1998) Structure and sizes of the genomes of the Archaea and Bacterial In Bruijn F *et al* (eds).*Bacterial Genomes: Physical Structure and Analysis*. (Chapman & Hall, New York)
Wallace D C and Morowitz H J (1973) genome size and evolution *Chromosome* **40**: 121-126
Zipkas, D. and Riley, M. (1975) Proposal concerning mechanism of evolution of the genome of *Escherichia coli Proc. Nat. Acad. Sci. USA* **72**(4):1354-1358

EXPERIMENTAL SIMULATION OF VOLCANIC LIGHTNING ON EARLY MARS

ANTÍGONA SEGURA AND RAFAEL NAVARRO-GONZÁLEZ
Laboratorio de Química de Plasmas y Estudios Planetarios, Instituto de Ciencias Nucleares, Universidad Nacional Autónoma de México. Circuito Exterior C.U., A. Postal 70-543, 04510. México, D.F., México.

Abstract

A mixture of probable volcanic gases was reproduced and irradiated by a high-energy infrared laser beam to simulate volcanic lightning on early Mars in order to determine the possible role of this phenomenon in prebiotic synthesis. Analysis of products was performed using a gas chromatograph interfaced in parallel with an infrared detector and a quadrupole mass spectrometer. Hydrogen cyanide, a key molecule for prebiotic synthesis, was detected among the products.

1. Introduction

Lightning produced in explosive volcanic eruption columns may had been the main source of fixed nitrogen and was therefore a favorable environment for prebiotic synthesis (Navarro-González, *et al.*, 1996) on early Earth (Navarro-González, *et al.*, 1998) and early Mars (Navarro-González and Basiuk, 1998).

Explosive eruption column of a volcano is relevant to prebiotic chemistry due to (1) the presence of reduced magmatic gases; (2) the production of volcanic lightning activity; and (3) the fast escape of the nascent molecules from the high temperature zone at sonic speeds (Navarro-González and Basiuk, 1998).

The exploration of Mars has shown that life may have developed about 3.8 billion years ago. Since volcanism was globally distributed (Mouginis-Mark *et al.*, 1992) during its early history, volcanic plumes could have been a favorable environment for the production of key molecules needed for chemical evolution and the origins of life.

2. Volcanism on Mars

Volcanic activity on Mars extended for a long period and volcanic surfaces cover more than half of Mars (Mouginis-Mark *et al.*, 1992). Numerical simulations have shown that explosive activity would occur if magma volatile contents exceed 0.01 weight percent on Mars (Wilson and Head, 1983). An explosive or plinian volcanic eruption is

J. Chela-Flores et al. (eds.), Astrobiology, 293–296.

characterized by magma that disrupts into very small fragments that become locked to an expanding gas plume rising buoyantly from the vent (Wilson and Head, 1994). Because of the lower martian gravity, nucleation of magma would occur at systematically greater depths than on Earth since volatile solubility is pressure-dependent, volatile release would likely be more efficient (Mouginis-Mark, *et al.*, 1992). More vigorous gas release and higher eruption rates in Martian magmas compared with those on Earth imply that it may have been common for basaltic magmas to produce plinian eruptions (Mouginis-Mark, *et al.*, 1992).

The oldest volcanic units identified on Mars are plateau plains that were formed by fissure eruptions (Greeley and Spudis, 1981). It has been shown that basaltic fissure eruptions can produce buoyant plumes of several kilometers when interaction with water is present (Woods, 1993). Then, because groundwater is an important part of Mars's upper crust (Wilson and Head, 1994) volcanic plumes may have been common on this planet.

2.1. VOLCANIC GASES ON EARLY MARS

Kuramoto and Matsui (1996) have developed a thermodynamic model considering a hot Earth which accreted homogeneously to calculate partitioning of H and C among fluid, silicate melt and molten metallic iron within a growing Earth at temperatures from 2000 to 2500 K and pressures from 0.2 to 5 GPa. The planetesimals accreted had the composition given by the two-component model slightly modified from Ringwood (1977) and Wänke (1981). On the two-component model is considered that accreting rock of a terrestrial planet is a mixture of a highly reduced, volatile free component A, and an oxidized, volatile rich component B (Kuramoto and Matsui, 1996). Kuramoto (1997) applied this model to Mars using a mixing fraction of 35% for the volatile-rich component. Because nitrogen was not considered in Kuramoto's model, for the experiment it is included considering the C/N ratio from the magmatic component measured on Nakhlitas and Chassigny meteorites (Wright, *et al.*, 1992).

The major volcanic provinces on Mars are due to upwelling mantle plumes, similar to hot spots on Earth (Schubert, *et al.*, 1992), this kind of volcanism comes from the deepest layer of the mantle. We choose the mixture formed at higher pressures because it has more probability to have been kept in the mantle and later degassed by volcanism.

3. Experimental

Volcanic lightning was simulated in the laboratory by focusing an infrared Nd-YAG laser that produce a Laser Induced Plasma (LIP). The laser delivers a beam of 1.06 μm photons with an energy of 480 mJ per pulse in 5-7 ns at 10 Hz. The beam was focused inside a closed Pyrex flask with a plano-convex optical glass lens coated with an anti-reflecting film and obtaining a focal spot size of ~9.7 μm. The energy deposited into the system was determined by the difference between the input laser energy and that transmitted by the LIP, and quantified with a power and energy measurement system. About 60% of the input laser energy was transmitted by the LIP at 194 mbar of 0.64 CH_4, 0.24 H_2, 0.10 H_2O, 0.02 N_2. The samples were irradiated 2.5, 5, 7.5, 10, 15, 20, 25 and 30 minutes to determine the energy yields. Each experiment was repeated three times in order to calculate errors.

The gases used for LIP irradiation were ultra-high purity (CH_4 =99.97%, H_2=99.99% and N_2=99.99%), supplied by Praxair, Inc. The anhydrous mixture was prepared using a Linde mass flow measuring and control gas blending console (FM4660) equipped with fast response mass flow control modules (FRC) of 20 cm^3 min^{-1} capacity. Water was added later as vapor to avoid condensation.

4. Results and Discussion

Analyses of the gases were performed using a Hewlett Packard (HP) gas chromatograph 5890 series interfaced in parallel with a HP FTIR-detector (model 5965) and a HP quadrupole mass spectrometer (5989B) equipped with an electron impact and chemical ionization modes. Table 1 lists the compounds identified. The main products have been identified as hydrocarbons and an uncharacterized yellow film deposit. It is especially relevant the presence of hydrogen cyanide (HCN) among the resultant compounds.

TABLE 1. Relative abundance of volcanic lightning products.

Compound	Identification technique	Relative Abundance
Acetylene,	MS, IR	1
Ethylene	MS, IR	1.0×10^{-2}
Ethane	MS, IR	3.5×10^{-2}
1-propene	MS, IR	1.0×10^{-2}
Hydrogen cyanide	MS, IR	3.0×10^{-3}
1,2-propadiene	MS, IR	1.5×10^{-2}
1-propyne	MS	4.5×10^{-2}
1-buten-3-yne	MS	3.0×10^{-2}
1-butyne	MS	5.0×10^{-3}
1,3-butadiyne	MS	1.0×10^{-1}
2-butyne	MS	5.0×10^{-3}
Benzene	MS, IR	1.0×10^{-2}

In order to explore the possibility that HCN could be formed in the volcanic plume due to its high temperature, a thermochemical model was developed. The model considers the formation of chemical species with one and two atoms of carbon from the original gas mixture used in the experiment, at temperatures between 1000 and 5000 K. According to the model, the HCN is formed above ~1600 K. Analysis of the orthopyroxene-silica assemblage in ALH84001, the oldest Martian meteorite, indicates a magmatic temperature of ~1700 K at 40 km depth in the planet (Kring and Gleason, 1997). But models of the Martian interior show that mantle temperatures were lower, about 1200 K (McSween, 1994); consequently HCN may have been a characteristic product of volcanic lightning. Therefore volcanic lightning on early Mars may have formed important quantities of HCN, a key molecule for origins of life and could have been an important source of fixed nitrogen.

References

Greeley R. and Spudis P.D. (1981) *Volcanism on Mars, Reviews of Geophysics and Space Physics* **19**, 13-41.

Kring, D.A. and Gleason, J.D. (1997) Magmatic temperatures and compositions on early Mars as inferred from the orthopyroxene-silica assemblage in Alan Hills 84001, *Meteoritics and Planetary Science Supplement* **32**, 4, A74.

Kuramoto, K. (1997) Accretion, core formation, H and C evolution of the Earth and Mars, *Physics of the Earth and Planetary Interiors* **100**, 3-20.

Kuramoto K. and Matsui T. (1996) Partitioning of H and C between the mantle and core during the core formation in the Earth: its implications for the atmospheric evolution and redox state of early mantle, *J. Geophysical Research* **101** (E6), 14, 909-14,932.

McSween, H.Y. (1994) What we have learned about Mars from SNC meteorites, *Meteoritics* **29**, 757-779.

Mouginis-Mark, P., Wilson, L., and Zuber, M. (1992) The physical volcanology of Mars in H.H. Kieffer, B.M. Yakosky, C.W. Snyder, M.S. Matthews (eds.), *Mars*, The University Arizona Press, pp. 425-452.

Navarro-González, R., Basiuk, V.A. and Rosembaum, M. (1996) Lightning associated to archean volcanic ash clouds in J. Chela-Flores and F. Raulin (eds.), *Chemical Evolution: Physics of the Origin and Evolution of Life*, Kluwer Academic Publishers, Netherlands, pp. 123-142.

Navarro-González, R. and Basiuk V.A. (1998) Prebiotic synthesis by lightning in martian volcanic plumes in J. Chela-Flores and F. Raulin (eds.), *Exobiology: Matter, Energy, and Information in the Origin and Evolution of Life in the Universe*, Kluwer Academic Publishers, Netherlands, pp. 255-260.

Navarro-González, R., Molina, M.J. and Molina L.T. (1998) Nitrogen fixation by volcanic lightning in the early Earth, *Geophysical Research Letters* **25**, 16, 3123-3126.

Ringwood, A.E. (1977) Composition of the Earth's core and implications for the origin of the Earth. *Geochemical J.*. **11**, 111-135.

Schubert, G., Solomon, S.C., Turcotte, D.L., Drake, M.J. and Sleep, N.H. (1992) Origin and thermal evolution of Mars in H.H. Kieffer, B.M. Yakosky, C.W. Snyder, M.S. Matthews (eds.), *Mars*, The University Arizona Press, pp.147-183.

Wänke, H. (1981) Constitution of the terrestrial planets, *Philosophical Transactions of the Royal Society of London* **A303**, 287-302.

Wilson, L. and Head, J.W. (1983) A comparison of the volcanic eruption processes on Earth, Moon, Io and Venus, *Nature* **302**, 663-669.

Wilson, L. and Head J.W. (1994) Mars: Review and analysis of volcanic eruption theory and relationships to observed landforms, *Reviews of Geophysics* **32**, 3, 221-263.

Woods, A.W. (1993) A model of the plumes above basaltic fissure eruptions, *Geophysical Research Letters* **20**, 12, 1115-1118.

Wright, I.P., Grady, M.M. and Pillinger C.T. (1992) Chassigny and the nakhlites: Carbon bearing components and their relationship to martian environmental conditions, *Geochimica et Cosmochimica Acta* **56**, 817-826.

TROPICAL ALPINE ENVIRONMENTS: A PLAUSIBLE ANALOG FOR ANCIENT AND FUTURE LIFE ON MARS

ITZEL PEREZ-CHAVEZ, RAFAEL NAVARRO-GONZALEZ
Laboratorio de Química de Plasmas y Estudios Planetarios
Instituto de Ciencias Nucleares, Universidad Nacional Autónoma de México, Circuito Exterior, Ciudad Universitaria, Apartado Postal 70-543, México D.F. 04510 México.

CHRISTOPHER P. MCKAY
Space Science Division, NASA Ames Research Center
Moffet Field, CA 94035 USA.

LUIS CRUZ-KURI
Instituto de Ciencias Básicas, Universidad Veracruzana
Carr. Xalapa-Veracruz Km 3.5, Las Trancas, Xalapa, Veracruz.

Abstract. El Pico de Orizaba, the highest mountain in Mexico has environmental conditions that could possibly have occurred in ancient Mars. Therefore, our interest is to determine why life is constrained at high altitudes, specifically what determines treelines at thermal gradients. Our working hypothesis is that diazotrophs can no longer fix nitrogen at low soil temperatures. To address this problem we have been monitoring the environmental conditions at different altitudes at the South and North faces of the volcano since March-1999 and have been studying the physical and chemical properties of soil.

1. Introduction

There is evidence that the Martian environmental conditions were more clement in the past (Kargel, 1996; Kasting, 1997; McKay, 1988; McKay, 1989); the estimated global temperature was near 0°C and was higher during certain seasons and at certain regions specifically near the equatorial zone (McKay, 1988; Squyres, 1994). Here on Earth, we find low latitude regions where temperature oscillates near 0°C at great altitudes; these are the tropical alpine environments. Nevertheless, as altitude increases environmental conditions such as temperature and pressure rapidly decrease and so organisms must adapt to them until they reach their physiological tolerance limits. A clear example of this is treeline at thermal gradients that represent an abrupt transition

J. Chela-Flores et al. (eds.), Astrobiology, 297–302.

in life form dominance; they are lines beyond which massive single stems and tall crowns either cannot be developed, become unaffordable, or are disadvantageous. Körner (1998) has hypothesized that treelines are limited by the potential investment: growth and development. The most sensitive range of temperatures for direct effects on growth and development appears to be higher than 3 and lower than 10°C, but the physiological and developmental mechanisms responsible for treeline position worldwide await clarification.

Nitrogen is a key constituent of living beings making up nucleic acids, proteins, energy-transfer molecules and coenzymes. Most of the organic nitrogen incorporated into Earth's biomass comes from atmospheric nitrogen that is fixed by prokaryotic organisms (Stacey, 1992). The enzymatic reaction involved in nitrogen fixation has a large activation energy and so the potential barrier increases with decreasing temperature. Therefore a plausible hypothesis for treeline position at thermal gradients is that bacteria can no longer fix atmospheric nitrogen due to low soil temperatures.

We are conducting a study at El Pico de Orizaba in the eastern Mexican region (19°01' N, 97°16' W); this mountain has an elevation of 5610 m and ranks in third place as the tallest in North America. It is one with the highest treeline positions in the world (Körner, 1998). The purpose of our work is to test if biological nitrogen fixation is the principal factor affected by soil temperature that determines treeline positions and their exobiological implications, specifically with Mars.

2. Methods

2.1. MONITORING OF ENVIRONMENTAL CONDITIONS

We have so far installed 12 meteorological stations at different altitudes (3 stations above treeline, one station at treeline and 2 stations below treelines) on the South and North Faces of the Pico de Orizaba. We are monitoring soil temperature in all stations at a depth of 10 cm. In one station at each face we are recording air temperature and relative humidity above treelines. Soil temperature gradient is being studied at various depths of 0, 20 and 40 cm in the highest stations at each face.

2.2. ORGANIC MATTER DETERMINATION

Soil samples were taken from each station at a depth of 5 cm on the South and North faces and were analyzed for their total organic carbon content by their loss of weight after heating 5 g of soil for 375°C for 8 hrs. Each sample was run in triplicate. In addition they were subjected to flash pyrolysis in a helium atmosphere and the resultant gas fragments were analyzed by gas chromatography coupled to mass spectrometry. This technique has been found to be useful for the characterization of soils (Stuczynsky *et al.*, 1997). About 0.02-0.04 g of freeze-dried, finely grounded soil was introduced into a quartz capillary tube and placed inside the pyroprobe. The temperature was initially set at 200°C for 1 sec and then was increased at a rate of 10°C s^{-1} up to 500°C, and then was held for 30 s at this temperature before injecting the gas mixture into the GC port. The volatilized compounds were analyzed using a HP 5890 Series II gas chromatograph (Column PoraPlot Q, 25m×0.32mm; program temperature: 4 min. at 30°C, ramp at 13°C min^{-1} up to 240°C, then held for 5.8 min) coupled to a HP 5989B mass spectrometer (electron impact at 70 eV, ion scan range: 45 to 300). Each sample was injected and analyzed at least 2 times.

3. Results and Discussion

Figure 1 shows a typical monthly plot of the variation of temperature with altitude for the South face of the Pico de Orizaba. Treeline is situated around 3900 to 4000 m and is dominated by *Pinus hartweggi*. It can be seen that the soil temperature is more stable under the forest and that oscillates more abruptly when the ground is directly exposed to meteorological changes.

Soil organic matter is more abundant in the forest, particularly around 3000 to 3500 m where temperature and humidity are probably optimal (see Fig. 2a). As altitude increases or decreases the total organic matter decreases. It is quite interesting that soils in the lower and upper treelines have similar organic content, about 1-2%. The lower treeline is limited by water availability while the upper one is constrained by inclement temperatures. As shown in Fig. 1a, the organic content in soil rapidly decreases to values ≤ 0.4% at altitudes ≥ 4200 m. Pyrolysis of soil samples produced a complex mixture of aliphatic and aromatic hydrocarbons. Among those positively identified by their mass spectra pattern are: Butune, 1,2-butadiene, 2-propenenitrile,

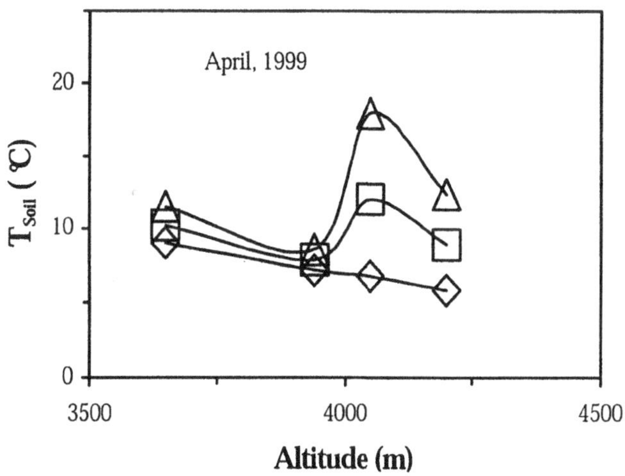

Figure 1. Soil temperature at 10 cm depth as a function of altitude for the South face of the Pico de Orizaba. Maximum (Δ), average (€) and minimum (◊).

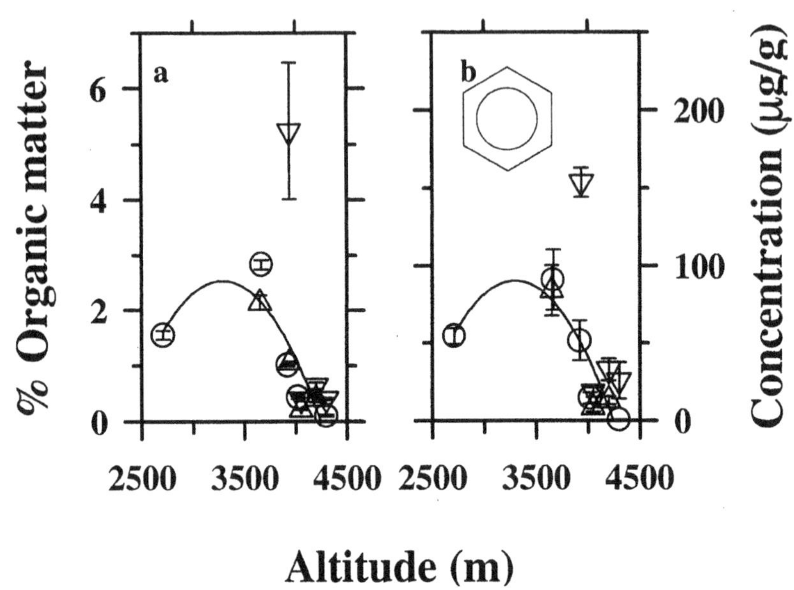

Figure 2. Organic matter content (a) and concentration of pyrolytic benzene (b) in soil at 5 cm depth from the Pico de Orizaba as a function of altitude. O North face, May 30, 1999; Δ South, March 27, 1999; and ∇ South face, May 29, 1999.

1,3-cyclopentadiene, 2-methyl-furan, benzene, 1-H-pyrrole, pyridine, toluene, 2-furan, carboxaldehyde, 4-methyl pyridine, 1,2-dimethyl benzene, 1,3-dimethyl benzene, styrene, phenol, benzaldehyde, benzonitrile, 2-methyl phenol and 4-methyl phenol. The most abundant pyrolytic product was benzene in all soil samples. Figure 2b shows the variation of its concentration as a function of altitude. Similar trends were obtained for other products and are consistent with that of total organic content shown in figure 2a.

The results presented here are just a summary of data collected during the first year of study. Further physical, chemical and biological studies are underway to determine treeline positions at thermal gradients.

4 Implications for Mars

4.1 POSSIBLE PAST BIOSPHERE.

Early Mars is thought to have had a dense atmosphere composed of CO_2 and N_2 (McKay, 1989). In such an atmosphere, abiotical N_2 fixing processes (i.e. lightning) would have provided a constant supply of nitrates as a nitrogen source for early life forms. In any case the most abundant source of nitrogen was the atmosphere where it was present as molecular nitrogen. If any organism were capable for using it, it would have an important advantage over the others. Consequently if biological nitrogen fixation originated on Mars, as the climate progressively changed and the surface temperature continued to drop, a point was eventually reached where the enzymatic reduction of N_2 was no longer possible limiting the flourishing of other organisms that depended on biological fixed nitrogen.

4.2. FUTURE COLONIZATION.

All the elements necessary for Earth-like life are present on Mars (McKay, 1988; Hiscox). To settle a human colony on the Red Planet, the first step would be to reach the minimum environmental conditions required for establishing extremophile microorganism communities (i.e. psychrophilic, radiation resistant life forms). In order to define these minimum conditions it is necessary to study microorganisms that inhabit regions on Earth that best approximate regions on Mars, like high altitude sites where it is cold and atmosphere is thinner. Any way, increasing the planet's temperature will be required. As Mars warms up CO_2 and H_2O will be released from the surface (McKay, 1999) and for preventing N_2 loss to space and make it available for pioneer organisms, biological nitrogen fixation is necessary so it is important to determine the environmental conditions that allows it to be carried out. The next step is to make the atmosphere breathable, this means to introduce photosynthetic organisms (Hiscox; McKay, 1999). As temperature rises, greening would be like hiking down a mountain (McKay, 1999) where there is a temperature gradient and different biological communities are adapted to the temperature at certain altitude., so we can find ideal organisms for each step in terraforming at the mountains.

Acknowledgements

This research has been supported by NASA Ames Research Center, Universidad Veracruzana en Xalapa and Universidad Nacional Autónoma de México (DGAPA-IN102796). We are indebted to Roberto Cruz-Kury, Nahum Castillo, José Mora Domínguez and Rafael Navarro-Aceves for their collaboration in the field work conducted at El Pico de Orizaba. One of us (IPC) acknowledges Prof. Julian Chela-Flores for the generous support to participate in the First Iberoamerican School on Astrobiology.

References

Hiscox, J. *http://spot.colorado.edu/~marscase/cfm/articles/biorev3.html*

Kargel J.S. and Strom R.G. (1996) Global Climatic Change on Mars, *Scientific American* **November,** 60-68.

Kasting J.F. (1997) Warming Early Earth and Mars, *Science* **276**, 1213-1215.

Körner C. (1998) A Re-assesment of High Elevation Treeline Positions and Their Explanation, *Oecologia* **115**, 445-459.

Lauer W. (1978) Timberline Studies in Central Mexico, *Arctic and Alpine Research* **10,** 383-396.

McKay C.P. (1988) Mars: A Reassesment of its Interest to Biology, in Carl G., D. Schwartz and Huntington J. (eds.), *Exobiology in Solar System Exploration.* NASA, pp. 67-81.

McKay C.P. (1999) Bringing Life to Mars, *Scientific American*, Special Issue, Spring 1999

Smith W.K. and Hinckley T.M. (1995) *Ecophysiology of Coniferous Forests*, Academic Press, USA.

Squyres S.W. and Kasting J.F. (1994) Early Mars: How Warm and How Wet?, *Science* **265**, 744-749.

Stacey G., Burris R.H. and Evans H.J.(1992) *Biological Nitrogen Fixation*, Chapman and Hall Routledge, USA.

Stuczynsky T.I., McCarty G.W., Reeves J.B. and Wright R.J. (1997) Use of Pyrolysis GC/MS for Assessing Changes in Soil Organic Matter Quality. *Soil Science* **162**, 97-105.

PLANETARY HABITABLE ZONES ON EARTH AND MARS

Biophysical Limits of Life in Planetary Environments

ABEL MÉNDEZ
University of Puerto Rico at Arecibo
Department of Physics and Chemistry, Arecibo, PR 00613, USA
email: a_mendez@cuta.upr.clu.edu

Abstract: The biosphere includes part of the atmosphere, hydrosphere and lithosphere. Microbial life is possible within this region due to the availability of liquid water, an energy source, and the right environment. This paper is focused in the necessary physical environment state for microbial growth and how that relates to Earth's and Mars' current environment.

1. Introduction

A Planetary Habitable Zone (PHZ) is a spatial and temporal region, within a planetary body, that is capable of supporting life (where microbial growth is possible). The biosphere, Earth's global PHZ, extends to the near-surface environment including part of the atmosphere, hydrosphere and lithosphere. It has a broad vertical extension as demonstrated by viable bacteria spores that have been collected from stratospheric air at altitudes up to 27 km (Greene *et al.*, 1964) and barophilic bacteria that have been isolated from a depth of 11 km below sea level (Kato *et al.*, 1998). The biosphere extension is also controlled by temporal variations due to daily and seasonal cycles. Life is limited in these regions by the temporal availability of liquid water.

The environment temperature, pressure, heat transfer (i.e. convection due to wind) and water content are some of the basic physical quantities that affects the microbial physiology. These environmental variables are very important as they control reaction and diffusion rates, viscosity, dielectric constant, stresses response and heat capacities of the cell's biochemicals and structures. Most known microorganisms require an environmental temperature for optimum growth of 310 K at standard atmospheric pressure (1.013 bars), but microbial growth is possible between 263 and 386 K (National Research Council, 1998). Although much has been studied about the effects of temperature, little is known about the combined effects of pressures and temperatures on microbial growth. Since the 1960's, studies conducted below standard pressure has been concentrated on the viability of spores in the upper atmosphere (Bruch, 1967) and the space environment (Halvorson and Srinivasan, 1964). However, there has been little interest about microbial growth in similar conditions. This may be due to the lack and difficulty of exploring natural Earth's environments with similar conditions.

J. Chela-Flores et al. (eds.), Astrobiology, 303-306.

Steady hydrostatic pressure changes have little or no effect on the growth and metabolism of most microorganisms. Growth of most terrestrial microorganisms is retarded between 300 and 400 bars at 303 K. At 600 bars most terrestrial microorganisms are sterilized, only a few species of marine bacteria grow at such pressure or higher at temperatures between 303 and 313 K. In general, lower temperatures accentuated the growth-retarding and sterilizing effects of pressure (ZoBell and Johnson, 1949). Some deep-sea obligately barophilic bacteria are not able to grow at pressures of less than 500 bars, but are able to grow well at higher pressures, even at 1000 bars (Kato *et al.*, 1998).

2. The PHZ-diagram

A Planetary Habitable Zone Diagram (PHZ-diagram) was constructed to compare Earth's and Mars' near-surface environment from a biophysical point of view (Méndez, 1999). It compares the current knowledge of Earth's and Mars' global near-surface environment, the water phase diagram, and known microbial physiology limits with two basic physical environment variables: pressure and temperature (see figure). The diagram shows the steady-state vertical PT-space environment of Earth's atmosphere, hydrosphere, and lithosphere compared to the Martian atmosphere and lithosphere (cryosphere).

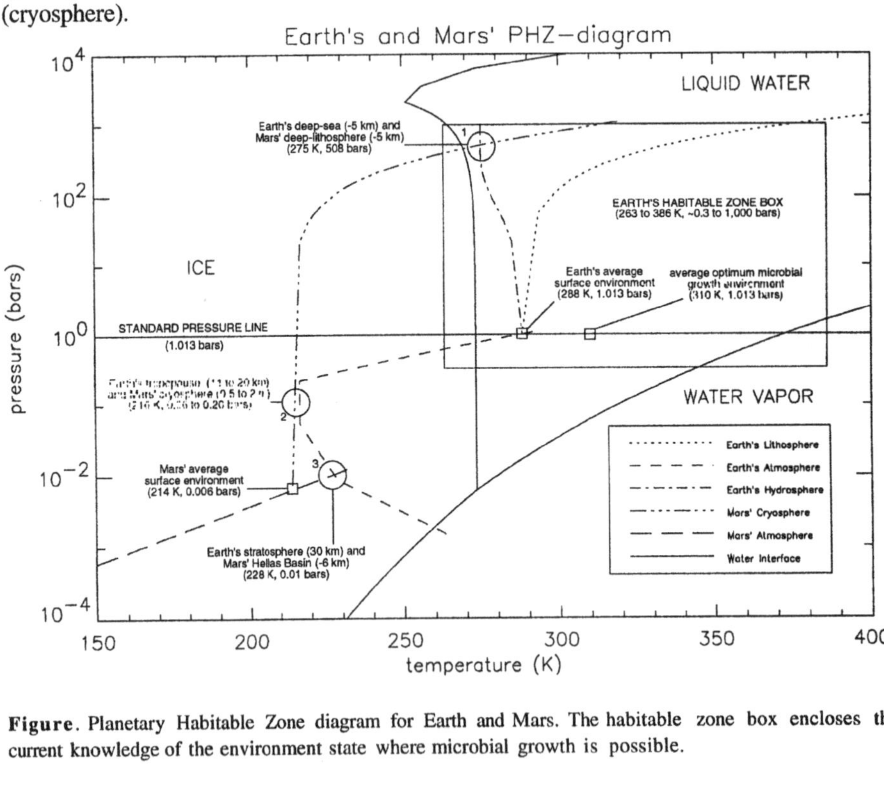

Figure. Planetary Habitable Zone diagram for Earth and Mars. The habitable zone box encloses the current knowledge of the environment state where microbial growth is possible.

The PHZ-diagram shows an Earth's atmosphere curve, derived from the 1976 U.S. Standard Atmosphere (Lide, 1999), from -0.5 to 30 km of mean sea level. The hydrosphere curve was derived from the hydrostatic equation, and a simplified model of the mixed layer and thermocline with boundary conditions fitted from global temperature data (Levitus and Boyer, 1994). The hydrosphere range is from 0 to -11 km (the depth of the Mariana Trench). The lithosphere curve was derived from the hydrostatic equation and Fourier's Law applied to a simple radioactive upper crustal layer model from 0 to −10 km. The Mars' atmosphere curve was constructed with a Mars Standard Atmosphere (Zurek, 1992) from −7.0 (the depth of Hellas impact basin) to 30 km of the geoid (Smith *et al.*, 1999). The lithosphere curve was derived from the current, but limited, understanding of the Mars' cryosphere.

The PHZ-diagram shows a water phase diagram constructed from temperature and pressure data for the vapor pressure of water, vapor pressure of ice and melting point of ice (Lide, 1999). The diagram shows phase transitions for pure water on a flat surface. Transitions are different for water solutions and around small particles. Also, water can be liquid below the freezing point in a supercooled state.

A database of some microbial physiology properties, such as optimum growth pH and temperature, was constructed from the scientific literature. This database shows that about 85% of the microbial diversity have an optimum growth temperature between 295 and 315 K, about 14% have an optimum growth temperature over 315 K and only 1% below 295 K. This is only a sample distribution of the current microbial diversity that have been cultured in the laboratory. However, it shows a strong tendency of microbial life toward mesophilic conditions. This may be due to evolutionary and physiological reasons or just a bias of the limited exploration of the biosphere.

3. Discussion

The Earth's and Mars' Planetary Habitable Zone diagram (PHZ-diagram) relates the global and average near-surface environment of both planets with the water state and microbial life. This type of diagram gives a general panorama of the limits of habitats as a function of the physical environment state described in terms of pressure and temperature. The diagram shows how different the Mars environment is when compared to the necessary environment for microbial growth. Similar diagrams could be constructed to compare the environment state of other planetary bodies such as Venus, Europa and Titan. Other interesting options are to plot diurnal and seasonal variations or to plot the environment state evolution of the primitive Earth and Mars.

The similarities and differences of Earth and Mars average environment states are evident in the PHZ-diagram. The diagram shows that in a global sense Earth and Mars share three average environments. From 0.5 to 2 meters below its surface, Mars has similar average PT-space conditions to Earth's lower stratosphere (tropopause) from 11 to 20 km above mean sea level. However, these regions are very different in terms of radiation levels (i.e. UV exposure) and substrate; but Mars' Hellas Basin at 6 km below the geoid has a similar environment to the stratosphere 30 km above. More interestingly, at 5 km below Mars' surface the planet has similar average PT-space environment to the deep-sea 5 km below. This is a very deep environment for future *in-situ* studies if localized energy sources, such as hydrothermal vents, are not considered.

It is interesting to note that about 85% of the microbial species have optimum growth temperatures between 295 to 315 K, just outside Earth's average surface environment of 288 K. Similar pressure limits have not been established because most studied microbial life is from the surface where atmospheric or hydrostatic pressure variations are minimal. Some near-surface regions on Mars are similar to Earth stratosphere and not much about life in similar environments is known, thus more field and laboratory studies of microbial life under low temperature and pressure environments are necessary. Knowledge of microbial growth and viability under such conditions might help in understanding and searching extinct or extant life in planetary bodies. It might also help in quantifying backward and forward contamination risks. Planetary environments could provide the necessary liquid water and energy source for life, but the physical environment state decides the biomass concentrations.

4. References

Bruch, C. W. (1967) Microbes in the Upper Atmosphere and Beyond, *Airborne Microbes*, Cambridge University Press, London, 345-373.

Greene, V. W., Pederson, P. D., Lundgren, D. A., and Hagberg, C. A. (1964) Microbial Exploration of Stratosphere: Results of Six Experimental Flights, in *Proceedings of the Atmospheric Biology Conference*, University of Minnesota, 199-212.

Halvorson, H. O., and Srinivasan, V. R. (1964) Can Spores Survive Space Travel? In *Proceedings of the Atmospheric Biology Conference*, University of Minnesota, 179-185.

Kato, C., Li, L., Nogi, Y., Nakamura, Y., Tamaoka, J., and Horikoshi, K. (1998) Extremely Barophilic Bacteria Isolated from the Mariana Trench, Challenger Deep, at a Depth of 11,000 Meters, *Appl. and Environm. Microbiol.*, **64**, 1510-1513.

Levitus, S., and Boyer, T. P. (1994) World Ocean Atlas 1994 Volume 4: Temperature, *U.S. Department of Commerce* Washington D. C., 177.

Lide, D. R., ed. (1999) *CRC Handbook of Chemistry and Physics*, CRC Press, Boca Raton, 6.9-6.12.

Méndez, A. (1999) Biophysical Comparison of the Habitable Zones on Earth and Mars. In the *Fifth International Conference on Mars*, Abstract #6197, LPI Contribution No. **972**, Lunar and Planetary Institute, Houston (CD-ROM).

National Research Council (1998) *Evaluating the Biological Potential in Samples Returned from Planetary Satellites and Small Solar System Bodies: Framework for Decision Making*. National Academy Press, Washington, D. C., 8-20.

Smith, D. E., Zuber, M. T., Solomon, S. C., Phillips, R. J., Head, J. W., Garvin, J. B. W., Banerdt, B., Muhleman, D. O., Pettengill, G. H., Neumann, G. A., Lemoine, F. G., Abshire, J. B., Aharonson, O., Brown, C. D., Hauck, S. A., Ivanov, A. B., McGovern, P. J., Zwally, H. J., and Duxbury, T. C., (1999) The Global Topography of Mars and Implications for Surface Evolution, *Science*, **284**, 1421-1576.

ZoBell, E., and Johnson, F. H. (1949) The Influence of Hydrostatic Pressure on the Growth and Viability of Terrestrial and Marine Bacteria, *J. of Bacteriol.*, **57**, 179-189.

Zurek, R. W., (1992) Comparative Aspects of the Climate of Mars: An Introduction to the Current Atmosphere, In *Mars*, H. H. Kieffer, B. M. Jakosky, C. W. Snyder and M. S. Matthews, eds., The University of Arizona Press, Tucson, 808-812.

QUANTITATIVE STUDY OF THE EFFECTS OF VARIOUS ENERGY SOURCES ON A TITAN'S SIMULATED ATMOSPHERE

SANDRA I. RAMIREZ AND RAFAEL NAVARRO-GONZALEZ
Laboratorio de Química de Plasmas y Estudios Planetarios
Instituto de Ciencias Nucleares, Universidad Nacional Autónoma de México Circuito Exterior, Ciudad Universitaria, Apartado Postal 70-543 México D.F. 04510 MEXICO

Abstract. A quantitative study of the production of organic species, mainly hydrocarbons and nitriles, is presented using different energy sources: gamma radiation, laser-induced plasma (LIP), arc and corona discharges. This is the first effort to derive energy yields of products arising under various conditions done by a single laboratory. Our results are discussed in terms of the dynamics of Titan's atmosphere and the possibility for the development of corona discharges within the tropospheric methane clouds. The energy yields derived by this mechanism are contrasted with those from other energy sources produced at different atmospheric levels. Our results can help to derive interesting conclusions for the Titan's dense atmosphere.

1. Introduction

The atmosphere of Titan, the largest Saturn's satellite, is mainly constituted by molecular nitrogen (90%), methane (0.5 - 3.4%), hydrogen (0.2%), (Coustenis, 1992) and traces of other molecules such as hydrocarbons, nitriles, and oxygen containing compounds. These minor compounds are essentially the products of an active N_2-CH_4 chemistry initiated in the upper atmosphere by solar ultraviolet radiation, Saturnian magnetospheric electrons and in the lower atmosphere by galactic cosmic rays and electric discharges. At present there is only one qualitative study carried out in the same laboratory of the products arising in a Titan's simulated atmosphere from ultraviolet light, electric discharge, gamma-, proton- and electron- beam irradiations (Gupta *et al.*, 1981). The results show significant differences among the different type of irradiations; however, with the lack of a quantitative knowledge of the amount of energy deposited in each experiment, it is quite difficult to decipher whether the differences are due to the type of irradiation and/or over exposure. The purpose of the present work is to provide such a comparison for the following energy sources: corona discharge, laser induced plasma, arc discharge and gamma radiation.

Gamma radiation allows us to simulate the effect caused by Saturnian electrons in the stratosphere and by galactic cosmic rays when they penetrate the atmosphere and reach the satellite's surface. Laser induced plasmas simulate the possible entrance of high velocity meteors into the atmosphere. Even when this event may be infrequent at

J. Chela-Flores et al. (eds.), Astrobiology, 307–310.

the present time, the information derived from the results can be useful to explain the actual composition of the atmosphere. The arc and corona discharges illustrate the role of electrical activity at tropospheric levels. As lightning has not been detected in Titan, the development, at tropospheric levels, of this alternative type of electrical activity is proposed.

2. Experimental

The atmosphere of Titan was simulated in the laboratory by mixing nitrogen (99.999%) and methane (99.97%) at 9/1 ratio using a Linde mass flow measuring and control gas blending console equipped with fast response flow modules. The resulted mixture was then introduced into appropriate reactors for irradiation at 670 mbar and 293 K with different energy sources. Corona discharges were generated in a Pyrex cylindrical reactor with a coaxial array: a central tungsten electrode was connected to high voltage, and a stainless steal plate, that surrounds the internal wall of the reactor, was grounded. The corona discharge was sustained at 1 mA using negative or positive high voltage from a regulated power source (100 W). The power deposited into the system was measured electrically according to the method described by Navarro-González *et al.* (1998). Arc discharges were produced across two tungsten electrodes separated by 1 cm in a cylindrical Pyrex reactor specially designed to measure the power deposited calorimetrically and described in detail by Navarro-González *et al.* (1998). Laser-induced plasmas were produced by focusing a 1.06 μm beam inside a 1-liter round flask with a pulse width of 7 ns from a high power Nd:YAG laser operating at 10 Hz using a 5 cm focal distance plano-convex lens with anti-reflection coating. The energy deposited into the system was determined by the difference between the incident pulse energy (300 mJ) and that transmitted after the formation of the plasma. This energy was measured with a power meter equipped with two-heads. Gamma ray experiments were conducted in the ^{60}Co source facility at ICN-UNAM. Dosimetry was done using methane as a chemical dosimeter at 1 mbar and taking $G(C_2H_6) = 2.1$ molecule per 100 eV. The power of the irradiated system was determined to be 1.23×10^{-4} W. Corrections were made for density and atomic composition to estimate the absorbed dose in Titan's simulated experiments.

After irradiation, the samples were immediately analyzed in a GC/FTIR/MS coupled system equipped with computerized data acquisition, analysis and database search capabilities. A PoraPLOTQ column was used. Mass and infrared spectral database matches were confirmed by visual inspection. Selective ion monitoring was used to increase sensitivity of the products. The ions used were 15, 16, 26, 27, 29, 30, 41, 43, 51, 52, 54, 77, 78, and 91. Calibration curves were done for C_2H_2, C_2H_4, C_2H_6 and CH_3CN. To estimate the yields of the other hydrocarbons and nitriles it was assumed that their responses were similar to those of C_2H_6 and CH_3CN, respectively.

3. Results and Discussion

The results from the laboratory analogy of Titan's atmospheric processes are summarized in Table 1. A total of 32 hydrocarbons and 6 nitriles were identified. We present here only a summary of the results. A remarkable feature is that all of the tested energy sources are good abiotic nitrogen fixers. Differences are observed with respect to the aliphatic and aromatic hydrocarbon production: corona discharges are likely to produce saturated and methyl-branched hydrocarbons. The arc discharge produces many conjugated hydrocarbons and aromatic compounds. The laser-induced plasma (LIP) produces mainly alkynes and benzene derivatives. Gamma rays seem to be the intermediate case between arc discharge, LIP and corona discharges producing saturated hydrocarbons and also unsaturated and aromatic compounds. The only detected unsaturated nitrile is produced exclusively by LIP and arc discharge. The products with the highest energy yields are C_2H_6 for corona discharges and gamma rays, and C_2H_2 for LIP and arc discharge. The nitrile production is dominated by HCN and CH_3CN. These facts contrast with the reported relative abundances of Gupta *et al.* (1981) where, except for the UV experiments, HCN is the main product in all their experiments.

The gas-phase compounds produced in Titan's upper atmosphere may condensate and travel to lower levels by action of gravitational forces. As they sediment, they form the haze layers that hide Titan's surface, and it is possible that they allow the formation of cloud particles. Navarro-González and Ramírez (1997) have presented a detailed description of these particles and the possible mechanisms for the formation of clouds in Titan. Observational evidence of titanian tropospheric methane clouds has been presented by Griffith *et al.* (1998). The search for terrestrial-like lightning discharges in Titan was performed during the Voyager 1 closest encounter in November 1980 (Desch and Kaiser, 1990). The instruments failed to detect any lightning-associated spherics. However, this result cannot completely dismiss the possibility of other types of electrical activity in Titan's clouds. It seems plausible that their electric fields are not strong enough to initiate lightning discharges and rather the atmospheric energy dissipation occurs by corona processes, which can occur in conditions of low current densities. Electrical activity in Titan's atmosphere is needed to explain the calculated abundance of some compounds like C_2H_4 and the dissociation of the nitrogen molecule in the lower atmosphere. From the results presented in Table 1, corona discharges seem to be an adequate option. Their occurrence in the satellite can be explained and they produce all of the observed species in Titan, except for benzene.

Knowledge of how a Titan's simulated atmosphere can be transformed by different types of energy sources increases the possibility of a better interpretation of the information that the Huygens probe will send from the atmosphere of Titan towards the end of the 2004 year. During the descend through the dense atmosphere, a new opportunity to detect atmospheric electricity and lightning activity will arrive (Grard, *et al.,* 1995). The possibility that some of the already identified compounds, typically produced by a single kind of energy, be detected and give us more insights of the energy dissipation mechanisms is open.

TABLE 1. Initial energy yields (P) of the main organic gases produced after irradiation by different energy sources of a N_2 and CH_4 (9/1) mixture at 670 mbar and 293 K*.

Compound	P (Molecule J^{-1})				
	LIP	Arc	Gamma rays	Positive Corona	Negative Corona
Ethane	7.9×10^{15}	8.6×10^{15}	9.2×10^{16}	3.0×10^{15}	6.6×10^{14}
Ethene	1.3×10^{16}	2.4×10^{15}	8.0×10^{14}	3.8×10^{14}	1.4×10^{14}
Ethyne	8.2×10^{16}	4.3×10^{16}	2.0×10^{14}	7.3×10^{14}	2.2×10^{14}
Propane	* *	4.8×10^{14}	1.1×10^{16}	6.2×10^{14}	5.7×10^{13}
n-Butane	* *	* *	9.0×10^{15}	4.3×10^{14}	9.0×10^{13}
2-Methylbutane	* *	* *	6.6×10^{14}	2.8×10^{14}	7.4×10^{13}
Butyne	6.5×10^{14}	4.1×10^{14}	6.0×10^{15}	* *	* *
Benzene	4.9×10^{14}	4.9×10^{14}	6.3×10^{15}	* *	* *
Toluene	1.9×10^{14}	1.9×10^{14}	2.8×10^{15}	* *	* *
Methanenitrile	3.6×10^{15}	2.6×10^{13}	5.5×10^{15}	1.9×10^{14}	5.6×10^{13}
Ethanedinitrile	7.3×10^{13}	3.8×10^{12}	6.6×10^{15}	3.8×10^{13}	6.4×10^{12}
Ethanenitrile	7.5×10^{13}	5.0×10^{13}	6.1×10^{14}	1.9×10^{14}	8.8×10^{13}
Propanenitrile	3.3×10^{13}	1.5×10^{13}	5.0×10^{14}	7.7×10^{13}	2.6×10^{13}
2-Propenenitrile	1.4×10^{12}	1.4×10^{12}	* *	* *	* *

*Energy yields were calculated in each case from the slope of number of molecule *vs.* energy deposition curves.
* *Not detected.
Molecules detected in Titan are in italics.

4. References

Coustenis, A. (1992) Titan's Atmosphere: Latitudinal Variations in Temperature and Composition, *Proceedings Symposium on Titan*, Toulouse, France, Sept. 1991, 53-58.

Desch, M.D. and Kaiser, M.L. (1990) Upper limit set for level of lightning activity on Titan, *Nature* **345,** 442-444.

Grard, R., Svedhem, H., Brown, V., Falkner, P., and Hamelin, M. (1995) An experimental Investigation of Atmospheric Electricity and Lightning Activity to be Performed During the Descend of the Huygens Probe onto Titan, *J. Atmos. Terr. Phys.* **57,** 575-585.

Griffith, C. A., Owen, T., Miller, G. A., and Geballe, T. (1998) Transient clouds in Titan's lower atmosphere, *Nature* **395**, 575-578.

Gupta, S., Ochiai, E. and Ponnamperuma, C. (1981) Organic synthesis in the atmosphere of Titan, Nature **293**, 725-726.

Navarro-González, R. and Ramírez, S.I. (1997) Corona Discharge of Titan's troposphere, *Adv. Space Res.* **19(7),** 1121-1133.

Navarro-González, R., Romero, A. and Honda, Y. (1998) Power Measurements of Spark Discharge Experiments, *Origins Life Evol. Biosphere* **28,** 131-153.

LIFE EXTINCTIONS AND THE GRAVITATIONAL COLLAPSE OF ONEMG ELECTRON–DEGENERATE OBJECTS

JORDI L. GUTIÉRREZ

Dept. de Física Aplicada, Universitat Politècnica de Catalunya, Pla de Palau, 18, 08003 – Barcelona, Spain

1. Introduction

Supernova explosions have been considered a possible (although improbable) thread to life on Earth. In 1987, a massive star exploded in the Large Magellanic Cloud and, coinciding with the arrival of the first photons of that event, several neutrino observatories detected a burst of neutrinos. This was the final evidence needed to validate the model of gravitational collapse supernovae.

In recent years, theorists have payed more attention to the effects of neutrinos on biological tissues. While they have an overwhelmingly small cross section (about 10^{-42} cm^2, slightly dependent on the neutron number of the atomic target), the large amount of energy emitted as neutrinos by a typical gravitational supernova (about 10^{53} erg) make them a suitable possibility to explain some (but not all, see Abbas & Abbas, 1996) mass extinctions.

Besides the canonical gravitational collapse model of supernovae, other objects can be destabilized to initiate such a process, as is the case of white dwarfs. Canal & Schatzman (1976) were the first in putting forward the idea that mass–accreting white dwarfs could reach the Chandrasekhar mass and, hence, collapse to form a neutron star. In this process, also known as a "silent supernova", some 10^{53} erg of energy are liberated, 99% of which in the form on neutrinos.

In this paper, I will show that stars between 8–11 $M_\odot$ and ONeMg white–dwarfs should be one of the main sources of these neutrino bursts.

J. Chela-Flores et al. (eds.), Astrobiology, 311–314.

2. The Final Evolution of 8–11 $M_\odot$ Stars

After the carbon burning in mildly degenerate conditions, the core of 8–11 $M_\odot$ stars fail to ignite the ^{20}Ne burning, contract and the structure becomes supported by the degenerate electrons. As the burning shells outside the core continue active, the core itself continues growing in mass. In this situation, the central density incrases and so does the Fermi Energy of the electrons. When it arrives to 5.514 MeV, the electrons have enough energy to start electron–captures on the nuclei of ^{24}Mg and its daughters (this at a slightly greater Fermi energy):

$$^{24}\mathrm{Mg}(\mathrm{e}^-,\gamma\,\nu_\mathrm{e})^{24}\mathrm{Na}(\mathrm{e}^-,\gamma\,\nu_\mathrm{e})^{24}\mathrm{Ne}$$

Current evolutionary models predict that the core of the star will have about a 3% of ^{24}Mg, not enough to give a prompt explosion at relatively low density ($\sim 4\times10^9$ g cm^{-3}). As the central density increases, the Fermi energy of the electrons reaches 7.029 MeV and the nuclei of ^{20}Ne start a similar chain of reactions:

$$^{20}\mathrm{Ne}(\mathrm{e}^-,\gamma\,\nu_\mathrm{e})^{20}\mathrm{F}(\mathrm{e}^-,\gamma\,\nu_\mathrm{e})^{20}\mathrm{O}$$

But now, the threshold energy for electron–capture of ^{20}F is *lower* (3.814 MeV) than that of ^{20}Ne, and the two captures proceed almost simultaneously. This gives a huge entropy release and makes the temperature rise up to the point on which material is processed to nuclear statistical equilibrium. However, the density is so high ($\sim 9.5\times10^9$ g cm^{-3}, see Gutiérrez et al., 1996 and Canal & Gutiérrez, 1997) that the electron captures on the NSE material ensures the collapse of the core to a neutron star (Timmes & Woosley, 1992, 1994).

Analogously, a white dwarf accreting hydrogen or helium at the correct rate, could grow in mass enough to be destabilized up to the accretion–induced collapse, roughly at the same density as the core of 8–11 $M_\odot$ stars. Then, both mechanisms produce sudden bursts of $\sim 10^{53}$ erg of neutrinos.

Quite oppositely, C+O white–dwarfs are more easily destabilized to a thermonuclear explosion that completely destroys the star (in what is seen as the standard model for type Ia supernovae).

The short lifespan of this stars makes them typical inhabitants of the spiral arms of the Galaxy; Leitch & Vasisht (1998) have presented evidences of a possible relationship between mass extinctions and the Sun's passage through the spiral arms. On the other hand, binary stars including an ONeMg white–dwarf should evolve in completely different timescales, and can be found, at least in principle, all around the Galaxy.

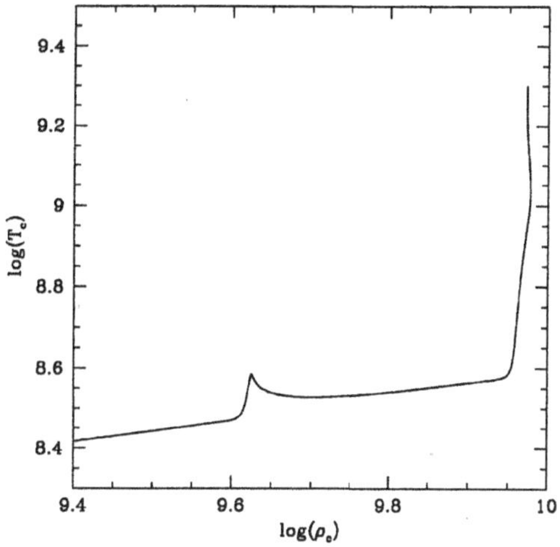

Figure 1. Final evolution of an ONeMg electron–degenerate core up to the begining of NSE.

Integrating a common form of the initial mass function for the mass range $m_1 = 8M_\odot$ and $m_2 = 11M_\odot$ it is possible to determine that about a 30% of all the stars that explode as a gravitational supernova belong to the 8–11 solar masses interval.

3. Biological Effects

In a recent paper, Collar (1996), the biological effects of MeV neutrinos were studied in the detail. The surprising conclusion of that work was that this ghostly particles interacted with the biological tissues and, via the recoil of nucleotides, triggered a number of malignant foci given approximately by the expression:

$$\langle MF \rangle \approx 490 \left(\frac{r}{1\ \mathrm{pc}} \right)^{-2} \text{malignant foci/kg} \tag{1}$$

where r is the distance in parsec.

So a high rate of pre-neoplasies consitute a serious threat to any organism, specially for the more massive ones. In fact, some extinction patterns seem to favour the disparition of large animals.

As Abbas & Abbas (1996) have noted, this mechanism is unlikely for many mass extinctions (as the K/T event). However, none of their arguments affected the main points of the mechanism proposed by Collar (1996).

4. Conclusions

The collapse of ONeMg electron–degenerate objects is a possible threat to life. Of course, given the relative scarcity of these objects, the hazard is far smaller than that associated with comets and asteroids, but must be taken into consideration. Stellar collapse neutrinos could be in the root of some mass extinctions not related to iridium rich strata or other causes, and given the relative abundance of ONeMg electron–degenerate objects, these should be one of the main sources for the neutrino bursts.

References

- Abbas, S. & A. Abbas, "Comments on Biological Effects of Stellar Collapse Neutrinos", astro–ph/9612003
- Canal, R. & J. Gutiérrez, "The Possible White Dwarf–Neutron Star Connection", *10th European Workshop on White Dwarfs*, ed. J. Isern, M. Hernanz & E. García–Berro, Dordrecht, Kluwer, pp. 49–55 (1997).
- Canal, R. & E. Schatzman, "Non Explosive Collapse of White Dwarfs", *A&Ap*, **46**, 229–235 (1976)
- Collar, J. I., "Biological Effects of Stellar Collapse Neutrinos", *Phys. Rev. Lett.*, **76**, 999-1002 (1996)
- Gutiérrez, J., E. García–Berro, I. Iben, J. Isern, J. Labay & R. Canal, "The Final Evolution of ONeMg Electron–Degenerate Cores", *ApJ*, **459**, 701–705 (1996)
- Gutiérrez, J., R. Canal, J. Labay, E. García–Berro, & J. Isern, "The Final Evolution of 8–10$M_\odot$ Stars", in *Thermonuclear Supernovae*, ed. P. Ruiz–Lapuente, R. Canal & J. Isern, Dordrecht, Kluwer, 303–311 (1997)
- Leitch, E. M. & G. Vasisht, "Mass Extinctions and the Sun's Encounters with Spiral Arms", *New Astronomy*, **3**, 51–56 (1998)
- Timmes, F. & S. Woosley, "The Conductive propagation of Nuclear Flames. 1 - Degenerate C+O and O+Ne+Mg White Dwarfs", *ApJ*, **396**, 649–667 (1992)
- Timmes, F. & S. Woosley, "The Conductive propagation of Nuclear Flames. 2 - Convectively Bounded Flames in C+O and O+Ne+Mg Cores", *ApJ*, **420**, 348–363 (1994)

NAME INDEX

SUBJECT INDEX

H

I

Participants

ACHONG, Claude
PDVSA
Caracas, VENEZUELA

ACOSTA, Jasmel
asmelacosta@hotmail.com
cuasaxi@telcel.net.ve
Caracas, VENEZUELA

ALDUARO, Verónica
UCAB, Caracas, VENEZUELA

ASCANIO, Sergio
Universidad Simón Bolívar,
Valle de Sartenejas, Baruta,
Caracas, VENEZUELA

BARRAL, José
Univ. de a Coruña, Dept. Biología Celular y Molecular, Fac. Ciencias,
Campus a Zapateeira s/n.
15071. A Coruña , ESPAÑA

BARROS PITA, José Carlos
Universidad Central de Venezuela, Instituto de Biología, Laboratorio de Zoofisiología, Colinas de Bello Monte. Caracas, VENEZUELA

BENÍTEZ, Pedro
Universidad de Los Andes, Centro de Microscopía Electrónica. Apartado Postal 163-175. Mérida, VENEZUELA

BOLIVAR, Nelson E.
Universidad Central de Venezuela, Departamento de Física, Facultad de Ciencias, Apartado Postal 47270. Los Chaguaramos, 1041-A. Caracas, VENEZUELA

BONYORNO, Rafael
Universidad Central de Venezuela. Departamento de Biología. Facultad de Ciencias. Apartado Postal 47270. Los Chaguaramos, 1041-A. Caracas, VENEZUELA

BRICEÑO, Carlos
Universidad Simón Bolívar, Valle de Sartenejas, Baruta, Caracas, VENEZUELA

CAICEDO SANDGREN, Mario
Universidad Simón Bolívar, Valle de Sartenejas – Baruta, Caracas, VENEZUELA

CALDEIRA SOUSA, María Elena
Universidad Central de Venezuela. Escuela de Física. Apartado Postal 47586. Los Chaguaramos, 1041-A. Caracas, VENEZUELA

CALVA-ALEJO, Leonel
Universidad Nacional Autónoma de México, Instituto de Ciencias Nucleares, 04510 México City, MEXICO

CAMPINS, Humberto
University of Arizona. Rm. 384, Life Sciences. North Bldg.
AZ 85724. Tucson
UNITED STATES OF AMERICA

CARDOZO, Damelis Mariela
Universidad Central de Venezuela. Departamento de Física. Apartado Postal 47270. Los Chaguaramos,
1041-A. Caracas, VENEZUELA

CEDEÑO, José Manuel
Universidad Central de Venezuela. Departamento de Física. Apartado Postal 47270. Los Chaguaramos, 1041-A, Caracas, VENEZUELA

CHACON BACA, Elizabeth
Universidad Nacional Autónoma de México. Institute of Geology, Department of Paleontology. A.P. 70-543, 04510 México City, MEXICO

CHANG ROMERO, Ricardo
nstituto de Estudios Avanzados (IDEA), Apartado 17606. Parque Central 1015-A, Caracas, VENEZUELA

CHAURIO, Ricardo Alfredo
Universidad Central de Venezuela. Departamento de Física. Facultad de Ciencias. Apartado Postal 47270. Los Chaguaramos, 1041-A. Caracas, VENEZUELA

CHELA-FLORES, Julián
The Abdus Salam International Centre for Theoretical Physics. Strada Costiera 11, P.O. Box 586, Miramare. 34100 Trieste, ITALIA

COLIN-GARCIA, Maria
Instituto de Ciencias Nucleares, Universidad Autónoma de México, Departamento de Química de Radiaciones y Radioquímica. Circuito Exterior. 04510 México City, MEXICO

COLMENARES, Valentina
Universidad Central de Venezuela. Departamento de Física. Facultad de Ciencias. Los Chaguaramos, 1041-A. Caracas, VENEZUELA

COLONNELLO, Claudia
Universidad Simón Bolívar, Valle de Sartenejas, Baruta, Caracas, VENEZUELA

CORDERO MORALES, Julio Francisco
Instituto de Biología Experimental (IBE), Calle Suapure, Colinas de Bello Monte, Caracas, VENEZUELA

DELAYE, Luis
Universidad Nacional Autónoma de México, Facultad de Ciencias, A.P. 70-407. Ciudad Universitaria, D.F. 04510 México City, MEXICO

DELGADO, Ronald Rodrigo
Universidad Central de Venezuela. Departamento de Física. Facultad de Ciencias. Apartado Postal 47270. Los Chaguaramos, 1041-A. Caracas, VENEZUELA

DIAZ, Luis Rafael
Universidad Central de Venezuela. Departamento de Física. Facultad de Ciencias. Apartado Postal 47270. Los Chaguaramos, 1041-A. Caracas, VENEZUELA

DIAZ, Naryttza Namelly
Universidad Central de Venezuela. Departamento de Biología. Facultad de Ciencias. Los Chaguaramos, 1041-A. Caracas, VENEZUELA

DOPAZO, Hernán
Universidad de Buenos Aires. Centro de Estudios Avanzados. Uriburu 950. 1114, Buenos Aires, ARGENTINA

DRAKE, Frank
SETI Institute. 2035 Landings Drive. 94043 CA Mountain View, UNITED STATES OF AMERICA

DRAYER, Gregorio
Universidad Simón Bolívar, Valle de Sartenejas. Baruta, Caracas, VENEZUELA

DRAYER, Roberto
rdrayer@cantv.net
Caracas, VENEZUELA

FAJARDO RODRÍGUEZ, Luis Arturo
Universidad de Los Andes, Centro de Microscopía Electrónica. Apartado Postal 163-175. Mérida, VENEZUELA

FALCON RODRIGUEZ, Jersys
Centro de Cibernética Aplicada a la Medicina (CECAM).
Calle 146, # 2504, E/25 Y 31. Cubanacán, La Habana,
CUBA

FALCON, Nelson Leonardo
Universidad de Carabobo, Departamento de Física, Facultad de Ciencias y Tecnología. Av. Los Colegios, Qta. Hilmar, Urb. Guaparo. Valencia,
Estado Carabobo, VENEZUELA

FIGUEROA, Daniela Alejandra
Universidad Central de Venezuela. Departamento de Biología. Facultad de Ciencias. Apartado Postal 47270. Los Chaguaramos, 1041-A. Caracas, VENEZUELA

GALAVIS, Martha Elena
Metromgalavis@unimet.edu.ve
Universidad politana, Caracas
Caracas, VENEZUELA

GALUÉ, Francisco
Sociedad Astronómica de Maracaibo, Diario "La Verdad". Maracaibo, VENEZUELA

GAMAZZA, Michele
Universidad Simón Bolívar, Valle de Sartenejas, Baruta, Caracas, VENEZUELA

GARCÍA PONCE, Luis Tomás
Centro de Estudios Astronómicos
Caracas, VENEZUELA

GARCÍA, Marco
Universidad Central de Venezuela,
Caracas, VENEZUELA

GÓMEZ LUGO, Emeterio
Universidad Simón Bolívar, Valle de Sartenejas, Baruta
Caracas, VENEZUELA

GOMEZ-CHABALA, Sandra María
Universidad Nacional de Colombia, Departamento de Física. Apartado Aéreo 3840, Medellín, COLOMBIA

GUTIERREZ-CABELLO, Jordi Luis
Universitat Politécnica de Catalunya, Dept. de Física Aplicada.
Campus Nord Modul B5. Pla de Palau, 18. 08003 Barcelona,
SPAIN

GUZMÁN MARMOLEJO, Andrés
Instituto de Ciencias Nucleares, Universidad Autónoma de México, Departamento de Química de Radiaciones y Radioquímica. Circuito Exterior. 04510 México City, MEXICO

HERNANDEZ COLMENARES, Javier A
Universidad Central de Venezuela. Departamento de Física. Facultad de Ciencias. Apartado Postal 47270. Los Chaguaramos, 1041-A. Caracas, VENEZUELA

HERNÁNDEZ, Laura
Universidad Central de Venezuela, Caracas, VENEZUELA

HOENICKA, Janet
Banco de Tejidos para Investigaciones Neurológicas, Universidad Complutense de Madrid. Facultad de Medicina. 28040, Madrid, SPAIN

ISLAS GRACIANO, Sara Ernestina
Universidad Nacional Autónoma de México, Facultad de Ciencias, Apdo. Postal 70-407, Ciudad Universitaria, D.F. 04510 México City, MEXICO

JIMÉNEZ, Douglas
Barquisimeto, VENEZUELA

KEMPIS FIGUEROA, Yanmir del Carmen
Universidad Central de Venezuela. Departamento de Física. Facultad de Ciencias. Apartado Postal 47270. Los Chaguaramos, 1041-A. Caracas, VENEZUELA

LAZO, Emilio
Universidad Central de Venezuela. Escuela de Física. Facultad de Ciencias. Apartado Postal 47270. Los Chaguaramos, 1041-A. Caracas, VENEZUELA

LAZCANO, Antonio
Universidad Nacional Autónoma de México, Facultad de Ciencias, Apdo. Postal 70-407, Ciudad Universitaria, D.F. 04510 México City, MEXICO

LEMARCHAND, Guillermo Andrés
Universidad de Buenos Aires, Centro de Estudios Avanzados, C.C.8, Sucursal 25. Uriburu 950, 1114 Buenos Aires, ARGENTINA

LEÓN, Jesús Alberto
Instituto de Zoologia Tropical, UCV Caracas, VENEZUELA

LEZAMA, José Rafael
Instituto de Estudios Avanzados (IDEA), Apartado 17606, Parque Central 1015-A, Caracas, VENEZUELA

LINARES JEREZ, Fedor Gabriel
Instituto de Estudios Avanzados (IDEA), Caracas, VENEZUELA

LONGART, Marinés
Instituto de Estudios Avanzados (IDEA), Caracas, VENEZUELA

LORENZONI, Laura
Universidad Simón Bolívar, Departamento de Biología, Valle de Sartenejas, Baruta,Caracas, VENEZUELA

LUCIANI TOLEDO, Christian Leonardo
Universidad Central de Venezuela. Departamento de Física. Facultad de Ciencias. Apartado Postal 47270. Los Chaguaramos, 1041-A. Caracas, VENEZUELA

MADURO MORA, Miguel Ángel
Universidad Simón Bolívar, Valle de Sartenejas, Baruta Caracas, VENEZUELA

MALAVÉ DE BILBAO, Caridad
Instituto de Estudios Avanzados (IDEA), Departamento de Biociencias, Caracas, VENEZUELA

MANNO, Carlo Danilo
nstituto Venezolano de Investigaciones Científicas (IVIC), Km. 11, Carretera Panamericana, Estado Miranda. Apartado Postal 21827, 1020-A, Caracas, VENEZUELA

MAQUEIRA, Braudel
Instituto de Estudios Avanzados (IDEA), Unidad de Neurobiología Molecular, Carretera Baruta, Hoyo de La Puerta, Caracas, VENEZUELA

MARCANO, Vicente
Universidad de Los Andes, Centro de Microscopía Electrónica, Apartado Postal 163-175, Mérida, VENEZUELA

MÁRQUEZ BRICEÑO, Jesús
jesuslalos@cantv.net lalos1@usa.net
Caracas, VENEZUELA

MARTIN-LANDROVE, Miguel
Universidad Central de Venezuela. Centro de Resonancia Magnética. Facultad de Ciencias. Apartado Postal 47586. Los Chaguaramos, 1041-A. Caracas, VENEZUELA

MASCAREÑO, Carlos
Universidad Central de Venezuela,
Caracas, VENEZUELA

MASSARINI, Alicia Isabel
Universidad de Buenos Aires, Departamento de Biología, Facultad de Ciencias Exactas y Naturales, Ciudad Universitaria, Pabellón I, 1417, Buenos Aires, ARGENTINA

MAYZ VALLENILLA, Ernesto
Instituto de Estudios Avanzados (IDEA), Apartado 89000, Valle de Sartenejas, Baruta 1080-A, Caracas, VENEZUELA

MEDINA DE CALLAROTTI, Elinor
Universidad Simón Bolívar, Departamento de Idiomas, Valle de Sartenejas, Baruta 1080-A, Caracas, VENEZUELA

MÉNDEZ, Abel
Universidad de Puerto Rico, Arecibo
Puerto Rico,
UNITED STATES OF AMERICA

MONSALVE DAM, Dorixa D.
Universidad Central de Venezuela. Departamento de Física. Facultad de Ciencias. Apartado Postal 47270. Los Chaguaramos, 1041-A. Caracas, VENEZUELA

MORA, Freddy
Universidad Metropolitana, Caracas
VENEZUELA

NAVARRO-GONZÁLEZ, Rafael
Universidad Nacional Autónoma de México, Instituto de Ciencias Nucleares, Laboratorio de Química de Plasmas y Estudios Planetarios. Ap. Postal 70-543, D.F. 04510 México City, MEXICO

NEGRÓN-MENDOZA, Alicia
Universidad Nacional Autónoma de México, Instituto de Ciencias Nucleares, Ap. Postal 70-543, Circuito Exterior, C.U., D.F. 04510 México City, MEXICO

OCAMPO, Adriana
NASA Headquarters, Office of Space Science, 300 e Street, S.W. 20546 Washington,
UNITED STATES OF AMERICA

ORÓ, Juan
University of Houston, Dept. of Biology and Biochemistry, TX, 77204-5934, Houston,
UNITED STATES OF AMERICA

PADRÓN, Carlos
Instituto de Estudios Avanzados (IDEA)
Unidad de Filosofía. Caracas,
VENEZUELA

PALACIOS PRU, Ernesto
Universidad de Los Andes, Centro de
Microscopía Electrónica, Apartado Postal
163-175. Mérida, VENEZUELA

PALERMO MAMMANA, Giuseppe A.
Universidad Central de Venezuela.
Departamento de Geofísica. Los
Chaguaramos, 1041-A. Caracas,
VENEZUELA

PAREDES, Pedro Pablo
Universidad del Zulia, Departamento de
Física, Facultad Experimental de Ciencias,
Módulo 1, Av. Universidad, Sector Grano
de Oro, Maracaibo, Estado Zulia,
VENEZUELA

PARJAN, Elena
Universidad Central de Venezuela,
Caracas, VENEZUELA

PEÑA, Francisco
Universidad Central de Venezuela. Escuela
de Física. Facultad de Ciencias. Apartado
Postal 47270. Los Chaguaramos, 1041-A.
Caracas, VENEZUELA

PERERA, Lucy
oldforest69@hotmail.com
Caracas, VENEZUELA

PEREZ CHAVEZ, Itzel
Universidad Nacional Autónoma de
México, Instituto de Ciencias Nucleares,
Circuito Exterior, Zona de Institutos,
Ciudad Universitaria. 04510 México City,
MEXICO

PEREZ DE VLADAR, Harold Paúl
Universidad Central de Venezuela. Escuela
de Física. Facultad de Ciencias. Apartado
Postal 47058. Los Chaguaramos, 1041-A.
Caracas, VENEZUELA

PETIT, Gustavo
Universidad Simón Bolívar, Valle de
Sartenejas, Baruta, Caracas,
VENEZUELA

PINTO, Ricardo Alberto
Universidad Central de Venezuela.
Departamento de Física. Facultad de
Ciencias.1041-A. Caracas, VENEZUELA

PUCCIARELLI, Gerardo
Universidad Simón Bolívar, Valle de
Sartenejas, Baruta, Caracas,
VENEZUELA

RAGO, Héctor Enrique
Universidad de Los Andes, Departamento
de Física, Facultad de Ciencias, La
Hechicera, 5101, Mérida, VENEZUELA

RAMIREZ JIMENEZ, Sandra Ignacia
Universidad Nacional Autónoma de
México, Instituto de Ciencias Nucleares,
Circuito Exterior, Zona de Institutos,
Ciudad Universitaria, 04510 México City,
MEXICO

RAMIREZ, César Ernesto
Universidad Central de Venezuela.
Departamento de Física. Facultad de
Ciencias. 1041-A. Caracas, VENEZUELA

RAMOS-BERNAL, Sergio
Universidad Nacional Autónoma de
México, Instituto de Ciencias Nucleares,
Apartado Postal 70-543, Circuito
Exterior, Ciudad Universitaria. D.F.
04510 México City, MEXICO

REISMAN, Garrett
NASA Headquarters
garrett.e.reisman1@jsc.nasa.gov
UNITED STATES OF AMERICA

REVILLA, Tomás Augusto
Universidad Central de Venezuela.
Departamento de Física. Facultad de
Ciencias. 1041-A. Caracas,
VENEZUELA

REYMONDIN, Christian
vaud@cantv.net
Caracas, VENEZUELA

RODRIGUEZ, Sandy Santiago
Universidad Central de Venezuela.
Departamento de Física. Facultad de
Ciencias. 1041-A, Caracas,
VENEZUELA

ROEDERER, Juan
University of Alaska, Geophysical
Institute, 99775-0800, Alaska, Fairbanks,
UNITED STATES OF AMERICA

ROJAS, Diego Rafael
Universidad Central de Venezuela.
Departamento de Biología. Facultad de
Ciencias. Apartado Postal 47270. Los
Chaguaramos, 1041-A. Caracas,
VENEZUELA

ROMERO, Gustavo-Esteban
Instituto Argentino de Radioastronomía
(IAR), C.C. Nº 5, 1894, Villa Elisa,
Buenos Aires, ARGENTINA

ROMERO, Jesús Guillermo
Instituto de Biología Experimental (IBE),
Universidad Central de Venezuela,
Laboratorio Fisiología Celular, Calle
Suapure, Colinas de Bello Monte,
Caracas, VENEZUELA

RON, Lupercio José
Universidad Central de Venezuela.
Departamento de Física. Facultad de
Ciencias. Apartado Postal 47270. Los
Chaguaramos, 1041-A. Caracas,
VENEZUELA

ROSALES, Pedro Pablo C.
Instituto de Estudios Avanzados (IDEA),
Unidad de Filosofía, Carretera Nacional
Hoyo de la Puerta, Edif. Bolívar, P.B.
Sartenejas, Baruta – Estado Miranda,
VENEZUELA

ROZENBAUM, Sami
Fundación Amigos del Planetario
Humboldt, Calle Negrín, edif.
DAVOLCA, Apto. 9, Sabana Grande,
Caracas, VENEZUELA

SALAZAR SORA, Estrella
Universidad Simón Bolívar, Edif.
Biblioteca Central, Nivel Jardín, Ofic. de
Relaciones Públicas, Sartenejas - Baruta,
Caracas, VENEZUELA

SALAZAR YAMARTE, Efraín Gerardo
CENAIP-IIA, Departamento de
Biotecnología, Zona Universitaria El
Limón, Edif. 08, 2101, Maracay, Estado
Aragua, VENEZUELA

SANCHEZ, Andrea Leticia
Universidad de la República de Uruguay,
Facultad de Ciencias, Instituto de Física,
Igua 4225 Y, Mataojo, 11400
Montevideo, URUGUAY

SANCHEZ, Rodrigo.
Universidad Central de Venezuela.
Departamento de Física. Facultad de
Ciencias. Apartado Postal 47270. Los
Chaguaramos, 1041-A. Caracas,
VENEZUELA

SANTANDER M., Sol D.
Universidad Central de Venezuela. Departamento de Biología. Facultad de Ciencias. Apartado Postal 47270. Los Chaguaramos, 1041-A. Caracas, VENEZUELA

SEGURA MOLINA, Antígona Peralta
Universidad Nacional Autónoma de México, Instituto de Ciencias Nucleares, Circuito Exterior, Zona de Institutos, Ciudad Universitaria,04510 México City, MEXICO

SOFIA, Sabatino
University of Yale, Department of Astronomy, P.O. Box 208101, New Haven, Connecticut 06520-8101,
UNITED STATES OF AMERICA

SUÁREZ MEZA, Luis
Universidad Centro Occidental Lisandro Alvarado, Barquisimeto, Estado Lara, VENEZUELA

SUBERO, Yolimar
Universidad Central de Venezuela. Escuela de Física. Facultad de Ciencias, A1041-A. Caracas, VENEZUELA

TAPIAS, Adonay José
Universidad Central de Venezuela. Departamento de Física. Facultad de Ciencias. 1041-A. Caracas, VENEZUELA

TORRES, Diego Fernando
Universidad Nacional de La Plata. Departamento de Física C.C. 67, Calle 49 Y 115, 1900 La Plata, ARGENTINA

TROMBINO, Giuseppe
Universidad Simón Bolívar, Valle de Sartenejas, Baruta, Caracas, VENEZUELA

URBINA PACHECO, Roymari
Universidad Central de Venezuela. Departamento de Física. Facultad de Ciencias. Apartado Postal 47270. Los Chaguaramos, 1041-A. Caracas, VENEZUELA

URRIOLA, Pedro
Instituto de Estudios Avanzados (IDEA), Baruta, Caracas, VENEZUELA

VALENCIA, Andrés
Observatorio ARVAL, Revista Astronomía Digital, Caracas, VENEZUELA

VEI SUNG, Ling
Universidad Metropolitana,
Caracas, VENEZUELA

VERRILLI, David Tony
Universidad Central de Venezuela. Departamento de Física. Facultad de Ciencias. A1041-A. Caracas, VENEZUELA

VILLEGAS, Gloria
Instituto de Estudios Avanzados (IDEA), Apartado 89000, Valle de Sartenejas, Baruta. 1080-A, Caracas, VENEZUELA

VILLEGAS, Raimundo
Instituto de Estudios Avanzados (IDEA), Apartado 89000, Valle de Sartenejas, Baruta. 1080-A, Caracas, VENEZUELA

WAMSTEKER, Willem
ESA/Vilspa, p.o. Box 50727, 28080, Madrid, SPAIN

ZAPATA, Fanny
fannyzapata@yahoo.com
VENEZUELA

FSC
www.fsc.org
MIX
Papier | Fördert
gute Waldnutzung
FSC® C083411

Kolibri 360 GmbH
Ferdinand-Jühlke-Straße 7
99095 Erfurt, Deutschland
produktsicherheit@kolibri360.de